CAMBRIDGE TRACTS IN MATHEMATICS

General Editors

B. BOLLOBAS, F. KIRWAN, P. SARNAK, C.T.C. WALL

128
An Introduction to Maximum Principles and Symmetry in Elliptic Problems

T0296936

An Introduction to Maximum Principles and Symmetry in Elliptic Problems

L.E. Fraenkel

School of Mathematical Sciences
University of Bath

CAMBRIDGE UNIVERSITY PRESS
Cambridge, New York, Melbourne, Madrid, Cape Town, Singapore,
São Paulo, Delhi, Dubai, Tokyo, Mexico City

Cambridge University Press
The Edinburgh Building, Cambridge CB2 8RU, UK

Published in the United States of America by Cambridge University Press, New York

www.cambridge.org
Information on this title: www.cambridge.org/9780521172783

© Cambridge University Press 2000

This publication is in copyright. Subject to statutory exception
and to the provisions of relevant collective licensing agreements,
no reproduction of any part may take place without the written
permission of Cambridge University Press.

First published 2000
First paperback edition 2010

A catalogue record for this publication is available from the British Library

ISBN 978-0-521-46195-5 Hardback
ISBN 978-0-521-17278-3 Paperback

Cambridge University Press has no responsibility for the persistence or
accuracy of URLs for external or third-party internet websites referred to in
this publication, and does not guarantee that any content on such websites is,
or will remain, accurate or appropriate.

Contents

v

Preface

During the academic year 1987–8 a group of young mathematicians at the University of Bath prepared (for the first time) a pamphlet, *Master of Science in nonlinear mathematics*, that contained the following entry.

PG14 Symmetry and the Maximum Principle

The maximum principle for elliptic operators will be proved from first principles and developed to the extent where the work on symmetry of positive solutions of semi-linear elliptic problems of Gidas, Ni, Nirenberg may be proved.

Naturally, the pamphlet did not state how this goal was to be reached in twenty lectures to students who could not be assumed to have any experience whatever of partial differential equations. Nor were detailed suggestions issued to me when, in the autumn of 1988, I joined the University of Bath and was ordered to give these lectures. What the authors of the pamphlet did do, however, was to attend the lectures themselves, to ask awkward questions, to imbue the course PG14 with their own youthful verve, and to appeal to my vanity by suggesting that I prepare something like the present book.

This explanation should indicate that the word *Introduction* in the title of the book is no gloss. I offer genuine apologies to B. Gidas, W.-M. Ni and L. Nirenberg for the extent to which I have used their paper *Symmetry and related properties via the maximum principle* (1979), to H. Berestycki and L. Nirenberg for my use of the easiest part of *On the method of moving planes and the sliding method* (1991), and to D. Gilbarg and N.S. Trudinger for the extent to which I have lifted theorems from their book *Elliptic partial differential equations of the second order* (1983). But I hope that it will be clear from the following pages that I have not merely copied, and that my aim has been to prepare beginners (and plodders like myself) for the more advanced and abbreviated treatment in those sources.

Theorems of 'Gidas, Ni and Nirenberg type' have their roots in work of Alexandrov (see H. Hopf 1956, p.147) and of Serrin (1971); such theorems establish monotonicity properties and, where the situation allows it, some type of symmetry for solutions of certain elliptic equations that may be slightly non-linear or highly non-linear. The symmetry may be invariance under reflection in a particular hyperplane or may be spherical symmetry. By now there is a profusion of such theorems. This book proceeds to simple, but perhaps basic, results of this kind, mainly for the equation $\triangle u + f(u) = 0$. Chapters 3 and 5 concern positive solutions in a bounded set $\Omega \subset \mathbb{R}^N$, with $u(x) = 0$ on all or part of the boundary $\partial\Omega$. In Chapter 3 the set Ω has either what will be called Steiner symmetry or spherical symmetry; once this has been specified, there is no need for worry about the form of $\partial\Omega$. In Chapter 5 no symmetry of Ω is required; then worry about $\partial\Omega$ is a necessary preliminary to the monotonicity results that are established there for solutions u. In both cases, our main task is patient analysis of the growth of $u(x)$ as the point x is moved inwards from the boundary $\partial\Omega$.

Chapter 4 deals with solutions of $\triangle u + f(u) = 0$ on the whole space $\mathbb{R}^N$; it precedes Chapter 5 because there is no boundary $\partial\Omega$ to cause grief. But neither is there a boundary condition to initiate the analysis; instead, one demands that $u(x)$ have what will be called admissible asymptotic behaviour as $|x| \rightarrow \infty$. Ultimately it emerges that there is a centre q of symmetry, that u depends only on $|x - q|$ and that u decreases as $|x - q|$ increases.

The book is organized as follows. Chapter 0 contains the terminology and apparatus from which later chapters proceed. Appendices A and B are intended to be a leisurely, elementary but serious introduction to methods and results that are basic for elliptic equations. This requires a naive but flexible divergence theorem, which I have tried to provide in Appendix D. Some (by no means all) results in Appendices A and B are used to build comparison functions and examples elsewhere in the book. Because the treatment is to be elementary, there is no whisper of distributions, weak derivatives and Sobolev spaces, but test functions and weak solutions play a central part. For the beginner, Appendices A, B and D may be at least as important as the rest of the book. This material is outside the main text only in deference to the reader who knows everything about the Newtonian potential but knows nothing about proofs of symmetry by means of the maximum principle.

Chapters 1 to 5 proceed from modest beginnings to results like Theorem 4.2 (described above) with only two digressions. The first is a sketch of symmetry, in §1.4, that is short but more general than later chapters demand. The second digression, in §§2.5 to 2.7, concerns Phragmén–Lindelöf theory for subharmonic functions. (The meaning of this phrase is explained at the beginning of §2.5.) These sections are included to show maximum principles at work on questions quite different from symmetry and of importance in their own right. However, §§2.5 *to* 2.7 *are not needed for Chapters* 3 *to* 5.

Appendices C and E contain proofs that would interrupt the story unduly if they were placed in the main text. The subject of Appendix C is construction of the first of three comparison functions of a type introduced by D. Siegel (1988); these are the main tools of the Phragmén–Lindelöf theory in §2.7. The *edge-point lemma* in Appendix E is a version of a maximum principle, previously called the *boundary-point lemma at a corner*, developed by Serrin (1971) and by Gidas, Ni and Nirenberg (1979). The lemma plays an important part in those papers, but has been circumvented in much recent work. It is not used in this book, but would be needed if Chapter 5 were taken further at our level. Moreover, it seems likely that, as a natural step beyond the boundary-point lemma, the edge-point lemma will find applications quite different from those for which it was devised.

The mnemonics C for construction, D for divergence and E for edge-point may be helpful where reference is made to the last three appendices.

Acknowledgements I am heavily indebted to a reviewer, of an early and incomplete draft of the book, who criticized it authoritatively and helpfully. The description of symmetry in §1.4 represents my attempt to meet one of his wishes. Points of detail throughout the text have been improved by his suggestions. However, I have rejected some of the reviewer's advice (for example, that Appendix A be moved into the main text, and I have failed to close the gap (which the reviewer deplored) between general theory and particular applications.

I am much indebted, also, to H. Brezis and to E.N. Dancer for tuition, at different times, on recent work in symmetry by way of the maximum principle; to B. Buffoni for suggesting various improvements, one of which has simplified greatly the proof of Theorem 2.11; to G.R. Burton for instruction in the material of §1.4, for suggesting the phrase *Steiner symmetry* (Definition 3.2) and for contributing Exercise D.15; to G. Keady for the essential idea in Exercise 5.30; to W. Reichel for various references and for attempts to bring me up to date; to G.C. Smith for correction of blunders in Chapter 0 and §1.4; and to J.F. Toland, who not only led me to Exercises 3.18 and 3.19, but also showed extraordinary patience with my questions as to whether I should do this or do that, and made valuable comments on first drafts of several sections.

Three conventions Square brackets [...] enclose explanations that will be obvious to some readers and, strictly speaking, are redundant. However, it may be that not all remarks in square brackets offend knowledgeable readers.

The abbreviated form iff (of if and only if) is used in definitions, despite its vulgarity, because repeated use of 'if and only if' causes undue length, while plain 'if' is dangerous.

The symbol □ marks the end of a proof, of a definition, of an exercise or of any statement that is not italicized and requires separation from the general commentary.

0

Some Notation, Terminology and Basic Calculus

Preamble Let Ω denote, now and henceforth, an open, non-empty subset of the real, N-dimensional, Euclidean space $\mathbb{R}^N$, and let $\overline{\Omega}$ denote its closure. The critical steps in this book will be of the following kind. An inequality

$$w(x) \leq 0 \quad \text{for all} \quad x \in \overline{\Omega},$$

which holds by hypothesis or for obvious reasons, will often be sharpened, by means of more nebulous properties of w and strenuous analysis, to

$$w(x) < 0 \quad \text{for all} \quad x \in \Omega.$$

This improvement may seem too small to be interesting or useful, but the ultimate result of many steps of this type can be astonishing. To prove such fine distinctions, we need careful language, precise notation and some classification of functions and of boundaries $\partial\Omega$ according to their smoothness properties.

It is appropriate, although unexciting, to begin with a list of such items. However, even the conscientious reader is advised merely to glance at the list now, referring to it more carefully later, as the need arises. Moreover, if this preliminary material should suggest that the text demands fluency in Lebesgue integration and sophisticated analysis, then that impression is false. Inspection of Chapter 1 and Appendix A, for example, should show that the reader is assumed to know only fragments of undergraduate material in calculus, linear algebra and analysis. However, a primitive grasp of inequalities between integrals is required for the basic material in Appendices A and B; the reader with no experience of the Hölder and Minkowski inequalities for integrals, and of the Lebesgue spaces $L_p(\Omega)$, should perhaps give special attention to sections (xii) to (xiv) of this preliminary chapter.

(i) **Basic usage** Some symbols in this list are not basic, but are included to prevent confusion with basic ones.

P or Q: at least one of the statements P and Q (possibly both).

Iff: if and only if.

Greater than, positive, increasing, ... have the *strict* meaning (strict inequality: $a > b$), unless inverted commas are added to the word. Thus a real-valued function f defined on the real line is decreasing iff $t > s \Rightarrow f(t) < f(s)$; it is non-increasing (alternatively, 'decreasing') iff $t > s \Rightarrow f(t) \leq f(s)$.

$x, y \in A$: abbreviation for $x \in A$ and $y \in A$.

$A \subset B$ iff $x \in A \Rightarrow x \in B$ (possibly $A = B$; we shall not use $\subseteq$). In words, A is a subset of B; it is a *proper* subset iff $A \subset B$ and $A \neq B$.

$A \setminus B := \{ x \mid x \in A \text{ and } x \notin B \}$.

$A - B := \{ a - b \mid a \in A \text{ and } b \in B \}$ when A and B are subsets of a linear space (or vector space).

$A \times B := \{ (a, b) \mid a \in A \text{ and } b \in B \}$, the *Cartesian product* of the sets A and B.

$\emptyset$: the empty set.

$\mathbb{N} := \{1, 2, 3, \ldots\}$, the set of positive integers or *natural* numbers.

$\mathbb{N}_0 := \{0\} \cup \mathbb{N} = \{0, 1, 2, 3, \ldots\}$, the set of non-negative integers.

$\mathbb{Z} := \{\ldots, -2, -1, 0, 1, 2, 3, \ldots\}$, the set of all integers.

$\mathbb{R}$: the set of real numbers, either as a field or as an inner-product space; for all $s, t \in \mathbb{R}$, the inner product is st and the norm is $|s|$.

$\mathbb{R}^{\#} := \mathbb{R} \cup \{-\infty\} \cup \{\infty\}$, the extended real-number system. The rules postulated for the symbols $-\infty$ and ∞ (such as $t \in \mathbb{R} \Rightarrow t + (-\infty) = -\infty$) are listed by Apostol (1974, p.14), by Rudin (1976, p.11) and in many other texts.

$\mathbb{C}$: the set of complex numbers.

Re, Im: respectively the real part, imaginary part of a complex number or complex-valued function.

(a, b): either an ordered pair (as in the definitions of $A \times B$ and $\mathbb{R}^2$), or the open interval $\{ t \in \mathbb{R} \mid a < t < b \}$; the context is supposed to imply which is meant.

$[a, b), [a, b]$: respectively the half-open interval $\{ t \in \mathbb{R} \mid a \leq t < b \}$, the closed interval $\{ t \in \mathbb{R} \mid a \leq t \leq b \}$.

$[a] := \{ x \mid x \sim a \}$, the equivalence class of a under a specified equivalence relation $\sim$.

(a_n): infinite sequence $(a_n)_{n=1}^{\infty} = (a_1, a_2, \ldots, a_n, \ldots)$, unless the contrary is indicated.

(ii) Real, N-dimensional, Euclidean space

$\mathbb{R}^N$: the Euclidean space of points $x = (x_1, \ldots, x_N)$ with each $x_j \in \mathbb{R}$ and $N \in \{1, 2, 3, \ldots\}$. The inner product is $x \cdot y = x_1 y_1 + \cdots + x_N y_N$ and the norm is $|x| = (x_1^2 + \cdots + x_N^2)^{1/2}$. We write $\mathbb{R}$ for $\mathbb{R}^1$.

$x' := (x_1, \ldots, x_{N-1})$, so that $x = (x', x_N)$.

$x'' := (x_2, \ldots, x_N)$, so that $x = (x_1, x'')$.

$(x^m) := \left((x_1^m, \ldots, x_N^m) \right)_{m=1}^{\infty}$. For a sequence of points in $\mathbb{R}^N$, superscripts are used as mere labels. Where there is a danger of confusion with exponents, or where subscripts label points of a sequence rather than co-ordinates, an explicit remark is made.

Ω: always (and often without explicit mention) an *open* non-empty subset of $\mathbb{R}^N$; we call it a *region* iff it is also *connected*.

$|\Omega|$: the volume, or N-dimensional Lebesgue measure, of Ω.

$|\partial\Omega|$: the surface area (when it exists) of the boundary $\partial\Omega$. See the definitions of ∂A in (iii), and of $dS(x)$ in (viii).

$\mathscr{B}_N(c, \rho) := \{ x \in \mathbb{R}^N \mid |x - c| < \rho \}$, $0 < \rho < \infty$, the ball in $\mathbb{R}^N$ having centre c and radius ρ. The subscript N is used only where dependence on N requires emphasis. See the definition of $\mathscr{B}(c, \rho)$ in (iii).

$\sigma_N := |\partial\mathscr{B}_N(0, 1)| = N\pi^{N/2}/(N/2)!$, the surface area of the unit sphere in $\mathbb{R}^N$. See Exercise 1.14.

(iii) Subsets of a metric space M Here $d : M \times M \to [0, \infty)$ denotes the distance function; A denotes any subset of M.

$\mathscr{B}(c, \rho) := \{ x \in M \mid d(x, c) < \rho \}$, $0 < \rho < \infty$, the *ball* having centre c and radius ρ in the metric space M implied by the context. Note that balls are defined to be *open*, not empty and of finite radius; nevertheless, we often repeat that $\rho > 0$.

int $A := \{ a \in A \mid \exists \, \rho = \rho(a) > 0 \text{ such that } \mathscr{B}(a, \rho) \subset A \}$, the interior of A.

ext $A := \text{int}(M \setminus A)$, the exterior of A.

$\overline{A} := A \cup \{\, y \mid y \text{ is a limit point of } A \,\}$, the *closure* of A. We call y a
   limit point of A iff every punctured ball $\mathscr{B}(y, \varepsilon) \setminus \{y\}$ contains a
   point of A.

$\partial A := \overline{A} \setminus \mathrm{int}\, A$, the *boundary* of A. It follows that $\partial A = M \setminus (\mathrm{int}\, A \cup \mathrm{ext}\, A)$
   and that $\partial A = \overline{A} \cap \overline{(M \setminus A)}$.

Let A and B be non-empty subsets of M.

$\mathrm{diam}\, A := \sup\{\, d(x, y) \mid x, y \in A \,\}$, the *diameter* of A.

$\mathrm{dist}(A, B) := \inf\{\, d(a, b) \mid a \in A \text{ and } b \in B \,\}$, the *distance* between A
   and B.

$\mathrm{dist}(p, B) := \mathrm{dist}(\{p\}, B)$ for any point $p \in M$.

The *characteristic function* χ_A of A is defined by $\chi_A(x) := \begin{cases} 1 & \text{if } x \in A, \\ 0 & \text{if } x \in M \setminus A. \end{cases}$

A and B are *separated* iff $\overline{A} \cap B = \emptyset$ and $A \cap \overline{B} = \emptyset$.

A subset C of M is *connected* iff it is not the union of two non-empty,
   separated subsets.

A *component* of a set $E \subset M$ is a maximal connected subset of E.

(iv) Sets associated with a given function Let A and B be arbitrary sets;
a function (or map or mapping) $f : A \to B$ is a rule that assigns to each
$x \in A$ exactly one value $f(x) \in B$. We call A the *domain* of f and B the
co-domain of f. (More often than not, A will be a subset of $\mathbb{R}^N$ and B
will be $\mathbb{R}$.)

$x \mapsto f(x)$ may be used to introduce particular functions. For example:
$$x \mapsto x^2 + 1 \quad (x \in \mathbb{R}).$$

$f(P) := \{\, f(x) \in B \mid x \in P \subset A \,\}$, the *image of P by f*.

$f^{-1}(Q) := \{\, x \in A \mid f(x) \in Q \subset B \,\}$, the (possibly empty) *inverse image
   of Q* under the mapping f.

$f(A)$ is the *range* of f.

$f^{-1}(\{b\})$, equivalently $\{\, x \in A \mid f(x) = b \in B \,\}$, is a (possibly empty)
   level set of f.

$\mathrm{supp}\, f := \overline{\{\, x \in A \mid f(x) \neq 0 \,\}}$, the *support* of f, provided that A is a
   subset of a metric space and B has a zero element. The overline
   denotes closure in the metric space containing A (closure in $\mathbb{R}^N$
   when $A \subset \mathbb{R}^N$). Thus $\mathrm{supp}\, f$ is the smallest closed set outside
   which f vanishes (equals zero).

$f|_P$, where $P \subset A$ and $P \neq A$, denotes the *restriction* of f to P [the function from P into B that is equal to f on P]. If $g := f|_P$, then $f : A \to B$ is an *extension* of $g : P \to B$. However, statements like 'f is continuous on P' are often preferred to '$f|_P$ is continuous'.

(v) Odds and ends

$\delta_{ij} := \begin{cases} 1 & \text{if } i = j, \\ 0 & \text{if } i \neq j, \end{cases}$ the Kronecker delta. Here $i, j \in \{1, 2, \ldots, N\}$.

$(\delta_{ij}) := (\delta_{ij})_{i,j=1}^N$, the $N \times N$ identity matrix.

Our rule for *matrix multiplication* is summation over adjacent subscripts:

$$(AB)_{ij} = \sum_{p=1}^N A_{ip} B_{pj},$$

$$Ax = \left(\sum_{j=1}^N A_{ij} x_j \right)_{i=1}^N,$$

$$yAx = \sum_{i,j=1}^N y_i A_{ij} x_j,$$

where A and B are $N \times N$ matrices and $x, y \in \mathbb{R}^N$. Thus row and column vectors need not be distinguished.

$t \downarrow a$ iff $t, a \in \mathbb{R}$, $t > a$ and $t \to a$. In words, t tends to a from above, or from the right.

$t \uparrow b$ iff $t, b \in \mathbb{R}$, $t < b$ and $t \to b$. In words, t tends to b from below, or from the left.

For any function f having values in $\mathbb{R}$,

$f^+(x) := \max_{[x \text{ fixed}]} \{f(x), 0\}$ defines the *non-negative* (or 'positive') *part* f^+ of the function f.

$f^-(x) := \min_{[x \text{ fixed}]} \{f(x), 0\}$ defines the *non-positive* (or 'negative') *part* f^- of f.

The *Landau symbols* $O(.)$ ['oh' or 'big oh'] and $o(.)$ ['little oh'] will not be used with great generality, so that the following examples of their meaning should suffice. For functions f, g from a subset of $\mathbb{R}^N$ to $\mathbb{R}^M$,

$f(x) = g(x) + O(r^2 \log r)$ as $r := |x| \to \infty$ iff there exist positive numbers C_1 and C_2 such that $r \geq C_1 \Rightarrow |f(x) - g(x)| \leq C_2 r^2 \log r$.

$f(x) = g(x) + o(r^2 \log r)$ as $r := |x| \to \infty$ iff $|f(x) - g(x)|/r^2 \log r \to 0$.

$f(x) = g(x) + o(1)$ as $x \to c$ iff $|f(x) - g(x)| \to 0$.

$f(x) = g(x) + O(|x - c|^{-1})$ as $x \to c$ iff there exist positive numbers δ and C such that $0 < |x - c| \le \delta \Rightarrow |f(x) - g(x)| \le C/|x - c|$.

$f_H(t) := \begin{cases} 1 & \text{if } t > 0, \\ 0 & \text{if } t \le 0, \end{cases}$ the characteristic function of $(0, \infty) \subset \mathbb{R}$, is the *Heaviside function*.

(vi) **Partial derivatives** Consider points $x \in \Omega$ [where Ω is an open non-empty subset of $\mathbb{R}^N$] and a function $f : \Omega \to \mathbb{R}$.

$e^j := (\delta_{ij})_{i=1}^N = (0, \dots, 0, 1, 0, \dots, 0)$, where 1 is the jth entry.

$\partial_j f(x) = \partial f(x_1, \dots, x_N)/\partial x_j := \lim_{t \to 0} \left(f(x + te^j) - f(x) \right)/t$ whenever the limit exists.

Here ∂_j means differentiation with respect to the jth argument, whatever that argument may be, and $\partial_j f$ is a function in its own right:

$$(\partial_j f)(y + z) = \left. \frac{\partial f(x)}{\partial x_j} \right|_{x = y + z}.$$

$\nabla = \text{grad} := (\partial_1, \dots, \partial_N)$, the *gradient* operator.

$\triangle = \nabla \cdot \nabla = \text{div grad} := \partial_1^2 + \cdots + \partial_N^2$, the *Laplace* operator.

$f_x, f_y, f_r, \dots := \partial f/\partial x, \partial f/\partial y, \partial f/\partial r, \dots$ respectively, when, for example, points of $\mathbb{R}^2$ are denoted by $\mathbf{x} = (x, y) = (r \cos \theta, r \sin \theta)$.

(vii) **Sets of continuous, and of continuously differentiable, functions** Let A be a non-empty subset of $\mathbb{R}^N$; let Ω have its standard meaning.

$C(A) := \{ f : A \to \mathbb{R} \mid f \text{ is continuous on } A \}$.

$C(A, \mathbb{R}^M) := \{ f = (f_1, \dots, f_M) : A \to \mathbb{R}^M \mid \text{each } f_j \text{ is continuous on } A \}$.

$C[a, b], C[a, b), \dots$: abbreviations for $C([a, b]), C([a, b)), \dots$ when $N = 1$.

$C^m(\Omega) := \{ f : \Omega \to \mathbb{R} \mid f \text{ and all partial derivatives } \partial_j f, \partial_i \partial_j f, \dots \text{ of order } \le m \text{ are continuous in } \Omega \}$.

$C^m(\overline{\Omega}) := \{ f : \overline{\Omega} \to \mathbb{R} \mid f \in C(\overline{\Omega}) \cap C^m(\Omega) \text{ and all partial derivatives of order } \le m \text{ have extensions continuous on } \overline{\Omega} \}$. It is necessary to introduce extensions; if $\Omega \subset \mathbb{R}^2$ is the triangular open set in Figure 0.1, and f is defined only on $\overline{\Omega}$, then $(\partial_1 f)(a), (\partial_2 f)(b)$ and $(\partial_2 f)(c)$ cannot be calculated even as the right-hand limit or left-hand limit of the relevant difference quotient, because this quotient is unknown.

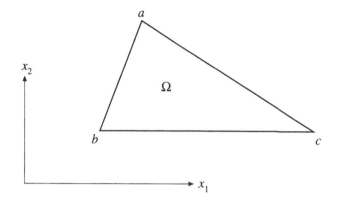

Fig. 0.1.

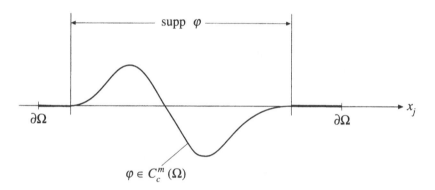

Fig. 0.2.

$C_c^m(\Omega) := \{ \varphi \in C^m(\Omega) \mid \operatorname{supp} \varphi \text{ is compact, } \operatorname{supp} \varphi \subset \Omega \}$. A basic theorem states that a subset of $\mathbb{R}^N$ is compact if and only if it is bounded and closed relative to $\mathbb{R}^N$. If $\varphi \in C_c^m(\Omega)$ and $\Omega \neq \mathbb{R}^N$, then $\operatorname{dist}(\operatorname{supp} \varphi, \partial\Omega) > 0$ [because $\operatorname{supp} \varphi$ is compact, $\partial\Omega$ is closed, and the two are disjoint], as is shown in Figure 0.2.

$C^0(\,.\,) := C(\,.\,)$ and $C_c(\,.\,) := C_c^0(\,.\,)$.

$C^\infty(\,.\,) := \bigcap_{m=0}^\infty C^m(\,.\,)$ and $C_c^\infty(\Omega) := \bigcap_{m=0}^\infty C_c^m(\Omega)$.

(viii) **Boundaries of class** C^k Let $N \in \{2,3,4,\dots\}$ and $k \in \mathbb{N}_0$. The boundary $\partial\Omega$ [of an open non-empty set $\Omega \subset \mathbb{R}^N$] is *of class* C^k iff

(a) $\partial\overline{\Omega} = \partial\Omega$;

(b) for each point $p \in \partial\Omega$, there exist a set $U = U(p)$ open in $\mathbb{R}^N$ and

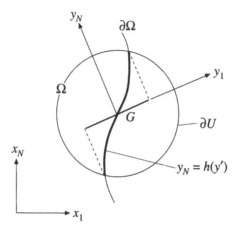

Fig. 0.3.

containing p, 'local' co-ordinates $y' := (y_1, \ldots, y_{N-1})$ and y_N, with $y = 0$ at $x = p$, and a function $h = h(., p)$ such that $\partial\Omega \cap U$ has a representation

$$y_N = h(y'), \qquad y' \in G, \qquad h \in C^k(\overline{G}), \qquad (0.1)$$

where $G = G(p)$ is open in $\mathbb{R}^{N-1}$ and convex (Figure 0.3).

The co-ordinate transformation is $y = A(p)(x - p)$, where $A(p)$ is an orthogonal $N \times N$ matrix depending on p, and $x_1, \ldots, x_N$ are the co-ordinates in terms of which Ω is defined.

Note that condition (a) rules out sets with isolated boundary points, such as $\mathscr{B}(0, 1) \setminus \{0\}$, and sets 'on both sides of their boundary', such as $\mathbb{R}^N \setminus \{x \mid x_N = 0\}$.

If $\partial\Omega$ is [at least] of class C^1, then the *outward unit normal* $n(x) = A(p)^{-1}v(y')$, and the *element* $dS(x) = d\sigma(y')$ *of surface area*, are defined as follows at points $x \in \partial\Omega \cap U(p)$. Let the y_N-axis point *into* Ω; then

$$v(y') := \frac{\big((\partial_1 h)(y'), \ldots, (\partial_{N-1} h)(y'), -1\big)}{\big\{ (\partial_1 h)(y')^2 + \cdots + (\partial_{N-1} h)(y')^2 + 1 \big\}^{1/2}}, \qquad (0.2)$$

$$d\sigma(y') := \frac{dy'}{|v_N(y')|} = \big\{ (\partial_1 h)(y')^2 + \cdots + (\partial_{N-1} h)(y')^2 + 1 \big\}^{1/2} \, dy', \quad (0.3)$$

where the components v_j of v are relative to the y_j-axes, and

$$dy' := dy_1 \, dy_2 \ldots dy_{N-1} .$$

(ix) **Integrals** For a function $f : A \to \mathbb{R}$, where $A \subset \mathbb{R}^N$, *integrable* will mean Lebesgue-integrable, or, in symbols, that $f \in L_1(A)$; the word *measurable*, applied to a subset of $\mathbb{R}^N$ or to a function, will mean Lebesgue-measurable. The notation

$$\int_A f := \int_A f(x)\, dx \qquad (dx := dx_1\, dx_2 \ldots dx_N) \qquad (0.4)$$

is often convenient.

In contrast to many texts, we do *not* admit integrals equal to $\pm\infty$. Therefore the statements

(a) the integral $\int_A f$ exists,
(b) f is integrable over A,
(c) $f \in L_1(A)$,

all have the same meaning. Each implies that

$$-\infty < \int_A f < \infty \quad \text{and} \quad 0 \le \int_A |f| < \infty.$$

For repeated integrals, both notations

$$\int_B g(y)\, dy \int_A f(x,y)\, dx \quad \text{and} \quad \int_B g(y)\left\{ \int_A f(x,y)\, dx \right\} dy$$

will appear. (The second is more logical but can lead, when several integral signs occur, to a plethora of braces, brackets and parentheses or, without them, to confusion.)

(x) **The divergence theorem** We shall often supplement (0.4) by a further abbreviation:

$$\int_\Omega g := \int_\Omega g(x)\, dx, \qquad dx := dx_1\, dx_2 \ldots dx_N,$$

$$\int_{\partial\Omega} g := \int_{\partial\Omega} g(x)\, dS(x), \qquad dS(x) \quad \text{as in (viii)}.$$

The following theorem is proved in Appendix D.

Let Ω be bounded and open in $\mathbb{R}^N$, $N \ge 2$, with $\partial\Omega$ of class C^1, and let $n = (n_1, \ldots, n_N)$ denote the outward unit normal on $\partial\Omega$. Then

$$\int_\Omega \partial_j f = \int_{\partial\Omega} n_j f \qquad (0.5)$$

whenever $f \in C^1(\overline{\Omega})$ and $j \in \{1, \ldots, N\}$.

(a) The theorem is called the *divergence* theorem because, for the specified set Ω, it is equivalent to

$$\int_\Omega \nabla \cdot V = \int_{\partial\Omega} n \cdot V \qquad (0.6)$$

whenever $V \in C^1(\overline{\Omega}, \mathbb{R}^N)$; here the divergence of the vector field V is

$$\operatorname{div} V := \nabla \cdot V = \sum_{j=1}^{N} \partial_j V_j. \tag{0.7}$$

[To derive (0.6) from (0.5), set $f = V_j$ and sum over j; to derive (0.5) from (0.6), set $V_i = f\delta_{ij}$ for fixed j and $i = 1$ to N.]

(b) That $\partial\Omega$ be of class C^1 is sufficient but by no means necessary. In Appendix D, the divergence theorem is extended to the sets listed in Remark D.4; the boundaries of these are not of class C^1. One can also weaken the condition $f \in C^1(\overline{\Omega})$ by means of the approximations Ω_m to a given set Ω that are constructed in Appendix D.

(xi) **Normed linear spaces** This heading occurs only because the norms of certain linear spaces are ingredients of inequalities that we need. No abstract result about normed linear spaces appears in this book. The norms to be encountered are those of $\mathbb{R}^N$ and $\mathscr{L}_p(\Omega)$, introduced in (ii) and (xv) of this chapter, and those of $C_b(\overline{\Omega})$ and $C_b^{0,\lambda}(\overline{\Omega})$, introduced by Definition A.10 (in Appendix A).

Let V be a linear space (or vector space) over the field $\mathbb{R}$ of real numbers. The zero elements of V and of $\mathbb{R}$ will both be denoted by 0. A *semi-norm on V* is a function $\|.\| : V \to \mathbb{R}$ such that, for all $u, v \in V$ and all $\alpha \in \mathbb{R}$,

$$\left.\begin{array}{ll} \|0\| = 0 \text{ and } \|u\| \geq 0 & \text{(non-negativity)}, \\ \|\alpha u\| = |\alpha| \, \|u\| & \text{(positive homogeneity)}, \\ \|u + v\| \leq \|u\| + \|v\| & \text{(triangle inequality)}. \end{array}\right\} \tag{0.8}$$

If also $\|u\| = 0 \Rightarrow u = 0$ (the zero element of V), then $\|.\|$ is a *norm on V*.

The notation $\|. \,|\, V\|$ is used to distinguish the semi-norm or norm of V from that of other linear spaces.

(xii) **The Lebesgue spaces $L_p(\Omega)$**

(a) An *open interval in $\mathbb{R}^N$* is a set of form

$$\begin{aligned} I \; &:= \; (a_1, b_1) \times (a_2, b_2) \times \cdots \times (a_N, b_N) \\ &= \; \{ x \in \mathbb{R}^N \mid a_j < x_j < b_j \quad \text{for} \quad j = 1, \ldots, N \}; \end{aligned}$$

its volume is $|I| := (b_1 - a_1)(b_2 - a_2) \cdots (b_N - a_N)$ if $a_j < b_j$ for each j. If $a_k \geq b_k$ for some k, then I is empty and its volume $|I| = 0$.

A subset S of $\mathbb{R}^N$ has *measure zero* iff, for each $\varepsilon > 0$, there is a sequence of open intervals $I_{\varepsilon,n}$ such that

$$S \subset \bigcup_{n=1}^{\infty} I_{\varepsilon,n} \quad \text{and} \quad \sum_{n=1}^{\infty} |I_{\varepsilon,n}| < \varepsilon.$$

[The nth point of a countable set of points can be covered by an open interval of volume $2^{-n-1}\varepsilon$.]

Almost everywhere (a.e.) in $\mathbb{R}^N$, or in Ω, means : except on a set of measure zero.

A description of the *supremum* of a real-valued function appears in Definition 1.3. If we re-define $f : \Omega \to \mathbb{R}$ on a set of measure zero, then its supremum may change. The *essential supremum* of $f : \Omega \to \mathbb{R}$ is, loosely speaking, the smallest supremum to result from re-definition on a set of measure zero. More precisely,

$$\text{ess sup}_{x \in \Omega} f(x) := \inf \{ \sup_{x \in \Omega} g(x) \mid g = f \text{ a.e. in } \Omega \}. \tag{0.9}$$

(b) Recall the notation (0.4) and the remark after (0.4). Given a set Ω and a number $p \in [1, \infty)$, we define

$$\left. \begin{array}{l} L_p := \{ u : \Omega \to \mathbb{R} \mid u \text{ is measurable}, \int_{\Omega} |u|^p \text{ exists} \}, \\[2mm] \|u\|_p := \left(\int_{\Omega} |u|^p \right)^{1/p} \text{ for all } u \in L_p. \end{array} \right\} \tag{0.10}$$

The semi-normed linear space $L_p(\Omega)$ is the pair $(L_p, \|.\|_p)$; here $p \in [1, \infty)$.

An explanation is in order. The function $\|.\|_p$ is not a norm because the condition $\|u\|_p = 0$ implies only that $u = 0$ a.e. in Ω, not that u is the zero function [which equals zero everywhere]. However, $\|.\|_p$ satisfies (0.8) for all $u, v \in L_p$ and all $\alpha \in \mathbb{R}$; that it satisfies the triangle inequality will be shown in (xiii). Therefore, if addition and scalar multiplication of elements of L_p are defined by

$$(u+v)(x) := u(x)+v(x) \text{ and } (\alpha u)(x) := \alpha u(x) \text{ for all } x \in \Omega \text{ and } \alpha \in \mathbb{R},$$

then L_p becomes a linear space over $\mathbb{R}$.

(c) *The semi-normed linear space $L_\infty(\Omega)$ is the pair $(L_\infty, \|.\|_\infty)$, where*

$$\left. \begin{array}{l} L_\infty := \{ u : \Omega \to \mathbb{R} \mid u \text{ is measurable}, \ \text{ess sup}_{x \in \Omega} |u(x)| < \infty \}, \\[2mm] \|u\|_\infty := \text{ess sup}_{x \in \Omega} |u(x)| \text{ for all } u \in L_\infty. \end{array} \right\}$$
$$\tag{0.11}$$

The explanation is much as before. The reason for the notation is that, if Ω has finite volume $(|\Omega| < \infty)$ and $u \in L_\infty(\Omega)$, then $\|u\|_p \to \|u\|_\infty$ as $p \to \infty$.

(d) *Caution* In (xv) we shall meet normed linear spaces $\mathscr{L}_p(\Omega)$ and

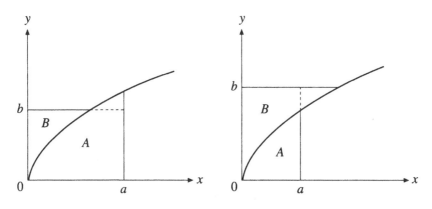

Fig. 0.4.

$\mathscr{L}_\infty(\Omega)$ that resemble $L_p(\Omega)$ and $L_\infty(\Omega)$, respectively. Even the most precise and respected sources (Rudin 1970, pp. 65–66; Weir 1973, pp. 165–166) blur the distinctions between $L_p(\Omega)$ and $\mathscr{L}_p(\Omega)$ and between $L_\infty(\Omega)$ and $\mathscr{L}_\infty(\Omega)$. We shall follow this practice: for example, $\|u\|_p$ (also written $\|u \mid L_p(\Omega)\|$) *will be called, quite incorrectly, the L_p norm of the function u.*

(xiii) **The Hölder and Minkowski inequalities for integrals**

The statement $1 \leq p \leq \infty$ or $p \in [1, \infty]$ is to mean that $p \in \mathbb{R}^{\#}$ and $p \geq 1$. If $p, q \in [1, \infty]$ and $1/p + 1/q = 1$, then q is the *Hölder conjugate* of p, and p is the Hölder conjugate of q. The pairs $(1, \infty), (\frac{5}{4}, 5), (\frac{3}{2}, 3)$ and $(2, 2)$ are examples of such conjugates.

By uv we mean here the pointwise product: $(uv)(x) := u(x)v(x)$.

Young's inequality For real numbers $a, b \in [0, \infty)$,

$$ ab \leq \frac{a^p}{p} + \frac{b^q}{q} \qquad \left(\frac{1}{p} + \frac{1}{q} = 1, \ \ 1 < p < \infty \right). \qquad (0.12) $$

Proof Let the curve in Figure 0.4 have equation $y = x^{p-1}$ or, equivalently, $x = y^{q-1}$; then

$$ ab \leq |A| + |B| = \int_0^a x^{p-1} \, \mathrm{d}x + \int_0^b y^{q-1} \, \mathrm{d}y = \frac{a^p}{p} + \frac{b^q}{q}. $$

$\square$

Hölder's inequality If $u \in L_p(\Omega)$ and $v \in L_q(\Omega)$, where p and q are Hölder conjugates, then $uv \in L_1(\Omega)$ and

$$\int_\Omega |uv| \leq \|u\|_p \|v\|_q. \tag{0.13}$$

Proof The function uv is measurable because u and v are measurable, by hypothesis; hence the inequality will establish that $uv \in L_1(\Omega)$. If $p = 1$, we have

$$\int_\Omega |uv| \leq \int_\Omega |u| \, \|v\|_\infty = \|v\|_\infty \int_\Omega |u|,$$

as desired. If $p = \infty$, we interchange u and v. Suppose then that $p \in (1, \infty)$; also that $\|u\|_p > 0$ and $\|v\|_q > 0$, otherwise the inequality is obvious. By Young's inequality,

$$\int_\Omega \left| \frac{u}{\|u\|_p} \frac{v}{\|v\|_q} \right| \leq \int_\Omega \left\{ \frac{1}{p} \frac{|u|^p}{(\|u\|_p)^p} + \frac{1}{q} \frac{|v|^q}{(\|v\|_q)^q} \right\} = \frac{1}{p} + \frac{1}{q} = 1.$$

$\square$

The case $p = q = 2$ of Hölder's inequality is often called the *Schwarz inequality* and sometimes *Buniakowsky's inequality*. We are now in a position to prove the *triangle inequality of* $L_p(\Omega)$; this is invariably attributed to Minkowski, who proved its counterpart for finite sums.

Minkowski's inequality Let $p \in [1, \infty]$. If $u, v \in L_p(\Omega)$, then $u+v \in L_p(\Omega)$ and

$$\|u + v\|_p \leq \|u\|_p + \|v\|_p. \tag{0.14}$$

Proof Since u and v are measurable, so is the function $u + v$. The cases $p = 1$ and $p = \infty$ of the inequality are immediate:

$$\int_\Omega |u + v| \leq \int_\Omega (|u| + |v|) = \int_\Omega |u| + \int_\Omega |v|,$$

and

$$\|u + v\|_\infty \leq \big\| \, |u| + \|v\|_\infty \, \big\|_\infty = \|u\|_\infty + \|v\|_\infty.$$

Therefore suppose that $p \in (1, \infty)$; also that $\|u+v\|_p > 0$, otherwise there is nothing to prove. We decompose $|u+v|^p$ and apply Hölder's inequality:

$$\int_\Omega |u + v|^p \leq \int_\Omega |u| \, |u + v|^{p-1} + \int_\Omega |v| \, |u + v|^{p-1}$$

$$\leq \|u\|_p \left(\int_\Omega |u + v|^p \right)^{1/q} + \|v\|_p \left(\int_\Omega |u + v|^p \right)^{1/q}, \quad q = \frac{p}{p - 1}.$$

Dividing both sides by $\left(\int_\Omega |u+v|^p\right)^{1/q}$, and observing that $1-1/q = 1/p$, we obtain (0.14).   □

(xiv) Three implications of the Hölder and Minkowski inequalities

(a) *An extended Minkowski inequality* It follows from (0.14) that, if $u_n \in L_p(\Omega)$ for $n = 1, \ldots, k$, then

$$\left\| \sum_{n=1}^{k} u_n \right\|_p \leq \sum_{n=1}^{k} \|u_n\|_p. \tag{0.15}$$

Now consider functions $u(.,y)$ in $L_p(\Omega)$ that depend on a parameter $y \in Y$, where Y is an open subset of $\mathbb{R}^M$. Replacing the summation over n in (0.15) by the operation

$$\int_Y . \, \alpha(y) \, \mathrm{d}y, \qquad \text{where } \alpha(y) \geq 0 \text{ for all } y \in Y,$$

we expect the result

$$\left\| \int_Y u(.,y) \, \alpha(y) \, \mathrm{d}y \right\|_p \leq \int_Y \|u(.,y)\|_p \, \alpha(y) \, \mathrm{d}y. \tag{0.16}$$

Claim If $u(.,y) \in L_p(\Omega)$ for all $y \in Y$ and for some $p \in [1,\infty]$, then (0.16) holds, provided that $u\alpha$ is measurable on $\Omega \times Y$ and the last integral in (0.16) exists.

Proof We note first that

$$\left\| \int_Y u(.,y) \, \alpha(y) \, \mathrm{d}y \right\|_p \leq \left\| \int_Y |u(.,y)| \, \alpha(y) \, \mathrm{d}y \right\|_p,$$

and define

$$U(x) := \int_Y |u(x,y)| \, \alpha(y) \, \mathrm{d}y, \qquad x \in \Omega.$$

Suppose that $p \in (1,\infty)$ and $\|U\|_p > 0$; the other cases follow readily once the inequality has been proved for this case. Since Fubini's theorem allows us to change the order of integration, we may imitate the proof of (0.14). In fact,

$$\int_\Omega \left\{ \int_Y |u(x,y)| \, \alpha(y) \, \mathrm{d}y \right\}^p \mathrm{d}x = \int_\Omega \left\{ \int_Y |u(x,y)| \, \alpha(y) \, \mathrm{d}y \right\} U(x)^{p-1} \, \mathrm{d}x$$

$$= \int_Y \left\{ \int_\Omega |u(x,y)| \, U(x)^{p-1} \, \mathrm{d}x \right\} \alpha(y) \, \mathrm{d}y$$

$$\leq \int_Y \|u(.,y)\|_p \, \|U^{p-1}\|_q \, \alpha(y) \, \mathrm{d}y \qquad \left(\frac{1}{p} + \frac{1}{q} = 1 \right),$$

by the Hölder inequality. The left-hand member is $\int_\Omega U^p$; on the right, we have

$$\| U^{p-1} \|_q = \left\{ \int_\Omega U^p \right\}^{1/q} ;$$

dividing both sides by this last, we obtain (0.16).   □

(b) *Interpolation between L_p spaces*   If $u \in L_a(\Omega)$ and $u \in L_c(\Omega)$, where $1 \leq a < c \leq \infty$, and if $1/b = (1-\theta)/a + \theta/c$ for some $\theta \in (0,1)$, then $u \in L_b(\Omega)$ and

$$\| u \|_b \leq (\| u \|_a)^{1-\theta} (\| u \|_c)^\theta. \tag{0.17}$$

Proof If $c = \infty$, we have

$$\left(\int_\Omega |u|^b \right)^{1/b} \leq \left(\int_\Omega |u|^a \, \| u \|_\infty^{b-a} \right)^{1/b} = \| u \|_a^{a/b} \, \| u \|_\infty^{1-a/b},$$

where $a/b = 1 - \theta$. If $c < \infty$, write $|u|^b$ as the product of $|u|^{b(1-\theta)}$ and $|u|^{b\theta}$; the Hölder inequality gives

$$\int_\Omega |u|^b \leq \left(\int_\Omega |u|^{pb(1-\theta)} \right)^{1/p} \left(\int_\Omega |u|^{qb\theta} \right)^{1/q} \quad \left(\frac{1}{p} + \frac{1}{q} = 1 \right).$$

Choose p so that $pb(1 - \theta) = a$; the definition of b shows that $p \in (1, \infty)$. A further calculation shows that $qb\theta = c$; then (0.17) follows.   □

(c) *The case of finite volume $|\Omega|$*   If $u \in L_p(\Omega)$ for some $p > 1$, and $|\Omega| < \infty$, then

$$\| u \|_1 \leq \| u \|_p \, |\Omega|^{1/q} \quad \left(\frac{1}{p} + \frac{1}{q} = 1 \right), \tag{0.18}$$

and $u \in L_s(\Omega)$ for $1 \leq s \leq p$.

Proof To derive (0.18), write u as the product of u and 1, then apply the Hölder inequality. The rest follows by interpolation between $L_1(\Omega)$ and $L_p(\Omega)$, as in (0.17).   □

(xv) **The normed linear spaces $\mathscr{L}_p(\Omega)$**   Let $p \in [1, \infty]$ and let L_p be as in (0.10) or (0.11). In order to convert $\| . \|_p$ into a norm, we proceed as follows. If $u \in L_p$, the *equivalence class* $[u]$ of u is the set of measurable functions $\tilde{u} : \Omega \to \mathbb{R}$ such that $\tilde{u} = u$ a.e. in Ω. Then $\| \tilde{u} \|_p = \| u \|_p$ whenever $\tilde{u} \in [u]$.

The elements of the *normed linear space* $\mathscr{L}_p(\Omega)$ are the equivalence classes $[u], [v], \ldots$ of all functions in L_p. The zero element is $[0]$, where 0 continues to denote the zero function, and

$$\alpha[u] + [v] := [\alpha u + v] \qquad \text{for all} \quad \alpha \in \mathbb{R} \quad \text{and all} \quad [u], [v] \in \mathcal{L}_p(\Omega).$$

The norm is defined by $\left\| [u] \mid \mathcal{L}_p(\Omega) \right\| := \|u\|_p$; then the condition $\left\| [u] \mid \mathcal{L}_p(\Omega) \right\| = 0$ implies that $[u] = [0]$.

1

Introduction

1.1 A glimpse of objectives

Our goal is to prove results like Theorems 3.3, 4.2, 5.12 and their various corollaries. The tools that we develop on the way have other uses, but it is desirable to work towards specific ends. Although the theorems in Chapters 3 to 5 are simple examples of their kind, they are too beset with ifs and buts to be described at this stage; here we observe two particular consequences of those theorems. The first concerns the Newtonian potential of constant density in a bounded subset of $\mathbb{R}^N$. Such potential functions, without the restriction to constant density, are important in physics and not without interest in mathematics.

Consider then the Newtonian potential u of constant density $\lambda > 0$ housed in a bounded open set $G \subset \mathbb{R}^N$; we write

$$u(x) := \begin{cases} -\dfrac{\lambda}{2} \displaystyle\int_G |x - \xi| \, \mathrm{d}\xi & \text{if } N = 1, \\[2ex] \dfrac{\lambda}{2\pi} \displaystyle\int_G \log \dfrac{1}{|x - \xi|} \, \mathrm{d}\xi & \text{if } N = 2, \\[2ex] \lambda \kappa_N \displaystyle\int_G \dfrac{1}{|x - \xi|^{N-2}} \, \mathrm{d}\xi & \text{if } N \geq 3, \end{cases} \tag{1.1}$$

where $x \in \mathbb{R}^N$, $\mathrm{d}\xi = \mathrm{d}\xi_1 \, \mathrm{d}\xi_2 \ldots \mathrm{d}\xi_N$ and κ_N is a positive normalization constant; in particular, $\kappa_3 = 1/4\pi$. The important facts about u are these.

(a) $u \in C^1(\mathbb{R}^N)$. For $N = 3$, we may interpret $m\nabla u(x)$ as proportional to the force on a test particle, having mass m and located at x, due to mass density λ (mass per volume) in the set G, which might be a naive model of some collection of heavenly bodies. Then the C^1 property of u implies that, if the test particle is moved a small distance from

17

the boundary ∂G into $\mathbb{R}^3 \setminus \overline{G}$, the gravitational force on the particle is changed only slightly.

(b) $u \in C^2(\mathbb{R}^N \setminus \partial G)$ and

$$\Delta u := (\partial_1^2 + \cdots + \partial_N^2)u = \begin{cases} -\lambda & \text{in } G, \\ 0 & \text{in } \mathbb{R}^N \setminus \overline{G}. \end{cases} \tag{1.2}$$

No relationship between the potential u and the density λ could be more pleasing than this. However, our first theorem concerns not a given set G, but the question of finding the shape of G when we are told that u is constant on ∂G; hence (1.2) can be used only indirectly in this search.

Note that G need not be connected and that ∂G need not be smooth; we are given only that G is bounded and open.

Theorem 1.1 *If the potential u in (1.1) is constant on the boundary ∂G, then G is a ball.*

For $N = 1$, this result follows directly from the definition (1.1) (Exercise 1.17), but for higher dimensions more subtle arguments are needed and the theorem seems a little surprising. For suppose that $N \geq 3$ and that G consists of two balls far apart: say $G = \mathscr{B}(c, 1) \cup \mathscr{B}(-c, 1)$, where $|c|$ is large and $\mathscr{B}(c, \rho)$ denotes the ball with centre c and radius ρ. Then a calculation shows that u is nearly constant on ∂G:

$$u\big|_{\partial G}(x) = \frac{\lambda}{N(N-2)} \left\{ 1 + (2|c|)^{-N+2} + |c|^{-N+1}v(x, c) \right\},$$

where $v(x, c)$ is bounded independently of $x \in \partial G$ and c for $|c| \geq 2$. Therefore it might seem likely that, by perturbing ∂G, one could construct a pair of nearly spherical bodies for which the potential is exactly constant on the boundary when $|c|$ is sufficiently large. The theorem shows that this cannot be done.

Although Theorem 1.1 establishes the shape of a set, we shall prove it by means of results for solutions on $\mathbb{R}^N$ of a non-linear elliptic equation of order two. A corresponding problem for the same equation on a *bounded* set Ω (open as always) is that of finding a function $u \in C^2(\overline{\Omega})$ such that

$$\left.\begin{aligned} \Delta u + f(u) &= 0 \quad \text{in } \Omega, \\ u\big|_{\partial\Omega} &= 0, \end{aligned}\right\} \tag{1.3}$$

where $f : \mathbb{R} \to \mathbb{R}$ is a given function in $C^1(\mathbb{R})$, say. Conclusions about the set Ω are again an important part of the results of Gidas, Ni and Nirenberg, but the theory also establishes properties of u when it is given that Ω is a ball. The next theorem is valuable because, for the general problem (1.3), existence of many solutions (or of none) tends to be the rule rather than the exception.

Theorem 1.2 *Suppose that in* (1.3) *the set* Ω *is a ball, say* $\Omega = \mathscr{B}(0,a) \subset \mathbb{R}^N$, *and that* u *is a positive solution:* $u(x) > 0$ *for all* $x \in \Omega$.

   Then u *is spherically symmetric: it depends only on* $r := |x|$, *and* $du/dr < 0$ *for* $0 < r < a$.

This theorem does not imply the existence or uniqueness of positive solutions in a ball. However, these questions can now be explored by seeking the *decreasing* positive solutions of an *ordinary* differential equation. Whether such exploration is easy, difficult or impossible depends on the nature of the function f.

If v is a negative solution of (1.3), with $\Omega = \mathscr{B}(0,a)$, then we apply Theorem 1.2 to $-v$. But if u is a solution that has both positive and negative values, then spherical symmetry of u cannot be inferred. This is shown by the eigenfunctions u_m, satisfying $u_m|_{\partial\Omega} = 0$, of the operator $-\triangle$ in the ball $\Omega = \mathscr{B}(0,a)$. For $N = 1$, these are

$$u_m(x) = c \sin \frac{m\pi(x+a)}{2a}, \qquad -a \le x \le a, \ c \in \mathbb{R} \setminus \{0\}, \atop m \in \{1,2,3,\ldots\}. \right\} \tag{1.4}$$

Thus $u_m \in C^{\infty}(\overline{\Omega})$ and

$$u_m'' + \left(\frac{m\pi}{2a}\right)^2 u_m = 0 \ \text{ in } \ \Omega, \qquad u_m|_{\partial\Omega} = 0.$$

Each of these differential equations has the form of that in (1.3) [with $f(t) = (m\pi/2a)^2 t$ for all $t \in \mathbb{R}$]. If m is even, then u_m is an odd function and hence not spherically symmetric.

The corresponding functions for $N = 2$ and $N = 3$ are displayed in Exercises 1.19 and 1.21.

1.2 What are maximum principles?

This question is relevant here because maximum principles form the subject of Chapter 2, and because these principles and the method of reflection in hyperplanes will be our main tools in Chapters 3 to 5. No short and accurate answer to the question seems possible; a vague answer will be given presently, but first there arises a more basic question. What do we mean by the maximum of a function?

Definition 1.3 Consider a function $f : A \to \mathbb{R}$, where A is a subset of a metric space.

(i) f has a (global) *maximum* iff there is a point $p \in A$ such that $f(x) \le f(p)$ whenever $x \in A$, and then $\max_{x \in A} f(x) := f(p)$.

(ii) f has a *local maximum* at $q \in A$ iff there is a number $\delta > 0$ such that $f(x) \le f(q)$ whenever $x \in \mathscr{B}(q, \delta) \cap A$.

(iii) The *supremum* $\sup_{x \in A} f(x)$ (which always exists) is the unique element s of the extended set $\mathbb{R}^{\#}$ such that

   (a) $f(x) \le s$ whenever $x \in A$,
   (b) if $t < s$, then there is a point $y \in A$ at which $f(y) > t$.

(iv) $\min_{x \in A} f(x) := -\max_{x \in A}\{-f(x)\}$, $\inf_{x \in A} f(x) := -\sup_{x \in A}\{-f(x)\}$.

   □

According to this definition, a maximum is a supremum that is *attained* at some point. To say that a maximum is attained (as is often said in the literature) states nothing beyond our definition; to say that a maximum is attained at a specified point, or in a specified subset, may be useful information; to say that a supremum is attained is always cause for rejoicing, because then the function has a maximum. [For example, the functions from $\mathbb{R}$ into $\mathbb{R}$ defined by

$$f(x) = x, \qquad g(x) = \begin{cases} 0 & \text{if } x \le 0, \\ e^{-x} & \text{if } x > 0, \end{cases}$$

have no maximum.]

Item (iv) of Definition 1.3 is somewhat indirect, but is useful for passing from a theorem about maxima to one about minima; no doubt the reader can supply an equivalent statement that is more direct.

Notation We shall often use abbreviations like

$$\max_A f, \quad \text{in place of} \quad \max_{x \in A} f(x), \tag{1.5a}$$

$$f < 0 \text{ in } A, \text{ in place of } f(x) < 0 \text{ for all } x \in A. \tag{1.5b}$$

If $B \subset A$, then $\max_B f$ and $\sup_B f$ refer to the restriction of f to B [and thus mean $\max_B(f|_B)$ and $\sup_B(f|_B)$, respectively].

Exercise 1.4 Prove the following for functions $g : \overline{W} \to \mathbb{R}$, where $W \subset \mathbb{R}^N$.

(a) If $U \subset V \subset \overline{W}$, then $\sup_U g \le \sup_V g$.
(b) If $g \in C(\overline{W})$, then $\sup_W g = \sup_{\overline{W}} g$.

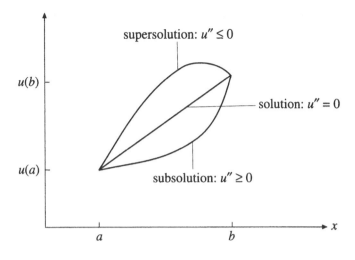

Fig. 1.1.

(c) If $g \in C(\overline{W})$ and W is bounded, then

$$\sup_W g = \sup_{\overline{W}} g = \max_{\overline{W}} g.$$

□

A loose description of maximum principles is that they are refinements and generalizations of the simple observations that we now make in Remarks 1.5 and 1.6. The extent and effectiveness of such improvements of our present remarks will emerge in later chapters.

Consider first the operator $(\,.\,)'' = d^2/dx^2$, which we regard as the Laplace operator in one dimension; an aspect of this operator is shown in Figure 1.1. The words *subsolution*, *supersolution* and *solution* will be defined precisely in Chapter 2 for a more general setting; here they refer to the differential equation $u'' = 0$, and the figure is sufficient indication. It is worth emphasis that, even in the more general situations to follow, *uncertainty about the direction of an inequality* ($\geq$ *or* $\leq$?) *can be resolved by visualizing Figure* 1.1. Remark 1.5 merely records three implications of the figure. Since these cannot possibly be false under the strong condition that $u \in C^2[a, b]$, the proof is indecently long; but it is a direct calculation, whereas later proofs will require tricks.

Remark 1.5 *If $u \in C^2[a, b]$ and $u'' \geq 0$ on $[a, b]$, then the following hold.*

(i) *Either $u(x) < \max\{u(a), u(b)\}$ for all $x \in (a, b)$, or u is constant on $[a, b]$.*

(ii) *If $u(b) \geq u(a)$, then either the outward derivative $u'(b) > 0$, or u is constant on $[a, b]$.*

(iii) *If $u(a) > u(b)$, then the outward derivative $-u'(a) > 0$.*

Proof Integration by parts gives

$$
\begin{aligned}
u(x) &= u(a) + \int_a^x u'(t)\, dt \\
&= u(a) + u'(a)(x - a) + \int_a^x u''(t)(x - t)\, dt, \quad a \leq x \leq b. \quad (1.6)
\end{aligned}
$$

To eliminate the 'unknown' $u'(a)$ in favour of the 'data' $u(a)$, $u(b)$ and $u'' \geq 0$, set $x = b$; then

$$
u'(a) = \frac{1}{b - a} \left\{ u(b) - u(a) - \int_a^b (b - t) u''(t)\, dt \right\}. \quad (1.7)
$$

Substituting (1.7) into (1.6), and tidying the result, one obtains

$$
u(x) = v(x) - \int_a^b G(x, t)\, u''(t)\, dt \quad (a \leq x \leq b), \quad (1.8)
$$

where v is the solution of $v'' = 0$ given by

$$
v(x) := u(a) + \frac{u(b) - u(a)}{b - a}(x - a),
$$

and, on the square $[a, b] \times [a, b]$ (Figure 1.2),

$$
G(x, t) := \begin{cases} (b - x)(t - a)/(b - a) & \text{if } x \geq t, \\ (x - a)(b - t)/(b - a) & \text{if } x \leq t. \end{cases}
$$

Thus $G > 0$ in $(a, b) \times (a, b)$ and $G = 0$ on the boundary of the square. Since also $u'' \geq 0$ on $[a, b]$, it follows from (1.8) that

$$
u(x) \leq v(x) \quad (a \leq x \leq b), \quad (1.9)
$$

with strict inequality for $x \in (a, b)$ if $u''(x_0) > 0$ at some point $x_0 \in [a, b]$, because then $u'' > 0$ in an interval of positive length, by the continuity of u''.

(i) To prove assertion (i) of the Remark, we consider three cases.

(1) If $u(a) \neq u(b)$, then $u(x) \leq v(x) < \max\{u(a), u(b)\}$ for $x \in (a, b)$.

(2) If $u(a) = u(b)$ and $u''(x_0) > 0$ at some $x_0 \in [a, b]$, then, by the remark following (1.9), $u(x) < v(x) = u(a)$ for $x \in (a, b)$.

(3) If $u(a) = u(b)$ and $u'' = 0$ on $[a, b]$, then $u(x) = v(x) = u(a)$ for $x \in [a, b]$.

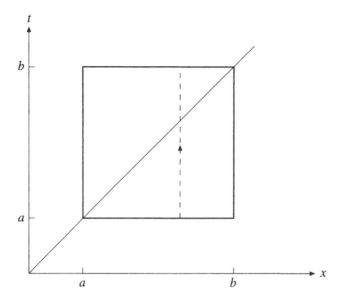

Fig. 1.2.

(ii) Differentiating (1.8) and setting $x = b$, or else forming the analogue of (1.7) for $u'(b)$, we obtain

$$u'(b) = \frac{1}{b-a}\left\{u(b) - u(a) + \int_a^b (t - a)\, u''(t)\, \mathrm{d}t\right\} \geq 0$$
$$\text{when } u(b) \geq u(a),$$

with equality only for case (3).

(iii) If $u(a) > u(b)$, then $u'(a) < 0$ by (1.7).

□

The next observation is almost trivial because it is the implication of a *strict* differential inequality. Nevertheless, a slight variant of this remark will play a part in the proof of Theorem 2.5.

Remark 1.6 *If $u \in C^2(\Omega)$ and $\triangle u > 0$ in Ω, then u cannot have a local (let alone a global) maximum at a point of Ω.*

Proof Assume (for contradiction) that u has a local maximum at $q \in \Omega$. Then for each $j \in \{1,\dots,N\}$ we have $(\partial_j u)(q) = 0$ and $(\partial_j^2 u)(q) \leq 0$. Hence $(\triangle u)(q) \leq 0$.

□

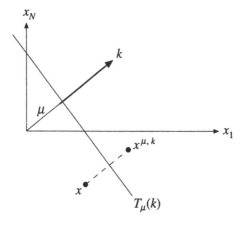

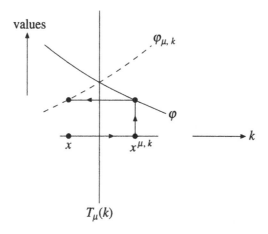

Fig. 1.3.

1.3 On reflection in hyperplanes

With this method, as with maximum principles, the proof of the pudding
will be in the eating. However, we can take a first step in this section, and
look ahead a little. Figure 1.3 illustrates the objects in the next definition;
the notation will be shortened when we come to use it in earnest.

Definition 1.7 For each unit vector $k \in \mathbb{R}^N$ and each number $\mu \in \mathbb{R}$,
define a hyperplane by

$$T_\mu(k) := \left\{ \xi \in \mathbb{R}^N \mid \xi \cdot k = \mu \right\} \qquad (|k| = 1).$$

The reflection in $T_\mu(k)$ of any point $x \in \mathbb{R}^N$ is

$$x^{\mu,k} := x + 2(\mu - x \cdot k)k,$$

and the reflection in $T_\mu(k)$ of any function $\varphi : \mathbb{R}^N \to \mathbb{R}$ is defined by

$$\varphi_{\mu,k}(x) := \varphi(x^{\mu,k}) \qquad \text{for all } x \in \mathbb{R}^N.$$

□

The reader should verify that $(x^{\mu,k})^{\mu,k} = x$.

There is no sign convention in the definition of hyperplanes, so that $T_{-\mu}(-k) = T_\mu(k)$. If φ is defined only on a proper subset A of $\mathbb{R}^N$, then we define $\varphi_{\mu,k}$ only on $\{x \in \mathbb{R}^N \mid x^{\mu,k} \in A\}$, which is the reflection in $T_\mu(k)$ of A.

Lemma 1.8 *If $v : \mathbb{R}^N \to \mathbb{R}$ has the property that $v(x^{0,k}) = v(x)$ for each unit vector k and all $x \in \mathbb{R}^N$, then v is spherically symmetric (depends only on $|x|$).*

Proof We need prove only that, for any two points y and z satisfying $|y| = |z|$ and $y \neq z$, we have $v(y) = v(z)$. Choose $k = (z - y)/|z - y|$. If $z = y^{0,k}$, as Figure 1.4 suggests, then

$$v(z) = v(y^{0,k}) = v(y),$$

as desired.

To check that $z = y^{0,k}$, we have

$$
\begin{aligned}
y^{0,k} &= y + 2(0 - y \cdot k)k = y - 2\left(y \cdot \frac{z-y}{|z-y|}\right)\frac{z-y}{|z-y|}, \\
y^{0,k} - z &= y - z + 2(y-z)\,y \cdot \frac{z-y}{|z-y|^2} \\
&= \frac{y-z}{|z-y|^2}\left\{\left(|z|^2 - 2z \cdot y + |y|^2\right) + 2(y \cdot z - |y|^2)\right\} \\
&= 0 \qquad \text{because } |y| = |z|.
\end{aligned}
$$

□

For a preliminary sketch of the method of reflection in hyperplanes, we return to Theorem 1.2 and consider a function u such that

$$\Delta u + f(u) = 0 \quad \text{in } \Omega = \mathscr{B}(0, a),$$

$$u\big|_{\partial\Omega} = 0, \qquad u\big|_\Omega > 0, \qquad u \in C^2(\overline{\Omega}).$$

Our aim is to prove the hypothesis of Lemma 1.8, adapted to the ball Ω: that $u(x^{0,k}) = u(x)$ for each unit vector k and all $x \in \Omega$. To this end,

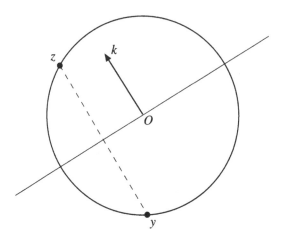

Fig. 1.4.

we *fix* the unit vector k and contemplate all positions of the hyperplane $T_\mu(k)$ from $\mu = a$ to $\mu = 0$ (Figure 1.5). Let

$$Z(\mu,k) := \left\{ z \in \Omega \mid z \cdot k > \mu \right\}.$$

The procedure is to prove that (for fixed but arbitrary k)

$$w(z,\mu) := u(z) - u\left(z^{\mu,k}\right) < 0 \qquad \text{if } z \in Z(\mu,k) \qquad (*)$$

whenever $\mu \in (0,a)$. Once this has been established, a continuity argument for $\mu \downarrow 0$ shows that $w(z,0) \le 0$ if $z \in Z(0,k)$; in other words, that

$$u(x) \le u\left(x^{0,k}\right) \quad \text{if } x \cdot k > 0 \quad \text{and } x \in \Omega.$$

But the same result for the direction $-k$ implies that

$$u\left(x^{0,k}\right) \le u(x) \quad \text{if } x \cdot k > 0 \quad \text{and } x \in \Omega.$$

These two inequalities yield the desired result that $u\left(x^{0,k}\right) = u(x)$ for all $x \in \Omega$ [since, if $x \cdot k = 0$, we have $x^{0,k} = x$, while points x with $x \cdot k < 0$ are points $y^{0,k}$ with $y \cdot k > 0$].

That $du/dr < 0$ for $0 < r < a$ is an easy consequence of $(*)$ for $0 < \mu < a$ and of a maximum principle called the *boundary-point lemma for balls*.

The proof of $(*)$ for all $\mu \in (0,a)$ proceeds in two stages. The first is for small, positive values $a - \mu$ and begins with application to w of a *maximum principle for thin sets*; this is possible because w satisfies an elliptic equation of order two and suitable boundary conditions. The

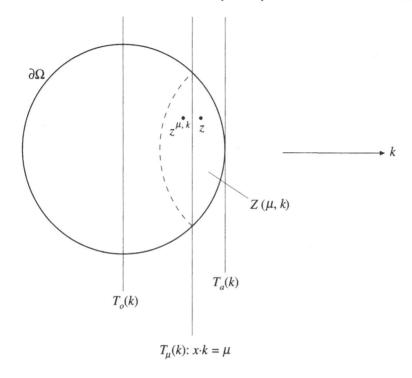

$$T_\mu(k): x \cdot k = \mu$$

Fig. 1.5.

second stage is a demonstration that, once (∗) holds in an interval (μ_0, a), it can fail only as $\mu \downarrow 0$. For equations less simple, or functions f significantly less smooth, than those in Theorem 1.2, the second stage requires tenacity and use of a substantial tool-kit.

This sketch of the method for the case of a ball serves also for any bounded set Ω that has appropriate symmetry under reflection in a hyperplane. But when the equation $\triangle u + f(u) = 0$ holds in the whole space $\mathbb{R}^N$, and no boundary condition is available, one must first locate a possible centre of symmetry by means of information about $u(x)$ for $|x| \to \infty$, and then pursue an analogue of (∗) in which $Z(\mu, k)$ is a half-space. Again there are two stages, but in the first stage one now labours to organize implications of the asymptotic behaviour of $u(x)$ for $|x| \to \infty$.

1.4 What is symmetry?

A small answer to this large question will be sufficient for our purpose, because the symmetries that will be established in Chapters 3 to 5 are simple enough to require no reference to any general theory. Nevertheless, some indication of a wider viewpoint seems in order.

By an *isometry of* $\mathbb{R}^N$ we mean a function $f : \mathbb{R}^N \to \mathbb{R}^N$ such that

$$|f(x) - f(y)| = |x - y| \qquad \text{for all} \quad x, y \in \mathbb{R}^N. \tag{1.10}$$

This definition ensures that f is injective [one-to-one: $f(x) = f(y) \Rightarrow x = y$], but does not demand *a priori* that f be surjective [that the range $f(\mathbb{R}^N) = \mathbb{R}^N$], nor that f be of any particular form. However, (1.10) has strong consequences, as follows.

Theorem 1.9 *If f is an isometry of $\mathbb{R}^N$, then it is affine [is a linear function plus a constant] and bijective [both injective and surjective]. In fact,*

$$f(x) = f(0) + Tx \qquad \text{for all} \quad x \in \mathbb{R}^N,$$

equivalently,

$$f_i(x) = f_i(0) + \sum_{j=1}^{N} T_{ij} x_j \qquad \text{for each} \quad i \in \{1, \ldots, N\} \quad \text{and all} \quad x \in \mathbb{R}^N,$$

where $T = (T_{ij})$ is an orthogonal $N \times N$ matrix independent of x.

The proof is outlined in the hint to Exercise 1.22.

Let A be a subset of $\mathbb{R}^N$ that contains $N + 1$ *affinely independent* points $a^0, a^1, \ldots, a^N$; this means that the N vectors $a^1 - a^0, \ldots, a^N - a^0$ are linearly independent. A *symmetry transformation* of A is an isometry f of $\mathbb{R}^N$ that maps A onto itself: $f(A) = A$.

For example, if $A \subset \mathbb{R}^2$ is a regular pentagon centred at the origin, or a good specimen of *carissa grandiflora* centred at the origin (Figure 1.6), then rotation through 72° is a symmetry transformation; if anti-clockwise, this rotation is the map $\theta \mapsto \theta + 2\pi/5$ for points $(r \cos \theta, r \sin \theta)$, or, more fully,

$$a(x) := \begin{bmatrix} \cos \alpha & -\sin \alpha \\ \sin \alpha & \cos \alpha \end{bmatrix} \begin{bmatrix} x_1 \\ x_2 \end{bmatrix}, \tag{1.11}$$

where $\alpha = 2\pi/5 = 72°$. If the regular pentagon has a vertex on the x_1-axis, then reflection in the x_1-axis is a symmetry transformation of the pentagon, but not of *carissa grandiflora* because each petal is asymmetric; this transformation is $(x_1, x_2) \mapsto (x_1, -x_2)$, or $\theta \mapsto -\theta$ for points $(r \cos \theta, r \sin \theta)$, or

$$b(x) := \begin{bmatrix} 1 & 0 \\ 0 & -1 \end{bmatrix} \begin{bmatrix} x_1 \\ x_2 \end{bmatrix}. \tag{1.12}$$

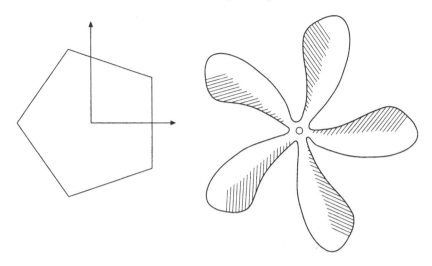

Fig. 1.6.

For a subset A of $\mathbb{R}^N$ that contains $N + 1$ affinely independent points, the set of all symmetry transformations of A forms a *group* if we define the group operation $\circ$ to be composition [so that $g \circ f$ is defined by $(g \circ f)(x) := g(f(x))$ for all $x \in \mathbb{R}^N$] and define the unit element to be the identity map I [such that $Ix := x$ for all $x \in \mathbb{R}^N$]. This group is the *symmetry group of A*; we denote it by $G(A)$.

The symmetry group of *carissa grandiflora* (centred at the origin) is

$$\{I, a, a^2, a^3, a^4\} =: \Gamma_5,$$

where a is as in (1.11) with $\alpha = 2\pi/5$, the symbol a^n means composition $a \circ a \circ \cdots \circ a$ with n factors a [so that a^3 is the map $\theta \mapsto \theta + 6\pi/5$ for points $(r\cos\theta, r\sin\theta)$], and it is understood that $a^5 = I$. The symmetry group of the regular pentagon (centred at the origin and having a vertex on the x_1-axis) is

$$\{I, a, \ldots, a^4, b, ba, \ldots, ba^4\} =: \Delta_5,$$

where a is as before, b is as in (1.12), $ba^n := b \circ a^n$, and it is understood that $b^2 = I$.

To extend these definitions of Γ_5 and Δ_5, we set

$$\alpha := 2\pi/n, \qquad n \in \mathbb{N},$$

in (1.11). Then

$$\Gamma_n := \{I, a, \ldots, a^{n-1}\} \tag{1.13}$$

is the symmetry group of a flower with n identical petals that are placed about the origin at equal angular intervals but lack individual symmetry. Also,

$$\Delta_n := \left\{ I, a, \ldots, a^{n-1}, b, ba, \ldots, ba^{n-1} \right\} \tag{1.14}$$

is the symmetry group, if $n \geq 3$, of a regular polygon centred at the origin, having n sides and with a vertex on the x_1-axis. The group Γ_n is a copy of the more abstract group usually written C_n and called the *cyclic group of order* n, the *order* of a group begin the number of its elements. Similarly, Δ_n is a copy of the *dihedral group of order* $2n$, often written D_n. [In D_n, the elements a and b are arbitary objects for which a group operation and a unit element can be defined.]

Let $\tilde{A}$ be an isometric image of A; that is, there is an isometry φ of $\mathbb{R}^N$ such that $\varphi(A) = \tilde{A}$. [For example, A might be the pentagon in Figure 1.6, and $\tilde{A}$ a translated and rotated version of it.] Then to each symmetry transformation g of A there corresponds a symmetry transformation $\tilde{g} := \varphi \circ g \circ \varphi^{-1}$ of $\tilde{A}$; conversely, to each $\tilde{g} \in G(\tilde{A})$ there corresponds $g := \varphi^{-1} \circ \tilde{g} \circ \varphi \in G(A)$. In other words, the map

$$g \mapsto \varphi \circ g \circ \varphi^{-1} \qquad \text{from} \ \ G(A) \ \ \text{onto} \ \ G(\tilde{A}) \tag{1.15}$$

is a *group isomorphism*; we regard $G(A)$ and $G(\tilde{A})$ as essentially the same and say that they are *equal up to isomorphism*.

The symmetry group $G(A)$ indicates the symmetry of a set A. If $G(A) \subset G(B)$ and $G(A) \neq G(B)$, then A is less symmetric than B; indeed, A is less symmetric than each isometric image $\tilde{B}$ of B, because $G(B)$ and $G(\tilde{B})$ are equal up to isomorphism. If $G(A)$ has many elements, then A has a large measure of symmetry; if $G(A) = \{I\}$, then A has one symmetry transformation but no symmetry in the everyday sense of the word. The group $G(\mathbb{R}^N)$ contains all isometries of $\mathbb{R}^N$ and is therefore the largest symmetry group for given N. The symmetry group $G(B)$ of a ball $B := \mathscr{B}(0, \rho)$ in $\mathbb{R}^N$ is almost as large: it contains all those isometries f of $\mathbb{R}^N$ for which $f(0) = 0$. The irregular polyhedron

$$P := \left\{ x \in \mathbb{R}^N \ \bigg| \ \sum_{j=1}^{N} jx_j \leq 1, \qquad \text{each} \ \ x_j \geq 0 \right\}, \ N \geq 2,$$

has symmetry group $G(P) = \{I\}$.

We turn now to the symmetry of functions. Again let A be a subset of $\mathbb{R}^N$ that contains $N + 1$ affinely independent points; an *invariance transformation of a function* $u := A \to \mathbb{R}^M$ is a symmetry transformation f of

A that leaves *u* unchanged : $u \circ f = u$. The invariance transformations of *u* form a group if the group operation is composition and the unit element is the identity map *I*; this group is the *invariance group of the function u* and we denote it by $H(u)$. (Some authors use the phrases *symmetry of u* and *symmetry group of u* for what are called here invariance transformation of *u* and invariance group of *u*.)

Here are two examples.

1. If $v : \mathbb{R} \to \mathbb{R}^2$ is defined by $v(t) := (\alpha \cos t, \beta \sin t)$ for $t \in \mathbb{R}$, the numbers α and β being positive constants, and if $f(x) := x + 2\pi$ for $x \in \mathbb{R}$, then the substitution $t = f(x)$ shows that $(v \circ f)(x) = v(x)$ for all $x \in \mathbb{R}$, so that *f* is an invariance transformation of *v*. The invariance group of this function *v* is

$$H(v) := \{f(.,k) \mid k \in \mathbb{Z}\}, \qquad \text{where } f(x,k) := x + 2k\pi \quad \text{for all } x \in \mathbb{R}.$$

Note that $H(v)$ is quite different from the symmetry group $G\big(v(\mathbb{R})\big)$ of the range of *v*; when $\alpha \neq \beta$, we have $G\big(v(\mathbb{R})\big) = \Delta_2$, a copy of the dihedral group of order 4.

2. The function $\psi : \mathbb{R}^2 \setminus \{0\} \to \mathbb{R}$ defined by

$$\psi(x) := -\frac{2x_1 x_2}{(x_1^2 + x_2^2)^2} = -\frac{\sin 2\theta}{r^2} \qquad \text{for all } x = (r \cos \theta, r \sin \theta) \in \mathbb{R}^2 \setminus \{0\}$$

is the stream function of a certain quadrupole at the origin, according to (A.55) and Exercises A.19 and A.21. The level sets of ψ are displayed in Figure A.10. We contemplate both ψ and the set

$$V := \big\{ x \in \mathbb{R}^2 \setminus \{0\} \mid |\psi(x)| > 1 \big\}.$$

This set has the symmetry group $G(V) = \Delta_4$. [Recall that, up to isomorphism, Δ_4 is the symmetry group of any square in $\mathbb{R}^2$.] The function ψ has a smaller invariance group, isomorphic to Δ_2; in fact, $H(\psi) = \{I, a, \tilde{b}, \tilde{b}a\}$, where *a* is the rotation $\theta \mapsto \theta + \pi$, while $\tilde{b}$ is the reflection $\theta \mapsto \pi/2 - \theta$. Four symmetry transformations of *V* are absent from $H(\psi)$ because they change $\sin 2\theta$ to $-\sin 2\theta$.

It seems desirable to link results like those in Chapters 3 to 5 to the theory of symmetry that has been sketched here, but so far there has been little work in this direction. We note one difficulty, and one encouraging feature, for any attempt to build a bridge between the two topics.

For the sake of definiteness, consider a positive solution *u* of (1.3). The difficulty is that the invariance group of this function *u* depends not only on the symmetry group of Ω but also on other properties of Ω,

such as its topological type, as well as on details of the function f in
(1.3). In particular, if the ball $B := \mathscr{B}(0,a)$ in Theorem 1.2 is replaced
by an annulus or spherical shell A centred at the origin, then these two
sets have the same symmetry group: $G(A) = G(B)$, but the first result in
Theorem 1.2, that $u(x)$ depends only on $|x|$, does not extend from $\Omega = B$
to $\Omega = A$. (Exercise 5.30 shows that (1.3), with $\Omega = A$, admits positive
solutions u which do not depend only on $|x|$.)

The encouraging feature is the fundamental role of reflection in both
theories. Let us rewrite a part of Definition 1.7 as

$$R(x) := x + 2(\mu - x \cdot k)k \qquad \text{for all } x \in \mathbb{R}^N; \qquad (1.16)$$

then the *reflection operator* R is an isometry of $\mathbb{R}^N$ that sends x to its
reflection in the hyperplane called $T_\mu(k)$ in Definition 1.7 and Figure 1.3.
In order to cast (1.16) into the form of an isometry displayed in Theorem
1.9, we write it as

$$R(x) = 2\mu k + Sx, \qquad \text{where } S_{ij} := \delta_{ij} - 2k_i k_j \qquad (|k| = 1); \qquad (1.17)$$

one checks without difficulty that the matrix S is orthogonal. It should be
clear from §1.3 that the operator R is basic in Chapters 3 to 5; maximum
principles and reflection in hyperplanes are the principal tools in those
chapters. It has not yet been stressed in this section that reflection is
basic to symmetry groups, but this is implied by the following theorem.

Theorem 1.10 *If f is an isometry of $\mathbb{R}^N$, then it can be written as a composition product*

$$f = R_0 \circ R_1 \circ \cdots \circ R_m, \qquad m \le N,$$

where each R_j is a reflection operator.

Proofs are given by Coxeter (1981, p.41 for $N = 2$; 1963, p.36 for
$N = 3$) and by Rees (1983, p.10).

1.5 Exercises

Notation, where not specified, is as in Chapter 0.

Exercise 1.11 For each of the following sets, state (with reasons) whether
the boundary is not of class C, or of class C^k but not of class C^{k+1} for
an integer $k \ge 0$ to be specified, or of class C^∞. [Refer to Chapter 0,
(viii).] The letters a, b, c denote positive constants.

(i) The elliptic disk $D := \{ x \in \mathbb{R}^2 \mid x_1^2/a^2 + x_2^2/b^2 < 1 \}$.

(ii) The elliptic cone $K := \{ x \in \mathbb{R}^3 \mid x_1 > 0,\ x_2^2/b^2 + x_3^2/c^2 < x_1^2 \}$.

(iii) The double cone $G := \{ x \in \mathbb{R}^3 \mid x_1 \in \mathbb{R} \setminus \{0\},\ x_2^2 + x_3^2 < x_1^2 \}$.

(iv) The hyperboloidal set $H := \{ x \in \mathbb{R}^3 \mid x_1^2 > a^2,\ x_2^2/b^2 + x_3^2/c^2 < x_1^2/a^2 - 1 \}$.

(v) The lens $L := \{ (x,y) \in \mathbb{R}^2 \mid 0 < x < 1,\ x^3 < y < x^2 \}$.

(vi) $\Omega := \left\{ (x,y) \in \mathbb{R}^2 \mid x \in \mathbb{R},\ y > 0 \text{ if } x \le 0,\ y > x^c \sin \dfrac{1}{x} \text{ if } x > 0 \right\}$.

Exercise 1.12 Given that in Theorem 1.1 the constant value of $u|_{\partial G}$ is β, where $\beta < 0$ if $N \le 2$, and $\beta > 0$ if $N \ge 3$, find the radius of the ball G in terms of λ, β and N.

[Refer to (A.23) in Appendix A.]

Exercise 1.13 The *factorial function* $(.)!$ is defined on $\mathbb{R} \setminus \{-1, -2, -3, \ldots\}$ by

$$x! = \int_0^\infty e^{-t} t^x \, dt \quad \text{if } x > -1, \qquad x! = \frac{(x+1)!}{x+1}.$$

The second equation is implied by the first for $x > -1$, and defines $x!$ for non-integer $x < -1$. The factorial function is related to the gamma function Γ by $x! = \Gamma(x+1)$. Given that $(-\tfrac{1}{2})! = \pi^{1/2}$, prove that

$$\int_0^{\pi/2} (\sin t)^n \, dt = \frac{1}{2}\pi^{1/2} \left(\frac{1}{2}n - \frac{1}{2} \right)! \Big/ \left(\frac{1}{2}n \right)! \quad \text{for } n \in \{0,1,2,\ldots\},$$

by showing that, if A_n denotes either expression, then $A_0 = \pi/2$, $A_1 = 1$ and $A_n/A_{n-2} = (n-1)/n$ for $n \in \{2,3,4,\ldots\}$.

Exercise 1.14 Use the result of Exercise 1.13 and induction over N to calculate the volume of a ball in $\mathbb{R}^N$:

$$|\mathscr{B}_N(0,\rho)| := \int \cdots \int_{|x|<\rho} dx_1 \cdots dx_N = \frac{\pi^{N/2}}{(N/2)!} \rho^N.$$

Infer that the surface area of the bounding sphere is

$$|\partial \mathscr{B}_N(0,\rho)| = \frac{\pi^{N/2}}{(N/2)!} N \rho^{N-1}.$$

Exercise 1.15 The function $\psi : \mathbb{R}^N \to \mathbb{R}$ is defined by

$$\psi(x) = \begin{cases} \exp\left(-\dfrac{1}{1-r^2}\right) & \text{if } r := |x| < 1, \\ 0 & \text{if } r \geq 1. \end{cases}$$

Prove that $\psi \in C_c^\infty(\mathbb{R}^N)$.

[It may be helpful to consider first all derivatives of

$$g(t) := \exp\left(-\frac{1}{1-t}\right), \qquad 0 \leq t < 1,$$

and ultimately to use multi-index notation (Definition A.3) for derivatives of ψ.]

Exercise 1.16 Let $f \in C(\Omega)$. Prove that, if $\int_\Omega \varphi f \geq 0$ whenever $\varphi \in C_c^\infty(\Omega)$ and $\varphi \geq 0$, then $f \geq 0$ in Ω; if $\int_\Omega \varphi f \leq 0$ whenever $\varphi \in C_c^\infty(\Omega)$ and $\varphi \geq 0$, then $f \leq 0$ in Ω; if $\int_\Omega \varphi f = 0$ whenever $\varphi \in C_c^\infty(\Omega)$ and $\varphi \geq 0$, then $f = 0$ in Ω.

[The abbreviation (1.5b) has been used. The function ψ in Exercise 1.15 has a useful variant defined by $\Phi(x) = \psi\left(\dfrac{x-c}{\delta}\right)$ for any fixed $c \in \mathbb{R}^N$ and $\delta > 0$.]

Exercise 1.17 Prove the one-dimensional case ($N = 1$) of Theorem 1.1.

[Any open set in $\mathbb{R}$ is a countable union of disjoint open intervals (Apostol 1974, p.51). It may be helpful to consider the case of finitely many intervals before proving the full result.]

Exercise 1.18 (i) Show that, when we operate on C^2 functions depending only on $r := |x|$ [say $u(x) = \tilde{u}(r)$],

$$\Delta = r^{-N+1}\frac{\mathrm{d}}{\mathrm{d}r}\left(r^{N-1}\frac{\mathrm{d}}{\mathrm{d}r}\right) \qquad \text{if } r > 0.$$

(ii) For $N \geq 3$, let $s := (x_2^2 + \cdots + x_N^2)^{1/2}$ and define $\theta \in [0, \pi]$ by $x_1 = r\cos\theta$, $s = r\sin\theta$. Show that, when we operate on C^2 functions depending only on x_1 and s,

$$\begin{aligned}
\Delta &= \left(\frac{\partial}{\partial x_1}\right)^2 + s^{-N+2}\frac{\partial}{\partial s}\left(s^{N-2}\frac{\partial}{\partial s}\right) \qquad (s > 0) \\
&= r^{-N+1}\frac{\partial}{\partial r}\left(r^{N-1}\frac{\partial}{\partial r}\right) + r^{-2}(\sin\theta)^{-N+2}\frac{\partial}{\partial\theta}\left\{(\sin\theta)^{N-2}\frac{\partial}{\partial\theta}\right\} \quad (s > 0).
\end{aligned}$$

(iii) Write $x = (r\cos\theta, r\sin\theta)$ for points of $\mathbb{R}^2$. By adapting the final result of (ii), or otherwise, show that

$$\Delta = \frac{1}{r}\frac{\partial}{\partial r}\left(r\frac{\partial}{\partial r}\right) + \frac{1}{r^2}\left(\frac{\partial}{\partial\theta}\right)^2 \qquad (N = 2,\ r > 0).$$

Exercise 1.19 The Bessel function J_ν, $\nu \geq 0$, is defined on $[0, \infty)$ by

$$J_\nu(t) = \sum_{j=0}^{\infty} \frac{(-1)^j}{j!\,(\nu + j)!}\left(\frac{1}{2}t\right)^{\nu+2j}, \qquad t \geq 0;$$

satisfies

$$\left\{\frac{1}{t}\frac{d}{dt}\left(t\frac{d}{dt}\right) + 1 - \frac{\nu^2}{t^2}\right\}J_\nu(t) = 0, \qquad t > 0;$$

and has denumerably many positive zeros, say $\beta_{\nu,p}$, ordered according to $0 < \beta_{\nu,1} < \beta_{\nu,2} < \cdots$. [See, for example, Burkill 1975, p.83; or Simmons 1972, p.121; or Whittaker & Watson 1927, p.367.] Referring to Exercise 1.18, (iii), show that the function $u_{n,p}$ defined by

$$\left.\begin{array}{l} u_{n,p}(x) = cJ_n\left(\beta_{n,p}\dfrac{r}{a}\right)\cos(n\theta + \kappa), \qquad 0 \leq r \leq a, \quad c \in \mathbb{R}\setminus\{0\}, \\ n \in \mathbb{N}_0,\ p \in \mathbb{N}, \qquad \kappa = 0 \text{ if } n = 0, \qquad \kappa \in \mathbb{R} \text{ if } n \geq 1, \end{array}\right\}$$

is in $C^\infty(\overline{B})$, where $B = \mathscr{B}(0, a) \subset \mathbb{R}^2$, and satisfies

$$\Delta u + (\beta_{n,p}/a)^2 u = 0 \quad \text{in } B, \qquad u|_{\partial B} = 0.$$

[Since $J_n(0) = 0$ for $n \geq 1$, it is to be understood that $u_{n,p}(x) = 0$ for $n \geq 1$ and $x = 0$. However, to establish the C^∞ property at $x = 0$, one must use co-ordinates x_1 and x_2.]

Exercise 1.20 Writing $x = (r\cos\theta, r\sin\theta\cos\varphi, r\sin\theta\sin\varphi)$, $0 \leq \theta \leq \pi$, for points of $\mathbb{R}^3$, show that for $N = 3$, and $r\sin\theta > 0$,

$$\Delta = \frac{1}{r^2}\left\{\frac{\partial}{\partial r}\left(r^2\frac{\partial}{\partial r}\right) + \frac{1}{\sin\theta}\frac{\partial}{\partial\theta}\left(\sin\theta\frac{\partial}{\partial\theta}\right) + \frac{1}{(\sin\theta)^2}\left(\frac{\partial}{\partial\varphi}\right)^2\right\}.$$

[It is not necessary to make six lengthy applications of the chain rule; see, for example, Kellogg 1929, p.183; or Sobolev 1964, p.393; or Spiegel 1959, p.151.]

Exercise 1.21 Let $n \in \mathbb{N}_0$ and $m \in \mathbb{N}_0$. The Legendre function P_n^m is defined on $[-1, 1]$ by

$$P_n^0(t) = P_n(t) = \frac{1}{2^n n!} \left(\frac{d}{dt} \right)^n (t^2 - 1)^n,$$

$$P_n^m(t) = (1 - t^2)^{m/2} \left(\frac{d}{dt} \right)^m P_n(t), \quad -1 \leq t \leq 1;$$

satisfies

$$\left\{ \frac{d}{dt} \left((1 - t^2) \frac{d}{dt} \right) + n(n+1) - \frac{m^2}{1 - t^2} \right\} P_n^m(t) = 0, \quad -1 < t < 1;$$

and has the property that, if r, θ, φ are spherical co-ordinates as in Exercise 1.20, then the function having values

$$r^n P_n^m(\cos \theta) \cos(m\varphi + \kappa) \qquad (\kappa \text{ as below}, \ r \sin \theta > 0)$$

is a homogenous polynomial of degree n in terms of x_1, x_2, x_3 [Sobolev 1964, p.404; Whittaker & Watson 1927, p.392].

Referring to Exercises 1.19 and 1.20, contemplate the function $u_{n,m,p}$ defined by

$$\left. \begin{array}{l} u_{n,m,p}(x) = cr^{-1/2} J_{n+1/2} \left(\beta_{n+1/2,p} \frac{r}{a} \right) P_n^m(\cos \theta) \cos(m\varphi + \kappa), \\[2mm] 0 < r \leq a, \quad c \in \mathbb{R} \setminus \{0\}, \quad \kappa = 0 \text{ if } m = 0, \quad \kappa \in \mathbb{R} \text{ if } m \geq 1, \end{array} \right\}$$

where p, n, m are as before, and $u_{n,m,p}(0)$ is defined by continuity. Establish results, analogous to those in Exercise 1.19, for $u_{n,m,p}$ on $\overline{\mathscr{B}(0, a)} \subset \mathbb{R}^3$. In particular, show that $u_{n,m,p}$ satisfies

$$\Delta u + k^2 u = 0 \text{ in } \mathscr{B}(0, a), \qquad \text{where } k := \beta_{n+1/2,p}/a.$$

[The Bessel functions $J_{n+1/2}$ have elementary representations: for $t > 0$,

$$J_{1/2}(t) = \left(\frac{2}{\pi} t \right)^{1/2} \frac{\sin t}{t}, \quad J_{3/2}(t) = \left(\frac{2}{\pi} t \right)^{1/2} \frac{\sin t - t \cos t}{t^2},$$

$$J_{5/2}(t) = \left(\frac{2}{\pi} t \right)^{1/2} \frac{3 \sin t - 3t \cos t - t^2 \sin t}{t^3}, \ \dots,$$

but this fact plays no part in the present exercise.]

Exercise 1.22 Prove Theorem 1.9.

[Let $g(x) := f(x) - f(0)$ for all $x \in \mathbb{R}^N$. Show that, for all $x, y \in \mathbb{R}^N$ and with each step following from earlier ones, $|g(x) - g(y)| = |x - y|$,

$|g(x)| = |x|$, $g(x) \cdot g(y) = x \cdot y$; moreover,

$$|g(x+y) - g(x) - g(y)|^2 = 0 \quad \text{and} \quad |g(\alpha x) - \alpha g(x)|^2 = 0 \quad \text{for all } \alpha \in \mathbb{R}.$$

Thus g has a finite-dimensional domain, is linear and injective, and conserves lengths and angles.]

Exercise 1.23 *The smoothing operation.* Let a function $k_1 : \mathbb{R}^N \to [0, \infty)$ have the properties

$$k_1 \in C_c^\infty(\mathbb{R}^N), \qquad \text{supp} \, k_1 \subset \overline{\mathscr{B}(0,1)}, \qquad \int_{\mathbb{R}^N} k_1 = \int_{\mathscr{B}(0,1)} k_1 = 1.$$

[For example, if ψ is as in Exercise 1.15, then a possible choice is $k_1(x) = \psi(x)/\int_{\mathscr{B}(0,1)} \psi$.] Set $k_\rho(x) := \rho^{-N} k_1(x/\rho)$ for any $\rho > 0$ and all $x \in \mathbb{R}^N$, so that $k_\rho \geq 0$ on $\mathbb{R}^N$ and

$$k_\rho \in C_c^\infty(\mathbb{R}^N), \quad \text{supp} \, k_\rho \subset \overline{\mathscr{B}(0,\rho)}, \quad \int_{\mathbb{R}^N} k_\rho = \int_{\mathscr{B}(0,\rho)} k_\rho = 1.$$

We call k_ρ a *smoothing kernel*, or *averaging* kernel, of *smoothing radius* ρ.

Now let $f : \mathbb{R}^N \to \mathbb{R}$ be locally integrable, that is, integrable on every compact subset of $\mathbb{R}^N$. Then we form

$$f_\rho(x) := \int_{\mathbb{R}^N} k_\rho(x-y) f(y) \, \mathrm{d}y = \int_{\mathbb{R}^N} k_\rho(z) f(x-z) \, \mathrm{d}z, \qquad x \in \mathbb{R}^N,$$

and call f_ρ a *regularization* of f, or a *mean function* of f.

Prove the following.

(i) $f_\rho \in C^\infty(\mathbb{R}^N)$.

(ii) If $f \in C(\mathbb{R}^N)$, then $f_\rho(x) \to f(x)$ as $\rho \to 0$, uniformly on each compact subset of $\mathbb{R}^N$.

(iii) If $f \in C^1(\mathbb{R}^N)$, then smoothing and differentiation commute:

$$\partial_j(f_\rho) = (\partial_j f)_\rho \qquad \text{for each} \quad j \in \{1, \ldots, N\},$$

so that $\partial_j f_\rho(x) \to \partial_j f(x)$ as $\rho \to 0$, uniformly on each compact subset of $\mathbb{R}^N$.

[For (i), multi-indices (Definition A.3) may be useful: so may a glance at the proof of Theorem A.5. For (ii), observe that

$$f(x) - f_\rho(x) = \int_{\mathscr{B}(0,\rho)} k_\rho(z) \big\{ f(x) - f(x-z) \big\} \, \mathrm{d}z \,.$$

For (iii), we have $(\partial/\partial x_j)k_\rho(x-y) = -(\partial/\partial y_j)k_\rho(x-y)$ and integration by parts.]

Exercise 1.24 *Continuity of translation for functions in $L_p(\mathbb{R}^N)$, $1 \le p < \infty$.*
Let $f \in L_p(\mathbb{R}^N)$ for some $p \in [1, \infty)$ and let $\|.\| = \|.|L_p(\mathbb{R}^N)|\|$ for that p.
Prove that

$$\|f(.-h) - f\| \to 0 \quad \text{as} \quad |h| \to 0,$$

where $h \in \mathbb{R}^N$ and $f(.-h)$ is the function with values $f(x-h)$.
    [There is a sequence (φ_n) in $C_c(\mathbb{R}^N)$ such that $\|f - \varphi_n\| \to 0$ as $n \to \infty$.
The symbol $C_c(.)$ refers to continuous functions of compact support; see
Chapter 0, (iv) and (vii).]

Exercise 1.25 *Averaging functions in $L_p(\mathbb{R}^N)$, $1 \le p < \infty$.* Let $a_1 : \mathbb{R}^N \to \mathbb{R}$
have the properties

$$a_1 \in L_1(\mathbb{R}^N) \qquad \text{and} \qquad \int_{\mathbb{R}^N} a_1 = 1.$$

As in Exercise 1.23, define an averaging kernel by $a_\rho(x) := \rho^{-N} a_1(x/\rho)$
for any $\rho > 0$ and all $x \in \mathbb{R}^N$. Given $p \in [1, \infty)$ and $f \in L_p(\mathbb{R}^N)$, define a
mean function, as before, by

$$f_\rho(x) := \int_{\mathbb{R}^N} a_\rho(z) f(x-z) \, dz, \qquad x \in \mathbb{R}^N.$$

Writing $\|.\| := \|.|L_p(\mathbb{R}^N)|\|$, prove that $\|f - f_\rho\| \to 0$ as $\rho \to 0$.
    [Observe that the substitution $z = \rho y$ yields

$$
\begin{aligned}
\|f - f_\rho\| &= \left\| \int_{\mathbb{R}^N} \{ f(.) - f(.-\rho y) \} \, a_1(y) \, dy \right\| \\
&\le \int_{\mathbb{R}^N} \| f(.) - f(.-\rho y) \| \, |a_1(y)| \, dy
\end{aligned}
$$

by the extended Minkowski inequality in (0.16); that, for given $\varepsilon > 0$,
there is a number $M = M(\varepsilon, f)$ such that

$$2\|f\| \int_{|y|>M} |a_1(y)| \, dy < \tfrac{1}{2}\varepsilon;$$

and that for $|y| \le M$ continuity of translation is relevant.]

2

Some Maximum Principles for Elliptic Equations

2.1 Linear elliptic operators of order two

As always, Ω denotes an open non-empty subset of $\mathbb{R}^N$.

Definition 2.1 (i) The operator L, defined by

$$Lu(x) := \left\{ \sum_{i,j=1}^{N} a_{ij}(x)\, \partial_i\partial_j + \sum_{j=1}^{N} b_j(x)\, \partial_j + c(x) \right\} u(x) \qquad (2.1)$$

whenever $u \in C^2(\Omega)$ and $x \in \Omega$, is a *linear partial differential operator*, of *order two*. Here

$$a = (a_{ij}) : \Omega \to \mathbb{R}^{N^2}, \quad b = (b_j) : \Omega \to \mathbb{R}^N, \quad c : \Omega \to \mathbb{R}$$

are given measurable functions. The $N \times N$ matrix a is *symmetric*: $a_{ji}(x) = a_{ij}(x)$ for all $i, j \in \{1,\dots,N\}$ and all $x \in \Omega$. [This involves no loss of generality because $\partial_j\partial_i u = \partial_i\partial_j u$.]

(ii) We say that L is *elliptic at* $x \in \Omega$ iff there is a number $\lambda(x) > 0$ such that

$$\sum_{i,j=1}^{N} a_{ij}(x)\, \xi_i\xi_j \geq \lambda(x)|\xi|^2 \quad \text{for all} \ \ \xi \in \mathbb{R}^N ; \qquad (2.2)$$

that L is *elliptic in* Ω iff it is elliptic at every $x \in \Omega$; and that L is *uniformly* elliptic in Ω iff there is a constant $\lambda_0 > 0$ such that $\lambda(x) \geq \lambda_0$ for all $x \in \Omega$. The best (largest) values $\lambda(x)$ and λ_0 are, respectively, the pointwise and uniform *moduli of ellipticity* of L. ☐

Here are three examples to which we can apply the definition with almost no calculation.

1. If $L = \triangle +$ lower order terms, then $a_{ij}(x) = \delta_{ij}$ [the Kronecker delta, Chapter 0, (v)], so that L is uniformly elliptic in every Ω, with $\lambda_0 = 1$.

2. Let $x_1, \ldots, x_{N-1}$ be space variables, while x_N denotes time. Then the operators $\partial_1^2 + \cdots + \partial_{N-1}^2 - \partial_N^2$ of the wave equation, and $\partial_1^2 + \cdots + \partial_{N-1}^2 - \partial_N$ of the heat equation, are not elliptic: choose $\xi_i = \delta_{Ni}$ in (2.2).

3. The Tricomi operator $\partial_1^2 + x_1 \partial_2^2$ is elliptic in the half-plane $\{ x \in \mathbb{R}^2 \mid x_1 > 0 \}$ but not uniformly so; the pointwise modulus of ellipticity is

$$
\lambda(x) = \begin{cases} x_1 & \text{if } 0 < x_1 \leq 1, \\ 1 & \text{if } x_1 > 1. \end{cases}
$$

Exercise 2.2 Given that $a(x)$ is symmetric and satisfies (2.2), prove the following.

(a) Ellipticity is invariant under rotation of co-ordinate axes. That is, if R is a constant, orthogonal $N \times N$ matrix, $y := Rx$ and $h(y) := Ra(x)R^{-1}$, then

$$
\sum_{i,j} a_{ij}(x) \frac{\partial}{\partial x_i} \frac{\partial}{\partial x_j} = \sum_{p,q} h_{pq}(y) \frac{\partial}{\partial y_p} \frac{\partial}{\partial y_q}
$$

and

$$
\sum_{p,q} h_{pq}(Rx) \eta_p \eta_q \geq \lambda(x) |\eta|^2 \quad \text{for all } \eta \in \mathbb{R}^N.
$$

(b) The pointwise modulus of ellipticity is the smallest eigenvalue of $a(x)$.

(c) Let $g(x)$ be a non-positive $N \times N$ matrix; we write $g(x) \leq 0$, meaning that $\xi g(x) \xi \leq 0$ for all $\xi \in \mathbb{R}^N$. Then

$$
\text{trace}\big(a(x)\, g(x)\big) := \sum_{i,j} a_{ij}(x)\, g_{ji}(x) \leq 0.
$$

(Here the rule for matrix multiplication is summation over adjacent subscripts:

$$
g(x)\, \zeta := \left(\sum_j g_{ij}(x)\, \zeta_j \right)^N_{i=1} , \quad \xi\, g(x)\, \zeta := \sum_{i,j} \xi_i\, g_{ij}(x)\, \zeta_j,
$$

so that row and column vectors need not be distinguished in such expressions.)   □

2.2 The weak maximum principle

Definition 2.3 The operators to be considered in this section and the next two are

$$L_0 := \sum_{i,j=1}^{N} a_{ij}(x)\, \partial_i \partial_j + \sum_{j=1}^{N} b_j(x)\, \partial_j,$$

$$L := L_0 + c(x), \qquad \text{with } c(x) \leq 0 \text{ for all } x \in \Omega,$$

$$L_1 := \sum_{i,j=1}^{N} a_{ij}\, \partial_i \partial_j + \sum_{j=1}^{N} b_j\, \partial_j + c, \qquad \text{with } c \leq 0;$$

in L_1 all coefficients a_{ij}, b_j and c are constants. Thus L_0 is the particular L with $c = 0$ (the zero function), while L_1 is the particular L with constant coefficients.

All three are *uniformly elliptic*: for all $x \in \Omega$ and $\xi \in \mathbb{R}^N$,

$$\sum_{i,j=1}^{N} a_{ij}(x)\, \xi_i \xi_j \geq \lambda_0 |\xi|^2, \quad \lambda_0 = \text{const.} > 0. \tag{2.3}$$

All coefficients are bounded and measurable: in L_0 and L, for all i and j,

$$\sup_{x \in \Omega} |a_{ij}(x)| < \infty, \quad \sup_{x \in \Omega} |b_j(x)| < \infty, \quad \sup_{x \in \Omega} |c(x)| < \infty.$$

□

Definition 2.4 We shall say that u is a C^2-*subsolution relative to L and Ω* iff $u \in C^2(\Omega)$ and $Lu \geq 0$ in Ω. (Here L may be replaced by L_0 or L_1.)

□

We distinguish L_0 from L because stronger conclusions are possible when $c = 0$, and L_1 from L because a different kind of subsolution will be used for L_1. However, in the following three versions of the weak maximum principle (which is not to be despised, relative to the strong maximum principle), hypothesis (a) is always the same; it ensures, as was noted in Exercise 1.4, that $\sup_\Omega u = \max_{\overline{\Omega}} u$.

Theorem 2.5 (the weak maximum principle for L_0). *Suppose that*
  (a) Ω *is bounded*, $u \in C(\overline{\Omega})$;
  (b) u *is a* C^2-*subsolution relative to* L_0 *and* Ω.
Then the supremum of u is attained on the boundary:

$$\max_{\overline{\Omega}} u = \max_{\partial\Omega} u.$$

Proof (i) Define, for arbitrary $\varepsilon > 0$ and for a constant K to be chosen presently,

$$v(x) := u(x) + \varepsilon e^{Kx_1}, \quad x \in \overline{\Omega}.$$

Now, for all $x \in \Omega$,

$$
\begin{aligned}
L_0\left(e^{Kx_1}\right) &= \left\{a_{11}(x)K^2 + b_1(x)K\right\} e^{Kx_1} \\
&\geq \left(\lambda_0 K^2 - \{\sup_\Omega |b_1|\} K\right) e^{Kx_1} \quad \text{[in (2.3), } \xi = (K,0,\ldots,0)] \\
&> 0 \quad \text{if we choose } K > \frac{1}{\lambda_0} \sup_\Omega |b_1|.
\end{aligned}
$$

Hence $L_0 v > 0$ in Ω.

(ii) Assume (for contradiction) that $\sup_\Omega v$ is attained at $x_0 \in \Omega$. Then $(\partial_j v)(x_0) = 0$ for all $j \in \{1, \ldots, N\}$, and the Hessian matrix

$$H(x_0) := \left((\partial_i \partial_j v)(x_0)\right) \leq 0.$$

[Otherwise $\zeta H(x_0)\zeta = \alpha > 0$, say, for some $\zeta \in \mathbb{R}^N$ with $|\zeta| = 1$, and the Taylor formula

$$v(x_0 + h) = v(x_0) + 0 + \tfrac{1}{2} \sum_{i,j}(\partial_i \partial_j v)(x_0)\, h_i h_j + o\left(|h|^2\right)$$

leads to a contradiction, because we can choose $h = \beta \zeta$ with $\beta > 0$ so small that $v(x_0 + h) > v(x_0)$.] The result of Exercise 2.2, (c), now shows that

$$
\begin{aligned}
(L_0 v)(x_0) &= \sum_{i,j} a_{ij}(x_0)\, (\partial_j \partial_i v)(x_0) + 0 \\
&= \text{trace}\left(a(x_0)\, H(x_0)\right) \leq 0,
\end{aligned}
$$

which contradicts the result of step (i).

(iii) Accordingly, for every $\varepsilon > 0$ and all $x \in \overline{\Omega}$,

$$u(x) < v(x) \leq \max_{\partial\Omega} v \leq \max_{\partial\Omega} u + \varepsilon K_1,$$

where

$$K_1 := \max_{x \in \partial\Omega} e^{Kx_1}.$$

It follows that $u(x) \leq \max_{\partial\Omega} u$ for all $x \in \overline{\Omega}$. [Otherwise $u(x_0) = \max_{\partial\Omega} u + \delta$ for some $x_0 \in \Omega$ and some $\delta > 0$; we obtain a contradiction by choosing $\varepsilon = \delta/2K_1$.] $\qquad \square$

The weak maximum principle for L involves the non-negative part u^+ of u [see Chapter 0, (v)] and states less than the theorem for L_0 when

$\max_{\partial\Omega} u < 0$. *However, if* $\max_{\partial\Omega} u \geq 0$, *then* $\max_{\overline{\Omega}} u = \max_{\partial\Omega} u$ *exactly as before*, because in that case $\max_{\partial\Omega} u^+ = \max_{\partial\Omega} u$, so that strict inequality in (2.4) is impossible.

Theorem 2.6 (the weak maximum principle for L). *Suppose that*
  (a) Ω *is bounded,* $u \in C(\overline{\Omega})$;
  (b) u *is a* C^2-*subsolution relative to* L *and* Ω.
Then

$$\max_{\overline{\Omega}} u \leq \max_{\partial\Omega} u^+. \tag{2.4}$$

Proof Let $\Omega^+ := \{ x \in \Omega \mid u(x) > 0 \}$. This set is open in $\mathbb{R}^N$: if $y \in \Omega^+$, say $u(y) = \alpha > 0$, then there is a number $\delta > 0$ such that both $\mathscr{B}(y, \delta) \subset \Omega$ [since Ω is open] and $u(x) > \alpha/2$ whenever $x \in \mathscr{B}(y, \delta)$ [since u is continuous], so that $\mathscr{B}(y, \delta)$ is in Ω^+.

If Ω^+ is empty, then $\max_{\overline{\Omega}} u \leq 0$ and the theorem is true.

Suppose then that Ω^+ is not empty. The hypotheses $L_0 u \geq -c(x)u$ in Ω and $c(x) \leq 0$ in Ω imply that $L_0 u \geq 0$ in Ω^+; by Theorem 2.5, the maximum of u over $\overline{\Omega^+}$ equals that over $\partial\Omega^+$; hence there is a point

$$x_0 \in \partial\Omega^+ \text{ such that } u(x_0) = \max_{\overline{\Omega^+}} u > 0.$$

If $x_0 \in \Omega$ (Figure 2.1) we have a contradiction: by continuity, $u > 0$ in $\mathscr{B}(x_0, \rho)$ for some $\rho > 0$; on the other hand, $\mathscr{B}(x_0, \rho)$ contains points of $\Omega \setminus \Omega^+$, because $x_0 \in \partial\Omega^+$, and $u \leq 0$ at such points. Therefore $x_0 \in \partial\Omega$. $\square$

Remark 2.7 If u is a C^2-*supersolution* relative to L and Ω, which means that $u \in C^2(\Omega)$ and $Lu \leq 0$ in Ω, then $-u$ is a C^2-subsolution. If also condition (a) holds, then

$$\max_{\overline{\Omega}}(-u) \leq \max_{\partial\Omega}(-u)^+,$$

where

$$(-u)^+(x) = \max_{[x \text{ fixed}]} \{ -u(x), 0 \} = -\min_{[x \text{ fixed}]} \{ u(x), 0 \}$$
$$= -u^-(x),$$

so that

$$\max_{\overline{\Omega}}(-u) \leq \max_{\partial\Omega}(-u^-).$$

Equivalently,

$$\min_{\overline{\Omega}} u \geq \min_{\partial\Omega} u^-. \tag{2.5}$$

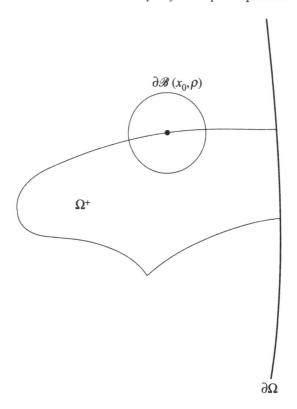

Fig. 2.1.

If u is a C^2-*solution* relative to L and Ω, which means that $u \in C^2(\Omega)$ and $Lu = 0$ in Ω, and condition (a) holds, then (2.4) and (2.5) imply that

$$\min_{\partial\Omega} u^- \le u(x) \le \max_{\partial\Omega} u^+ \quad \text{for all } x \in \overline{\Omega}. \tag{2.6}$$

Similarly, all our results for subsolutions have implications for supersolutions and solutions. □

Remark 2.8 The *Dirichlet problem* for L in a bounded set Ω is to find v such that

$$\left.\begin{array}{l} Lv = f \quad \text{in } \Omega, \\ v\big|_{\partial\Omega} = g, \quad v \in C(\overline{\Omega}) \cap C^2(\Omega), \end{array}\right\} \tag{2.7}$$

where f and g are given functions. This problem has *at most one solution*, because the difference $u := v_1 - v_2$ of two solutions satisfies $Lu = 0$ in Ω,

$u = 0$ on $\partial\Omega$, and has the smoothness required for (2.6); therefore $u = 0$ on $\overline{\Omega}$. □

Remark 2.9 (i) *The condition $c \leq 0$ in Ω (Definition 2.3) cannot be omitted from Theorem 2.6.* Once again this is illustrated by eigenfunctions of the Laplace operator. For example, let Ω be the rectangle $(0, \alpha) \times (0, \beta)$ in $\mathbb{R}^2$, and let

$$u(x) = \sin\frac{m\pi x_1}{\alpha} \sin\frac{n\pi x_2}{\beta}, \qquad m, n \in \mathbb{N}. \tag{2.8}$$

Calculating $\triangle u$, we see that

$$\triangle u + cu = 0 \quad \text{in} \quad \Omega, \quad \text{where} \quad c = \left(\frac{m\pi}{\alpha}\right)^2 + \left(\frac{n\pi}{\beta}\right)^2 > 0,$$

and, in contrast to (2.4), $\max_{\overline{\Omega}} u = 1$ while $\max_{\partial\Omega} u^+ = 0$.

(ii) *We cannot replace u^+ by u in (2.4).* [As was noted earlier, this would give $\max_{\overline{\Omega}} u = \max_{\partial\Omega} u$.] For, let Ω be the unit ball $\mathscr{B}(0,1)$ in $\mathbb{R}^N$, let $L = \triangle - 1$, and let $u(x) = -3N - |x|^2$ on $\overline{\Omega}$. Then $\triangle u = -2N$, so that

$$Lu(x) = \triangle u(x) - u(x) = N + |x|^2 > 0 \quad \text{in} \quad \Omega,$$

and

$$\max_{\overline{\Omega}} u = -3N > -3N - 1 = \max_{\partial\Omega} u.$$

□

The requirement in Definition 2.4 that subsolutions be in $C^2(\Omega)$ can cause embarrassment. For example, the Newtonian potential of constant density in a bounded open set G is not twice differentiable at points of ∂G; when ∂G is unknown *a priori* and may be unpleasant, a need to consider second derivatives of the potential would be a source of difficulty. We now define subsolutions for which membership of $C^1(\Omega)$ is ample smoothness. However, we do this only for the operator L_1, because a proof of something like Theorem 2.11 for an operator with variable coefficients requires (I believe) considerably more machinery.

Definition 2.10 We shall say
(a) that u is a *generalized subsolution relative to L_1 and Ω* iff $u \in C^1(\Omega)$ and

$$\Lambda_1(\varphi, u; \Omega) := \int_\Omega \left\{ -\sum_{i,j=1}^N a_{ij}(\partial_i\varphi)(\partial_j u) + \sum_{j=1}^N b_j \varphi \partial_j u + c\varphi u \right\}$$

$$\geq 0 \quad \text{whenever} \quad \varphi \in C_c^\infty(\Omega) \text{ and } \varphi \geq 0;$$

(b) that u is a *distributional subsolution relative to* L_1 *and* Ω iff u is locally integrable in Ω (integrable on each compact subset of Ω) and

$$\Lambda_{1d}(\varphi, u; \Omega) \quad := \quad \int_\Omega \left\{ \sum_{i,j=1}^N a_{ij} (\partial_j \partial_i \varphi) u - \sum_{j=1}^N b_j (\partial_j \varphi) u + c\varphi u \right\}$$

$$\geq \quad 0 \ \text{ whenever } \ \varphi \in C_c^\infty(\Omega) \ \text{ and } \ \varphi \geq 0.$$

Then u is a *generalized supersolution* iff $-u$ is a generalized subsolution; u is a *generalized solution* iff it is both a generalized subsolution and a generalized supersolution (cf. Remark 2.7). *Distributional supersolutions* and *distributional solutions* are defined similarly. $\qquad\square$

Evidently the key to this definition is integration by parts:

$$\Lambda_1(\varphi, u; \Omega) = \int_\Omega \varphi L_1 u \ \text{ if } \ \varphi \in C_c^\infty(\Omega) \ \text{ and } \ u \in C^2(\Omega); \qquad (2.9)$$

$$\Lambda_{1d}(\varphi, u; \Omega) = \Lambda_1(\varphi, u; \Omega) \ \text{ if } \ \varphi \in C_c^\infty(\Omega) \ \text{ and } \ u \in C^1(\Omega). \qquad (2.10)$$

Since a C^2-subsolution u satisfies $L_1 u \geq 0$ in Ω, we see from (2.9) that *a C^2-subsolution* (relative to L_1 and Ω) *is a generalized subsolution*, and from (2.10) that *a generalized subsolution is a distributional subsolution*. On the other hand, a distributional subsolution is a generalized subsolution only if it is also in $C^1(\Omega)$, and a generalized subsolution is a C^2-subsolution only if it is also in $C^2(\Omega)$. [In this last case, we use (2.9) and Exercise 1.16 to deduce that $L_1 u \geq 0$ in Ω.]

Note that, in the following theorem, hypothesis (a) swamps the condition of local integrability demanded in Definition 2.10, (b).

Theorem 2.11 (the weak maximum principle for L_1). *Suppose that*
(a) Ω *is bounded,* $u \in C(\overline{\Omega})$;
(b) u *is a distributional subsolution relative to* L_1 *and* Ω.
Then the previous conclusions hold:

$$\max{}_{\overline{\Omega}} \, u = \max{}_{\partial\Omega} \, u \quad \textit{if } c = 0, \qquad (2.11a)$$

$$\max{}_{\overline{\Omega}} \, u \leq \max{}_{\partial\Omega} \, u^+ \quad \textit{if } c < 0. \qquad (2.11b)$$

Proof (i) Let an arbitrary point $\xi \in \Omega$ be given; we shall prove the theorem by showing that

$$u(\xi) \leq \begin{cases} \max_{\partial\Omega} u & \text{if } c = 0, & (2.12a) \\ \max_{\partial\Omega} u^+ & \text{if } c < 0. & (2.12b) \end{cases}$$

Adopting a standard trick, we choose the following test function φ in the definition of distributional subsolution.

$$\varphi(y) = k_\rho(x - y) \quad \text{for all } y \in \Omega, \tag{2.13a}$$

where k_ρ is a smoothing kernel as in Exercise 1.23; ρ and x are parameters satisfying

$$0 < \rho \le \tfrac{1}{3} \operatorname{dist}(\xi, \partial\Omega), \tag{2.13b}$$

$$x \in \overline{G(\rho)}, \qquad \text{where } G(\rho) := \left\{ z \in \Omega \mid \operatorname{dist}(z, \partial\Omega) > 2\rho \right\}, \tag{2.13c}$$

as is illustrated in Figure 2.2. This choice of φ is legitimate because $k_\rho(x - y) = 0$ when $|y - x| \ge \rho$, so that $\operatorname{supp} k_\rho(x - .) \subset \Omega$, and certainly $k_\rho(x - .)$ is infinitely differentiable and non-negative in Ω.

(ii) Now let

$$u_\rho(x) = \int_\Omega k_\rho(x - y)\, u(y) \, \mathrm{d}y, \quad x \in \overline{G(\rho)}, \tag{2.14}$$

where, equally well, the integral could be written as one over $\mathscr{B}(x, \rho)$. Then $u_\rho \in C^\infty(\overline{G(\rho)})$ by Exercise 1.23; the present boundary $\partial\Omega$ plays no part when $x \in \overline{G(\rho)}$. The definition of distributional subsolution states that

$$
\begin{aligned}
0 \;\le\; & \Lambda_{1d}(k_\rho(x - .), \, u; \, \Omega) \\
= \; & \int_\Omega \left\{ \sum_{i,j=1}^N a_{ij} \left[\frac{\partial^2}{\partial y_i \partial y_j} k_\rho(x - y) \right] u(y) \right. \\
& \left. - \sum_{j=1}^N b_j \left[\frac{\partial}{\partial y_j} k_\rho(x - y) \right] u(y) + c k_\rho(x - y) u(y) \right\} \mathrm{d}y \\
= \; & \sum_{i,j=1}^N a_{ij} \int_\Omega \left[\frac{\partial^2}{\partial x_i \partial x_j} k_\rho(x - y) \right] u(y) \, \mathrm{d}y \\
& + \sum_{j=1}^N b_j \int_\Omega \left[\frac{\partial}{\partial x_j} k_\rho(x - y) \right] u(y) \, \mathrm{d}y + c \int_\Omega k_\rho(x - y) u(y) \, \mathrm{d}y \\
= \; & L_1 u_\rho(x).
\end{aligned}
$$

Thus u_ρ is a C^2-subsolution relative to L_1 and $G(\rho)$; by the weak maximum principle for L_0 and for L,

$$u_\rho(\xi) \le \begin{cases} \max_{\partial G(\rho)} u_\rho & \text{if } c = 0, \\ \max_{\partial G(\rho)} (u_\rho)^+ & \text{if } c < 0. \end{cases}$$

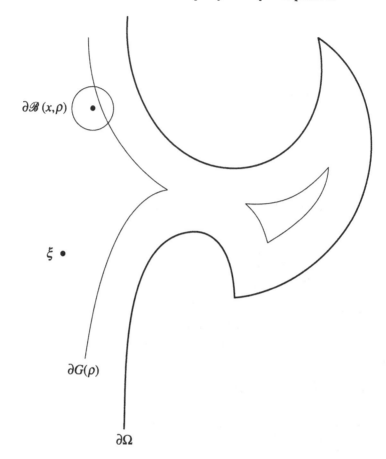

$\partial \mathscr{B}(x, \rho)$

$\xi \, \bullet$

$\partial G(\rho)$

$\partial \Omega$

Fig. 2.2.

Consequently, if

$$\lim_{\rho \to 0} u_\rho(\xi) = u(\xi), \tag{2.15}$$

$$\lim\sup_{\rho \to 0} \left\{ \max_{\partial G(\rho)} u_\rho \right\} \leq \max_{\partial \Omega} u, \tag{2.16a}$$

$$\lim\sup_{\rho \to 0} \left\{ \max_{\partial G(\rho)} (u_\rho)^+ \right\} \leq \max_{\partial \Omega} u^+, \tag{2.16b}$$

then (2.12) will follow [because for $c = 0$ we shall be able to contradict $u(\xi) = \max_{\partial \Omega} u + \mu$, $\mu > 0$, by choosing ρ sufficiently small; similarly for $c < 0$].

(iii) Consider in passing the statements

$$\lim_{\rho \to 0} \left\{ \max_{\partial G(\rho)} u_\rho \right\} = \max_{\partial \Omega} u, \tag{2.17a}$$

$$\lim_{\rho \to 0} \left\{ \max_{\partial G(\rho)} (u_\rho)^+ \right\} = \max_{\partial \Omega} u^+. \tag{2.17b}$$

These may seem simpler than (2.16a,b) and, with (2.15), they certainly imply (2.12). Moreover, (2.17a,b) are true. However, their proof is longer, and slightly harder, than that of (2.16a,b) because a *lower* bound for $\max_{\partial G(\rho)} u_\rho$ emerges less easily than the upper bound that we shall find.

(iv) Since $u \in C(\overline{\Omega})$ and $\overline{\Omega}$ is compact, u is uniformly continuous: for every $\varepsilon > 0$ there is a number $\delta_\varepsilon > 0$ such that

$$y, z \in \overline{\Omega} \text{ and } |y - z| < \delta_\varepsilon \Rightarrow |u(y) - u(z)| < \varepsilon; \tag{2.18}$$

we reduce δ_ε, if necessary, in order that $\delta_\varepsilon \leq \frac{1}{3} \operatorname{dist}(\xi, \partial\Omega)$.

To prove (2.15), we observe that, for every $\varepsilon > 0$,

$$
\begin{aligned}
|u(\xi) - u_\rho(\xi)| &= \left| \int_{\mathscr{B}(\xi,\rho)} k_\rho(\xi - y)\{u(\xi) - u(y)\} \, dy \right| \\
&< \int_{\mathscr{B}(\xi,\rho)} k_\rho(\xi - y) \, \varepsilon \, dy \quad \text{if } \rho < \delta_\varepsilon \\
&= \varepsilon.
\end{aligned}
$$

To prove (2.16a), we write

$$M := \max_{\partial\Omega} u, \quad v_\rho := u_\rho \big|_{\partial G(\rho)}.$$

Now, if $x \in \partial G(\rho)$ and $y \in \mathscr{B}(x, \rho)$, then $\operatorname{dist}(y, \partial\Omega) < 3\rho$ [because $\operatorname{dist}(x, \partial\Omega) = 2\rho$ and $|y - x| < \rho$]; if also $3\rho < \delta_\varepsilon$, then $u(y) < M + \varepsilon$ [because $\operatorname{dist}(y, \partial\Omega) < \delta_\varepsilon$ and by (2.18)]. Accordingly, for all $x \in \partial G(\rho)$ and every $\varepsilon > 0$,

$$
\begin{aligned}
v_\rho(x) &= \int_{\mathscr{B}(x,\rho)} k_\rho(x - y) u(y) \, dy \\
&< \int_{\mathscr{B}(x,\rho)} k_\rho(x - y) \, (M + \varepsilon) \, dy \quad \text{if } 3\rho < \delta_\varepsilon \\
&= M + \varepsilon, \tag{2.19}
\end{aligned}
$$

which proves (2.16a).

It remains to prove (2.16b). If $M < 0$, then (2.19) shows that, for $3\rho < \delta_{-M}$ and for all $x \in \partial G(\rho)$, we have $v_\rho(x) < 0$ and hence $(v_\rho)^+(x) = 0$; therefore, both sides of (2.16b) are zero. If $M = 0$, then (2.19) shows that $v_\rho(x) < \varepsilon$ for every $\varepsilon > 0$ and for all $x \in \partial G(\rho)$, if $3\rho < \delta_\varepsilon$; again

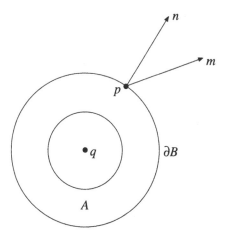

Fig. 2.3.

both sides of (2.16b) are zero. If $M > 0$, then $\max_{\partial\Omega} u^+ = M$, and (2.19) implies (2.16b) once more.   □

2.3 The boundary-point lemma and the strong maximum principle

Lemma 2.12 (the boundary-point lemma for balls). *Suppose that*
(a) $B \subset \Omega$ is a ball, $u \in C(\overline{B})$;
(b) u is a C^2-subsolution relative to L_0 or L and B, or a distributional subsolution relative to L_1 and B;
(c) $u(x) < u(p)$ for all $x \in B$ and some $p \in \partial B$, with $u(p) \geq 0$ when the coefficient c is not the zero function.
Let m be an outward unit vector at p $(m \cdot n > 0$ and $|m| = 1$, where n denotes the outward unit normal to ∂B at p). Then

$$\liminf\nolimits_{t \downarrow 0} \frac{u(p) - u(p - tm)}{t} > 0, \tag{2.20}$$

which implies that

$$\frac{\partial u}{\partial m}(p) := \lim\nolimits_{t \downarrow 0} \frac{u(p) - u(p - tm)}{t} > 0 \tag{2.21}$$

whenever this one-sided directional derivative exists.

Proof (i) As in Figure 2.3, let $B =: \mathscr{B}(q, \rho)$ and $A := \mathscr{B}(q, \rho) \setminus \overline{\mathscr{B}(q, \tfrac{1}{2}\rho)}$; it will suffice to consider the annular set $\overline{A}$. Also, let $M := u(p) = \sup_B u$.

If we can find a function $v \in C^2(\overline{A})$ such that

$$v(p) = 0, \qquad \qquad \text{(I)}$$

$$\frac{\partial v}{\partial m}(p) < 0, \qquad \qquad \text{(II)}$$

$$u + v \le M \quad \text{on} \quad \overline{A}, \qquad \qquad \text{(III)}$$

then we can prove (2.20) as follows. Let $w := u + v$. For $0 < t < \frac{1}{2}\rho$,

$$\frac{w(p) - w(p - tm)}{t} = \frac{M - w(p - tm)}{t} \ge 0 \qquad \text{[by (I) and (III)]},$$

whence

$$\lim \inf_{t \downarrow 0} \frac{u(p) - u(p - tm)}{t}$$

$$= \lim \inf_{t \downarrow 0} \frac{\{w(p) - w(p - tm)\} - \{v(p) - v(p - tm)\}}{t}$$

$$\ge \lim \inf_{t \downarrow 0} \frac{-v(p) + v(p - tm)}{t}$$

$$= -\frac{\partial v}{\partial m}(p) > 0 \qquad \text{[by (II)]}.$$

(ii) Consider the function defined on $\overline{A}$ by

$$v(x) := \delta \left(e^{-Kr^2} - e^{-K\rho^2} \right), \qquad r := |x - q|,$$

and shown in Figure 2.4; both positive constants δ and K are still to be chosen.

Certainly $v \in C^2(\overline{A})$; also (I) and (II) hold, since

$$\frac{\partial v}{\partial m}(p) = (m \cdot n) \frac{dv}{dr}\bigg|_{r=\rho} < 0.$$

For (III), we shall use the weak maximum principle, first considering the values of $u + v$ on ∂A. For $r = \rho$ we have $u \le M$, $v = 0$ and hence $u + v \le M$, with equality at p. For $r = \frac{1}{2}\rho$, we have $u < M$ by hypothesis (c); if $M - \alpha$ denotes the maximum of u for $r = \frac{1}{2}\rho$ [the supremum of a continuous function on a compact set is attained], then $\alpha > 0$. Choose $\delta = \alpha$; then $u \le M - \alpha$ and $v < \alpha$ for $r = \frac{1}{2}\rho$. Accordingly,

$$\max_{\partial A} (u + v) = M.$$

(iii) If we can choose K so that $Lv \ge 0$ in A (hence so that $L_0 v \ge 0$ in A, or $L_1 v \ge 0$ in A), then condition (III) will follow from one of

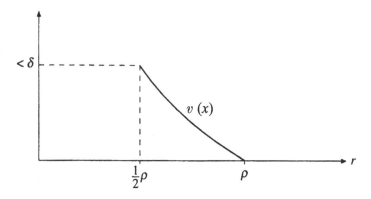

Fig. 2.4.

our three versions of the weak maximum principle, applied to $u + v$ and A. The condition $u(p) \geq 0$ when $c \neq 0$ banishes one difference in these versions. To deal with a distributional subsolution u, we add to the given condition,

$$\Lambda_{1d}(\varphi, u; A) \geq 0 \quad \text{whenever} \quad \varphi \in C_c^{\infty}(A) \quad \text{and} \quad \varphi \geq 0,$$

the condition $\Lambda_{1d}(\varphi, v; A) \geq 0$ for the same φ; this property of v will be implied by integration by parts [as in (2.9) and (2.10)] once we have $L_1 v \geq 0$ in A. Then, with $u + v \in C(\overline{A})$ and $\Lambda_{1d}(\varphi, u + v; A) \geq 0$, condition (III) will follow from Theorem 2.11.

(iv) It remains to calculate Lv and choose K. Since

$$\partial_j e^{-Kr^2} = e^{-Kr^2}(-K)2(x_j - q_j),$$
$$\partial_i \partial_j e^{-Kr^2} = e^{-Kr^2}\left\{4K^2(x_i - q_i)(x_j - q_j) - 2K\delta_{ij}\right\},$$

we have

$$\frac{1}{\delta}Lv(x) = e^{-Kr^2}\left\{4K^2 \sum_{i,j} a_{ij}(x)\,(x_i - q_i)(x_j - q_j)\right.$$

$$- 2K \sum_j a_{jj}(x) - 2K \sum_j b_j(x)\,(x_j - q_j)\Bigg\}$$

$$+ c(x)\left\{e^{-Kr^2} - e^{-K\rho^2}\right\}.$$

By the condition (2.3) of uniform ellipticity,

$$\frac{1}{\delta} Lv(x) \;\geq\; e^{-Kr^2} \left\{ 4K^2 \lambda_0 r^2 - 2K \sup_A \left(\sum_j |a_{jj}| + |b|\rho \right) - \sup_A |c| \right\}$$

$$> \; 0 \qquad \text{for } x \in \overline{A}$$

if we choose K sufficiently large, because $r^2 \geq (\tfrac{1}{2}\rho)^2$. □

Note a change of direction in the statement of the next theorem: there is no mention of $\overline{\Omega}$ or of $\partial\Omega$.

Theorem 2.13 (the strong maximum principle). *Suppose that*
(a) Ω *is a region (open and connected, possibly unbounded);*
(b) u *is a C^2-subsolution relative to L_0 or L and Ω, or a generalized subsolution relative to L_1 and Ω;*
(c) $\sup_\Omega u \geq 0$ *when the coefficient c is not the zero function.*
Under these hypotheses, if $\sup_\Omega u$ is attained at a point of Ω, then u is constant in Ω.

Proof Let $M := \sup_\Omega u$, and assume that this supremum is attained at $\hat{x} \in \Omega$. Define

$$F := \left\{ x \in \Omega \mid u(x) = M \right\}, \qquad G := \left\{ x \in \Omega \mid u(x) < M \right\};$$

then F is closed in the metric space Ω, and not empty because $\hat{x} \in F$; the set G is open in the metric space Ω. If G is empty, the theorem is true.

Suppose then that there is a point $x_0 \in G$. We shall obtain a contradiction by means of Lemma 2.12, first using the result that, because Ω is open and connected in $\mathbb{R}^N$, it is pathwise connected (Burkill & Burkill 1970, p.44; Cartan 1971, p.42). This implies existence of a continuous arc

$$\gamma := \left\{ \xi(t) \mid 0 \leq t \leq 1 \right\} \subset \Omega \quad \text{with } \xi(0) = x_0, \; \xi(1) = \hat{x},$$

as shown in Figure 2.5. Here $\xi \in C([0,1], \mathbb{R}^N)$, so that γ is compact; if Ω has a boundary, then $\text{dist}(\gamma, \partial\Omega) > 0$ because $\partial\Omega$ is closed in $\mathbb{R}^N$ and disjoint from γ.

Let $\tilde{x}$ be the first point of γ at which $u(x) = M$; here 'first' means 'with smallest t'. Possibly $\tilde{x} = \hat{x}$. Let q be any point of γ that is strictly between x_0 and $\tilde{x}$, and is such that $|q - \tilde{x}| < \text{dist}(\gamma, \partial\Omega)$ when Ω has a boundary. Now consider the ball $B := \mathcal{B}(q, \rho)$ with $\rho := \text{dist}(q, F)$. Then $\rho \leq |q - \tilde{x}| < \text{dist}(\gamma, \partial\Omega)$, so that $B \subset \Omega$; also, $B \subset G$ by construction. There exists a point $p \in F \cap \partial B$ because F is closed (possibly $p = \tilde{x}$). All

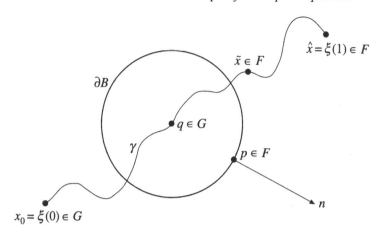

Fig. 2.5.

the hypotheses of Lemma 2.12 hold, so that at p the outward normal derivative

$$\frac{\partial u}{\partial n}(p) = n \cdot (\nabla u)(p) > 0.$$

But since $p \in F$, it is an interior maximum point of $u \in C^1(\Omega)$. Hence $(\nabla u)(p) = 0$ and we have our contradiction.   □

There are many boundary-point lemmas for elliptic operators and sets other than balls, but Lemma 2.12 is probably the heart of the matter. Theorem 2.15 is a consequence of that lemma, seasoned by a touch of the strong maximum principle. First, we need a definition.

Definition 2.14 A set Ω has the *interior-ball property* at a point $p \in \partial\Omega$ iff there exists a ball $B_0 \subset \Omega$ such that $p \in \partial B_0$; it has the *exterior-ball property* at p iff there exists a ball $B_1 \subset \mathbb{R}^N \setminus \overline{\Omega}$ such that $p \in \partial B_1$.   □

Figure 2.6 shows two cases of the interior-ball property for Ω, and therefore two cases of the exterior-ball property for $\mathbb{R}^N \setminus \overline{\Omega}$. Note that a unit vector m at p, outward from an interior ball B_0, need not be outward from Ω.

Theorem 2.15 (a boundary-point theorem for Ω). *Suppose that*
   (a) Ω *is a region*;
   (b) u *is a C^2-subsolution relative to L_0 or L and Ω, or a generalized subsolution relative to L_1 and Ω;*

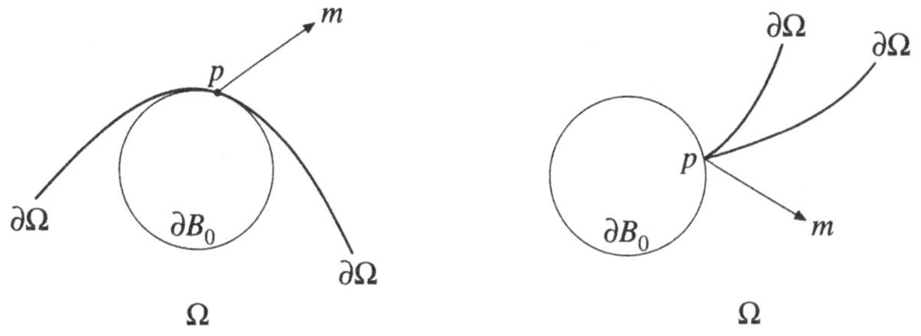

Fig. 2.6.

(c) *there is a point $p \in \partial\Omega$ such that $u \in C(\Omega \cup \{p\})$ and $u(p) = \sup_\Omega u$, with $u(p) \geq 0$ when the coefficient c is not the zero function;*

(d) *Ω has the interior-ball property at p.*

Let m be a unit vector at p, outward from an interior ball B_0 at p. Then either

$$\liminf_{t \downarrow 0} \frac{u(p) - u(p - tm)}{t} > 0 \tag{2.22}$$

(which implies that $(\partial u/\partial m)(p) > 0$ whenever this derivative exists), or u is constant in Ω.

Proof Let x_0 be the centre of the ball B_0 and let $\rho_0 := |p - x_0|$, so that $B_0 = \mathcal{B}(x_0, \rho_0)$. Now consider the smaller ball $B := \mathcal{B}(q, \frac{1}{2}\rho_0)$ with $q := \frac{1}{2}(p + x_0)$. Since $\overline{B} \subset B_0 \cup \{p\}$, we have $\overline{B} \subset \Omega \cup \{p\}$ and hence $u \in C(\overline{B})$.

If $u(x) < u(p)$ for all $x \in B$, then Lemma 2.12 implies (2.22). If $u(\hat{x}) = u(p)$ for some $\hat{x} \in B$, then $u(\hat{x}) = \sup_\Omega u$ and the strong maximum principle implies that u is constant in Ω. □

Suppose that p is what may be called an *edge point*; for example, $\Omega = (0,1)^2 \subset \mathbb{R}^2$ and $p = (0,0)$, or $\Omega = (0,1)^3 \subset \mathbb{R}^3$ and $p = (\frac{1}{2}, 0, 0)$. Then Ω lacks the interior-ball property at p, but something can still be said, for a subsolution, about an outward derivative or difference quotient at p. This is the subject of Appendix E.

Remark 2.16 (on the condition $c \leq 0$ in Ω). *For a subsolution u, if $\sup_\Omega u = 0$ in an application of the weak or strong maximum principle, or if $\sup_B u = 0$ in an application of the boundary-point lemma for a ball B, then the condition $c \leq 0$ in Ω (imposed in Definition 2.3) can be omitted.*

Proof We use the decomposition $c(x) = c^+(x) + c^-(x)$ [defined in Chapter 0, (v)]. The foregoing theorems and lemma are valid for the operator L^- and bilinear form Λ_1^- defined by

$$L^- u := Lu - c^+(x)u \ \left(= L_0 u + c^-(x)u \right),$$

$$\Lambda_1^-(\varphi, u; \Omega) := \Lambda_1(\varphi, u; \Omega) - c^+ \int_\Omega \varphi u \qquad (c^+ = c > 0).$$

When $\sup_\Omega u = 0$, we can use L^- in place of L because $Lu \geq 0$ and $u \leq 0$ imply that $L^- u \geq 0$. Again, when $\sup_\Omega u = 0$, we can use Λ_1^- in place of Λ_1 because $\Lambda_1(\varphi, u; \Omega) \geq 0$, $\varphi \geq 0$ and $u \leq 0$ imply that $\Lambda_1^-(\varphi, u; \Omega) \geq 0$. $\qquad \square$

2.4 A maximum principle for thin sets Ω

All our maximum principles so far have required that the coefficient $c(x) \leq 0$ for all $x \in \Omega$, unless it happens to be known for a subsolution u that $\sup_\Omega u = 0$ (Remark 2.16), or for a supersolution v that $\inf_\Omega v = 0$. In this section we proceed to a weak maximum principle for thin sets Ω in which both $c(x)$ and $u(x)$ are unrestricted in sign *a priori*. By a *thin set* Ω we mean one of specified diameter and small volume: $|\Omega| < \delta$, where the positive number δ depends only on diam Ω and on constants independent of Ω. To derive this maximum principle, we need some form of the basic estimate for elliptic equations that is presented here as Theorem 2.18. This estimate, in turn, is a consequence of elementary results in Appendix A for the Newtonian potential and of the weak maximum principle in Theorem 2.11.

Given a bounded open subset G of $\mathbb{R}^N$, we define a *modified Newtonian kernel* $\tilde{K}$ by

$$\tilde{K}(x) := \begin{cases} -\frac{1}{2}|x| + \frac{1}{2} \operatorname{diam} G & \text{if } N = 1, \\[2mm] \dfrac{1}{2\pi} \log \dfrac{\operatorname{diam} G}{|x|} & \text{if } N = 2, \\[2mm] \kappa_N \dfrac{1}{|x|^{N-2}} & \text{if } N \geq 3, \end{cases} \qquad (2.23)$$

where $x \neq 0$ if $N \geq 2$ and where κ_N is as in (A.18b) of Appendix A. This function differs from the Newtonian kernel K introduced by (A.18) only for $N = 1$ or 2, and then only by the addition of a constant which ensures that $\tilde{K}(x_0 - x) \geq 0$ whenever $x_0, x \in \overline{G}$. The corresponding *modified Newtonian potential* of a suitable density function $g : G \to \mathbb{R}$ is defined by

$$v(x_0) := \int_G \tilde{K}(x_0 - x)\, g(x)\, dx, \quad x_0 \in \mathbb{R}^N, \tag{2.24}$$

but here we restrict attention to field points $x_0 \in \overline{G}$.

Lemma 2.17 *Let G be a bounded open subset of $\mathbb{R}^N$ and let v be the modified Newtonian potential defined by (2.23) and (2.24). If $g \in L_p(G)$ with $1 \le p < \infty$ for $N = 1$, or with $N/2 < p < \infty$ for $N \ge 2$, then $v \in C(\overline{G})$ and v is a distributional solution (Definition A.7) of $-\triangle v = g$ in G. Moreover,*

$$|v(x_0)| \le \Gamma(N,p)\,(\mathrm{diam}\,G)^{2-(N/p)}\, \| g\,|L_p(G)\| \quad \text{for all} \;\; x_0 \in \overline{G}, \tag{2.25}$$

where, with the notation $1/p + 1/q = 1$ and $\sigma_N := |\partial \mathscr{B}_N(0,1)|$,

$$\left.\begin{aligned}
&\Gamma(1,1) = \frac{1}{2}, \quad \Gamma(1,p) = \left(\frac{1}{2}\right)^{1/p} \left(\frac{1}{q+1}\right)^{1/q} \quad \text{for} \;\; 1 < p < \infty, \\
&\Gamma(2,p) = \left(\frac{1}{2\pi}\right)^{1/p} \left\{ \int_0^1 \left(\log\frac{1}{\rho}\right)^q \rho \, d\rho \right\}^{1/q} \quad (1 < p < \infty), \\
&\text{for } N \ge 3, \quad \Gamma(N,p) = \frac{1}{N-2} \left(\frac{1}{\sigma_N}\right)^{1/p} \left(\frac{1}{N - Nq + 2q}\right)^{1/q} \\
&\qquad\qquad\qquad\qquad \left(\frac{N}{2} < p < \infty\right).
\end{aligned}\right\} \tag{2.26}$$

Proof For $N \ge 2$, it follows from Theorem A.6 that $v \in C(\overline{G})$, and from Theorem A.8 that v is a distributional solution of $-\triangle v = g$ not merely in G but in $\mathbb{R}^N$; for $N = 1$, the proofs are similar in strategy but much easier in detail. The bound (2.25) results from the Hölder inequality; we integrate $\tilde{K}(x_0 - x)^q$ over the ball $\mathscr{B}(x_0, \mathrm{diam}\,G)$, using $R := |x - x_0|$ as variable of integration. $\quad\square$

Notation and terminology The next theorem involves both the constant-coefficient operator L_1 (Definition 2.3) and the Lebesgue space $L_1(\Omega)$. Confusion will be avoided by unfailing display of Ω in the symbol $L_p(\Omega)$. Extending slightly the terminology in Definition 2.10, we shall say that $L_1 u + f \ge 0$ in Ω *in the distributional sense* iff u and f are locally integrable in Ω and

$$\int_\Omega \left\{ \sum_{i,j=1}^N a_{ij} \left(\partial_j \partial_i \varphi \right) u - \sum_{j=1}^N b_j \left(\partial_j \varphi \right) u + c\varphi u + \varphi f \right\} \geq 0$$

$$\text{whenever} \quad \varphi \in C_c^\infty(\Omega) \quad \text{and} \quad \varphi \geq 0. \qquad (2.27)$$

Theorem 2.18 (a basic estimate for the operator L_1). *Suppose that*

(a) Ω *is bounded,* $u \in C(\overline{\Omega})$;

(b) $L_1 u + f \geq 0$ *in* Ω *in the distributional sense, where* $f \in L_p(\Omega)$ *with* $1 \leq p < \infty$ *if* $N = 1$, *or with* $N/2 < p < \infty$ *if* $N \geq 2$;

(c) $u|_{\partial\Omega} \leq 0$.

Then

$$\max_{\overline{\Omega}} u \leq A \left\| f \mid L_p(\Omega) \right\|, \qquad (2.28)$$

where A is independent of u and $|\Omega|$ (but depends on $\operatorname{diam}\Omega$). *In fact, coarse inequalities give*

$$A = \frac{\Gamma(N,p)}{\lambda_0} \exp\left(\frac{|b| \operatorname{diam}\Omega}{\lambda_0} \right) (\operatorname{diam}\Omega)^{2-(N/p)}, \qquad (2.29)$$

where $\Gamma(N,p)$ is as in Lemma 2.17, λ_0 is the (positive) smallest eigenvalue of the matrix (a_{ij}) and $b = (b_1,\ldots,b_N)$ is the vector of coefficients in the term $b \cdot \nabla$ of L_1.

Proof (i) We make two co-ordinate transformations. First, let P be an orthogonal $N \times N$ matrix such that the transformation $y = Px$ makes the y_j-axes principal axes of the matrix a; say $\left(PaP^{-1} \right)_{ij} =: \lambda_i \delta_{ij}$ for $i, j \in \{1,\ldots,N\}$, where $\lambda_0 := \min_j \lambda_j > 0$. Second, we make a dilatation $z = Ey$, where $E_{ij} = \lambda_i^{-1/2} \delta_{ij}$, in order to transform L_1 to $\triangle + \cdots$. Writing

$$b^* := EPb, \quad G := EP(\Omega), \quad u^*(z) := u\left(P^{-1}E^{-1}z \right) = u(x),$$

and transforming f and φ like u, we obtain from hypothesis (b) that

$$\left\{ \triangle + b^* \cdot \nabla + c \right\} u^*(z) + f^*(z) \geq 0 \qquad \text{in } G \text{ in the d.s.,}$$

where $\triangle$ and ∇ are with respect to z, and d.s. means 'distributional sense'. More explicitly,

$$\int_G \left\{ \left(\triangle\varphi^* - b^* \cdot \nabla\varphi^* + c\varphi^* \right) u^* + \varphi^* f^* \right\} dz \geq 0$$

$$\text{whenever} \quad \varphi^* \in C_c^\infty(G) \quad \text{and} \quad \varphi^* \geq 0.$$

(ii) Next, first derivatives are removed by the transformation

$$u^*(z) =: \eta(z)\,\hat{u}(z), \quad f^*(z) =: \eta(z)\,\hat{f}(z), \quad \varphi^*(z) =: \frac{1}{\eta(z)}\check{\varphi}(z), \left.\begin{array}{}\\\\\end{array}\right\} \quad (2.30)$$
$$\text{where } \eta(z) := \exp\left(-\tfrac{1}{2}b^* \cdot z\right).$$

Hypothesis (b) now becomes

$$\left(\triangle - k^2\right)\hat{u}(z) + \hat{f}(z) \geq 0 \quad \text{in } G \text{ in the d.s.,} \qquad (2.31\text{a})$$

where $-k^2 := c - \tfrac{1}{4}|b^*|^2 \leq 0$; more explicitly,

$$\int_G \left\{ \left(\triangle\check{\varphi} - k^2\check{\varphi}\right)\hat{u} + \check{\varphi}\hat{f} \right\} dz \geq 0 \quad \text{whenever } \check{\varphi} \in C_c^\infty(G) \text{ and } \check{\varphi} \geq 0.$$

Conditions (a) to (c) also imply that $\hat{u} \in C(\overline{G})$, that $\hat{f} \in L_p(G)$ for the same p as in (b), and that

$$\hat{u}\big|_{\partial G} \leq 0. \qquad (2.31\text{b})$$

(iii) We compare the function $\hat{u}$ with the modified Newtonian potential v of $\left(\hat{f}\right)^+$. (Note that $\left(\hat{f}\right)^+ = (f^+)\hat{\,}$.) In other words,

$$v(z) := \int_G \widetilde{K}(z - \zeta)\left(\hat{f}\right)^+(\zeta)\,d\zeta, \qquad z \in \overline{G},$$

whence $v \in C(\overline{G})$ and

$$\left(\triangle - k^2\right)v(z) + \left(\hat{f}\right)^+(z) \;=\; -k^2 v(z) \quad \text{in } G \text{ in the d.s.,} \quad (2.32\text{a})$$
$$v(z) \;\geq\; 0 \quad \text{on } \overline{G}, \qquad\qquad\qquad\qquad\qquad (2.32\text{b})$$

by Lemma 2.17 and because $\widetilde{K}(z - \zeta) \geq 0$ and $\left(\hat{f}\right)^+(\zeta) \geq 0$.

Let $w := \hat{u} - v$; then (2.31a), (2.31b) and (2.32a), (2.32b) imply that

$$\left(\triangle - k^2\right)w \geq k^2 v - \left(\hat{f}\right)^- \geq 0 \quad \text{in } G \text{ in the d.s.,}$$

$$w\big|_{\partial G} \leq 0,$$

and the weak maximum principle (Theorem 2.11) ensures that

$$\max_{\overline{G}} w \;\leq\; \max_{\partial G} w^+ \;=\; 0.$$

The inequality (2.25) now yields

$$\max_{\overline{G}} \hat{u} \;\leq\; \max_{\overline{G}} v \;\leq\; \Gamma(N,p)(\text{diam } G)^{2-(N/p)}\|\hat{f} \mid L_p(G)\|. \qquad (2.33)$$

Returning to Ω, u and f, we note first that

$$\left|b^*\right| \le \lambda_0^{-1/2} |b|, \qquad \text{diam } G \le \lambda_0^{-1/2} \text{ diam } \Omega.$$

We may suppose that $0 \in \Omega$, because the operator L_1 and the norm $\| f \mid L_p(\Omega) \|$ are invariant under translation of co-ordinate axes; then (2.30) implies that

$$\text{max}_{\overline{\Omega}} u \le \exp\left(\frac{1}{2} \frac{|b| \text{ diam } \Omega}{\lambda_0}\right) \text{max}_{\overline{G}} \hat{u},$$

$$\| \hat{f} \mid L_p(G) \| \le \exp\left(\frac{1}{2} \frac{|b| \text{ diam } \Omega}{\lambda_0}\right) \lambda_0^{-N/2p} \| f \mid L_p(\Omega) \|.$$

The result (2.28) now follows from (2.33) and these inequalities.   □

As was mentioned earlier, the virtue of the following maximum principle is that the signs of the coefficient $\gamma(x)$ and of the subsolution u are both unrestricted in Ω.

Theorem 2.19 (a maximum principle for thin sets Ω). *Suppose that*

  (a) Ω *is bounded,* $u \in C(\overline{\Omega})$;
  (b) $L_{10}u + \gamma(x)u \ge 0$ *in* Ω *in the distributional sense, where* L_{10} *is the operator* L_1 *with coefficient* $c = 0$ *and* $\gamma \in L_\infty(\Omega)$;
  (c) $u\big|_{\partial\Omega} \le 0$.

Then

$$u \le 0 \quad \text{on} \quad \overline{\Omega} \qquad \text{whenever} \quad |\Omega| < \delta, \tag{2.34}$$

where the positive number δ *is independent of* u *and* $|\Omega|$ *(but depends on* diam Ω*). In fact, we may take*

$$\delta^{1/N} = \frac{\lambda_0}{2\Gamma(N,N) \text{ diam } \Omega \, \|\gamma \mid L_\infty(\Omega)\|} \exp\left(-\frac{|b| \text{ diam } \Omega}{\lambda_0}\right), \tag{2.35}$$

where the notation is that explained after (2.29).

Proof (i) The first step is to write $L_{10} + \gamma(x) \ge 0$ in a more tractable form. We introduce a constant $c \le 0$ with $|c|$ so large that

$$g(x) := -c + \gamma(x) \ge 0 \quad \text{almost everywhere in} \quad \Omega;$$

this can be done with $\| g \mid L_\infty(\Omega) \| \le 2\| \gamma \mid L_\infty(\Omega) \|$. Then

$$L_1 u + g(x)u = L_{10}u + \gamma(x)u \ge 0 \quad \text{in} \quad \Omega \quad \text{in the d.s.},$$

where d.s. means 'distributional sense', as before.

(ii) The second step is to decompose $g(x)u$:

$$L_1 u + g(x)u^+ \geq -g(x)u^- \geq 0 \quad \text{in } \Omega \text{ in the d.s.,}$$

to recall that $u|_{\partial\Omega} \leq 0$, and to apply Theorem 2.18 with $f = gu^+$; the choice $p = N$ is admissible for all $N \in \mathbb{N}$. This yields

$$\begin{aligned}
\max_{\overline{\Omega}} u &\leq A \| gu^+ \mid L_N(\Omega) \| \\
&\leq A \| g \mid L_\infty(\Omega) \| \max_{\overline{\Omega}} u^+ |\Omega|^{1/N}.
\end{aligned} \tag{2.36}$$

If $\max_{\overline{\Omega}} u < 0$, then (2.34) holds. If $\max_{\overline{\Omega}} u \geq 0$, then (2.36) and our bound for $\| g \mid L_\infty(\Omega) \|$ imply that

$$\left(\max_{\overline{\Omega}} u^+ \right) \left\{ 1 - 2A \| \gamma \mid L_\infty(\Omega) \| |\Omega|^{1/N} \right\} \leq 0,$$

from which (2.34) and (2.35) follow if we choose $|\Omega|$ to be so small that the expression in braces is positive. $\qquad\square$

2.5 Steps towards Phragmén–Lindelöf theory

All three versions of the weak maximum principle in §2.2 require Ω to be bounded and u to be continuous on $\overline{\Omega}$. If we relax one or other of these conditions, what other hypotheses will ensure that a subsolution, relative to L and Ω, can be bounded above in terms of its values on $\partial\Omega$? This is the theme of the remainder of this chapter, but, as it stands, the question is much too wide; we narrow it as follows.

(a) Among the many unbounded, proper subsets of $\mathbb{R}^N$ that might be considered, our favourite will be the half-space $D := \{ x \in \mathbb{R}^N \mid x_N > 0 \}$.

(b) In the rest of this chapter, the dimension $N \geq 2$.

(c) The condition $u \in C(\overline{\Omega})$ will be relaxed at only one or two boundary points; typically to $u \in C(\overline{\Omega} \setminus \{p\})$, where p is a specified point of $\partial\Omega$.

(d) Only the Laplace operator $\triangle$ will be considered. There is no disgrace in this restriction; good answers to our question are sensitive to details of the differential operator L, and each proof involves a comparison function tailored rather closely to the task in hand. To launch here into the more general theory initiated by Gilbarg (1952) and E. Hopf (1952a) would be a catastrophic attempt to run before we have learned to walk.

Definition 2.20 We shall say that u is *subharmonic in Ω* iff it is a distributional subsolution relative to $\triangle$ and Ω; that is, iff u is locally integrable in Ω and

$$\int_\Omega (\triangle \varphi) u \geq 0 \quad \text{whenever} \quad \varphi \in C_c^\infty(\Omega) \quad \text{and} \quad \varphi \geq 0.$$

Then u is *superharmonic in Ω* iff $-u$ is subharmonic there; u is *harmonic in Ω* iff it is both subharmonic and superharmonic in Ω. $\square$

This definition of 'harmonic' (which follows inevitably from the useful definition of 'subharmonic' that we have adopted) scarcely does justice to harmonic functions. Theorems B.6 and B.10 show that, if u is harmonic in Ω according to Definition 2.20, then, after re-definition on a set of measure zero, not only is u a C^2-solution of $\triangle u = 0$, but also u is real-analytic in Ω.

Our opening question can now be replaced by the following. If u is continuous on $\overline{D}$ and subharmonic in D, to what rate of growth, as $|x| \to \infty$, must $u(x)$ be restricted in order that $\sup_D u = \sup_{\partial D} u$? If u is continuous merely on $\overline{\Omega} \setminus \{p\}$ and subharmonic in Ω, to what rate of growth, as $x \to p$, must $u(x)$ be restricted in order that $\sup_\Omega u = \sup_{\partial \Omega \setminus \{p\}} u$?

We begin by inspecting some simple and explicit harmonic functions that *vanish* on the boundary of the half-space D or on a punctured boundary $\partial \Omega \setminus \{p\}$; these functions indicate rates of growth that are too large in the context of our questions.

Examples 1. The harmonic polynomials

$$p_1(x) = x_N, \ p_2(x) = 2x_1 x_N, \ p_3(x) = 3x_1^2 x_N - x_N^3, \ldots,$$
$$p_m(x) = \operatorname{Im}(x_1 + ix_N)^m, \ldots \qquad (2.37)$$

all vanish on ∂D; if $N \geq 3$, there are many more such polynomials.

But, apart from the zero function, no function springs to mind that is continuous on $\overline{D}$, is harmonic in D, vanishes on ∂D and is $o(r)$ as $r := |x| \to \infty$. This is significant: the critical rate of growth for the result $\sup_D u = \sup_{\partial D} u$ will turn out to be close to growth like r as $r \to \infty$.

2. If we seek functions that are continuous on $\overline{D} \setminus \{0\}$, tend to zero at infinity, are harmonic in D and vanish on $\partial D \setminus \{0\}$, then the prototype is

$$q_1(x) = x_N / r^N, \quad x \in \mathbb{R}^N \setminus \{0\}, \quad r := |x|. \qquad (2.38)$$

This is the potential of a particular *dipole* (§A.4); more precisely, the potential of a multipole of type $(0, \ldots, 0, 1)$. Differentiating this repeatedly in horizontal directions (with respect to x_j, $j \leq N - 1$), we generate multipole potentials like

$$\left. \begin{aligned} q_2(x) &= \partial_1 q_1(x) = -N x_1 x_N r^{-N-2}, \\ q_3(x) &= \partial_1^2 q_1(x) = N x_N \{ (N+2) x_1^2 - r^2 \} r^{-N-4}, \end{aligned} \right\} \tag{2.39}$$

which retain the properties listed before (2.38), but have a stronger singularity at the origin, relative to q_1, and a more rapid decay at infinity.

However, no non-trivial function springs to mind that has the properties listed before (2.38) and is $o(r^{-N+1})$ as $x \to 0$ with $x \in D$. Again this is significant: the critical rate of growth for the result $\sup_\Omega u = \sup_{\partial\Omega \setminus \{p\}} u$ will turn out to be close to growth like $|x - p|^{-N+1}$ as $x \to p$, when $\partial\Omega$ is smooth.

3. We now allow a singularity at the south pole $p := (0, \ldots, 0, -a)$ of the ball $B := \mathscr{B}(0, a)$ in $\mathbb{R}^N$. The *Poisson kernel* (§B.5) gives an example of a function continuous on $\overline{B} \setminus \{p\}$, harmonic in B and equal to zero on $\partial B \setminus \{p\}$:

$$P(x, p) = \text{const.} (a^2 - r^2) |x - p|^{-N}, \quad x \in \overline{B} \setminus \{p\}, \ r := |x|. \tag{2.40}$$

Again the singularity at p is of dipole type, and again appropriate differentiation generates a stronger singularity at p, while conserving the value zero on $\partial B \setminus \{p\}$. Thus the harmonic function

$$\begin{aligned} Q(x) &= (x_N \partial_1 - x_1 \partial_N) P(x, p) \\ &= \text{const.} \, x_1 (a^2 - r^2) |x - p|^{-N-2}, \quad x \in \overline{B} \setminus \{p\}, \tag{2.41} \end{aligned}$$

has a quadrupole singularity at p.

Rather as in Example 2, no non-trivial function springs to mind that is continuous on $\overline{B} \setminus \{p\}$, harmonic in B, equal to zero on $\partial B \setminus \{p\}$ and is $o(|x - p|^{-N+1})$ as $x \to p$; this is significant in the same way as before.

The Phragmén–Lindelöf theory that follows *must be distinguished from Phragmén–Lindelöf theory for holomorphic functions* (complex-analytic functions). In that theory one supposes that, for example, $\sup_{\partial D} |u + iv|$ is known, where D is the upper half of the complex plane $\mathbb{C}$; the analogous situation for us, when $N = 2$, is that only $\sup_{\partial D} |u|$ (or only $\sup_{\partial D} |v|$) is known. In the case of holomorphic functions (Hille 1973, Chapter 18; Titchmarsh 1932, §§5.6–5.8) much more can be inferred because much more is given.

Definition 2.21 Let $D := \{ x \in \mathbb{R}^N \mid x_N > 0 \}$, $N \geq 2$; let $D_a = D \cap \mathscr{B}(0, a)$

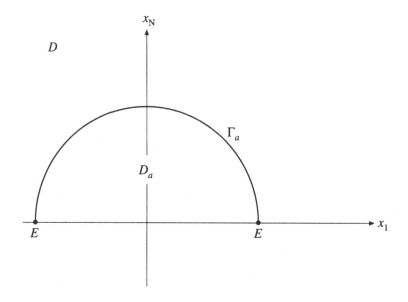

Fig. 2.7.

and $\Gamma_a := D \cap \partial \mathscr{B}(0, a)$ (Figure 2.7); denote the equator of $\mathscr{B}(0, a)$ by $E := \partial D \cap \partial \mathscr{B}(0, a)$.

A function with values $V(x, a)$ will be called a *comparison function of the first kind* iff

(a) for each $a \in (0, \infty)$,

$$V(., a) \in C(\overline{D}_a \setminus E) \cap C^2(D_a) \quad \text{and} \quad V(., a) \geq 0 \text{ on } \overline{D}_a \setminus E, \quad (2.42)$$

$$(\triangle V)(., a) = 0 \quad \text{in} \quad D_a, \quad (2.43)$$

$$V(., a) > 0 \quad \text{on} \quad \Gamma_a; \quad (2.44)$$

(b) $\inf \left\{ V(x, a) \mid x \in \Gamma_a \right\} \to \infty$ as $a \to \infty$;

(c) there is a function $\lambda : D \to (0, \infty)$ such that $V(x_0, a) \leq \lambda(x_0)$ whenever $x_0 \in D$ and $a \geq 2|x_0|$.   □

Functions having these properties will be displayed in due course; first, we prove the lemma for which the definition has been designed. This lemma shows that, if a function u is continuous on $\overline{D}$, is subharmonic in D and is smaller, in order of magnitude, than $V(., a(n))$ on some sequence $(\Gamma_{a(n)})$ of hemispheres marching to infinity, then we retain the result $\sup_D u = \sup_{\partial D} u$.

The proof will show that the important case of the growth condition (2.45) is that in which the limit inferior *equals* zero.

Lemma 2.22 *Let V be a comparison function of the first kind. If $u \in C(\overline{D})$, if u is subharmonic in D and if*

$$\liminf_{a\to\infty} \sup\left\{ \frac{u(x)}{V(x,a)} \;\middle|\; x \in \Gamma_a \right\} \leq 0, \qquad (2.45)$$

then

$$\sup_D u = \sup_{\partial D} u.$$

Proof (i) We may suppose that $\sup_{\partial D} u < \infty$, otherwise the result is trivial. Let $\tilde{u} := u - \sup_{\partial D} u$. Then $\sup_{\partial D} \tilde{u} = 0$ and $\tilde{u}$ also satisfies the growth condition (2.45), because the definition of $\tilde{u}$ and hypothesis (b) imply that

$$\sup\left\{ \frac{|u(x) - \tilde{u}(x)|}{V(x,a)} \;\middle|\; x \in \Gamma_a \right\} \to 0 \quad \text{as } a \to \infty.$$

Let both $x_0 \in D$ and $\varepsilon > 0$ be given; we shall prove the lemma by showing that $\tilde{u}(x_0) < \varepsilon$.

(ii) If

$$\liminf_{a\to\infty} \sup\left\{ \tilde{u}(x) \;\middle|\; x \in \Gamma_a \right\} \leq 0,$$

then no comparison function is needed. For, there is a sequence $(a(n))$ tending to infinity for which the supremum of $\tilde{u}(x)$ over $\Gamma_{a(n)}$ tends to a non-positive limit. We choose $a(k)$ so large that $x_0 \in D_{a(k)}$ and so large that $\tilde{u}(x) < \varepsilon$ on $\Gamma_{a(k)}$. Then the weak maximum principle, Theorem D.11, applied to $\tilde{u}$ on $\overline{D}_{a(k)}$ shows that $\tilde{u}(x_0) < \varepsilon$ as desired.

(iii) It remains to consider the following case: there is a number $A \geq 0$ such that

$$\sup\left\{ \tilde{u}(x) \;\middle|\; x \in \Gamma_a \right\} > 0 \quad \text{whenever} \quad a > A,$$

and

$$\liminf_{a\to\infty} \sup\left\{ \frac{\tilde{u}(x)}{V(x,a)} \;\middle|\; x \in \Gamma_a \right\} = 0. \qquad (2.46)$$

Let

$$s(a) := \sup\left\{ \frac{\tilde{u}(x)}{V(x,a)} \;\middle|\; x \in \Gamma_a \right\} \quad \text{for} \quad a > A;$$

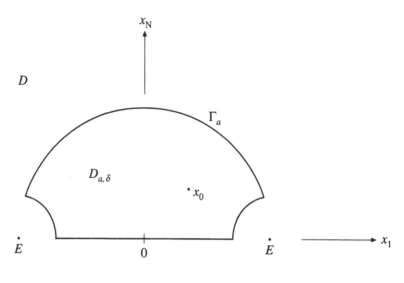

Fig. 2.8.

then $s(a) > 0$. We so choose a that, at the given point x_0,

$$s(a)\, V(x_0, a) < \tfrac{1}{2}\varepsilon; \tag{2.47}$$

this can be done because (2.46) states that there is a sequence $(a(n))$ for which $s(a(n)) \to 0$ as $n \to \infty$ and $a(n) \to \infty$, while hypothesis (c) ensures that $V(x_0, a(n)) \le \lambda(x_0)$ whenever $a(n) \ge 2|x_0|$. With a now fixed at this value, define

$$\begin{aligned}
\varphi(x) &:= \tilde{u}(x) - s(a)\, V(x, a) \qquad \text{for } x \in \overline{D}_a \setminus E, \\
D_{a,\delta} &:= \{x \in D \mid |x| < a,\ \operatorname{dist}(x, E) > \delta > 0\}
\end{aligned}$$

(see Figure 2.8). Choose δ so small that $x_0 \in D_{a,\delta}$ and so small that

$$\operatorname{dist}(x, E) = \delta \quad \text{and} \quad x \in \overline{D} \;\Rightarrow\; \tilde{u}(x) < \tfrac{1}{2}\varepsilon;$$

this choice is possible because $\tilde{u}\big|_E \le 0$ and $\tilde{u} \in C(\overline{D})$.

(iv) Finally, we apply the weak maximum principle (Theorem 2.11) to φ on $\overline{D}_{a,\delta}$. Certainly $\varphi \in C(\overline{D}_{a,\delta})$, and φ is subharmonic in $D_{a,\delta}$, because $\tilde{u}$ is subharmonic there and $V(.,a)$ is harmonic.

The boundary values of φ are as follows. On $\partial D_{a,\delta} \cap \partial D$ we have $\tilde{u}(x) \le 0$ and $V(x, a) \ge 0$, hence $\varphi(x) \le 0$. On the part of $\partial D_{a,\delta}$ distant δ from E we have $\tilde{u}(x) < \tfrac{1}{2}\varepsilon$ and $V(x, a) \ge 0$, hence $\varphi(x) < \tfrac{1}{2}\varepsilon$. On

$\partial D_{a,\delta} \cap \Gamma_a$ we have $\varphi(x) \leq 0$ by the definition of $s(a)$:

$$x \in \Gamma_a \Rightarrow \varphi(x) = V(x,a)\left\{ \frac{\tilde{u}(x)}{V(x,a)} - \sup_{y\in\Gamma_a} \frac{\tilde{u}(y)}{V(y,a)} \right\} \leq 0.$$

Therefore the weak maximum principle implies that $\varphi(x) < \frac{1}{2}\varepsilon$ on $\overline{D}_{a,\delta}$; it follows from (2.47) that

$$\tilde{u}(x_0) = \varphi(x_0) + s(a)V(x_0,a) < \varepsilon,$$

as desired. $\qquad\qquad\qquad\qquad\qquad\qquad\qquad\qquad\qquad\qquad\qquad\square$

The next item is a naive application of Lemma 2.22, based on a simple comparison function and intended to make Definition 2.21 less mysterious. In this example, V is independent of a, and does not have a discontinuity on the equator E. The full force of Lemma 2.22 will emerge only in §2.7, after more elaborate comparison functions have been constructed.

Example 2.23 *Let $D := \{ x \in \mathbb{R}^2 \mid x_2 > 0 \}$. If $u \in C(\overline{D})$, if u is subharmonic in D and if, for some constant $\beta \in (0,1)$,*

$$\liminf_{a\to\infty} \sup\{ a^{-\beta} u(x) \mid x \in \Gamma_a \} \leq 0 \qquad (2.48)$$

(in particular, if $u(x) = o(r^\beta)$ for some $\beta \in (0,1)$ as $r := |x| \to \infty$), then

$$\sup_D u = \sup_{\partial D} u.$$

Proof Denote points of $\overline{D}$ by $x = (r\cos\theta, r\sin\theta)$, $0 \leq \theta \leq \pi$. We claim that the formula

$$V(x) := r^\beta \sin(\beta\theta + k), \qquad k := (1-\beta)\frac{\pi}{2}, \qquad x \in \overline{D},$$

defines a comparison function of the first kind. For, referring to Definition 2.21, we observe that $V \subset C(\overline{D}) \cap C^2(D)$; that $V \geq 0$ on $\overline{D}$, with $V > 0$ on $\overline{D} \setminus \{0\}$, because

$$\sin(\beta\theta + k) \geq \sin k \quad \text{for } 0 \leq \theta \leq \pi;$$

and that $\triangle V = 0$ in D because

$$V(x) = \operatorname{Im} e^{ik} z^\beta \qquad (z := x_1 + ix_2 = re^{i\theta}).$$

Thus V satisfies condition (a) of Definition 2.21; it satisfies (b) because $V(x) > a^\beta \sin k$ when $x \in \Gamma_a$; for (c), we may choose $\lambda(x_0) := V(x_0)$ or $\lambda(x_0) := r_0^\beta$.

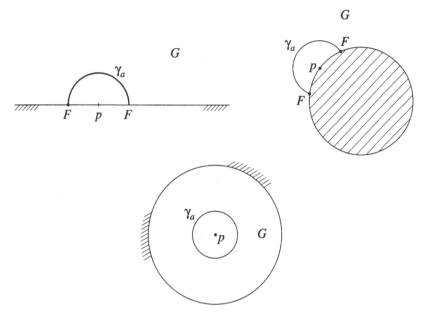

Fig. 2.9.

If the growth condition (2.48) implies (2.45) for the present function V, then Lemma 2.22 implies the present result. Now,

$$\sin k < \frac{V(x)}{a^\beta} \le 1 \quad \text{when } x \in \Gamma_a,$$

so that the two growth conditions are equivalent. $\square$

Definition 2.24 Let G be a connected open set in $\mathbb{R}^N$, $N \ge 2$, let $p \in \partial G$ be given, let $G_a := G \setminus \overline{\mathscr{B}(p,a)}$, let $\gamma_a := G \cap \partial \mathscr{B}(p,a)$ and let $F := \partial G \cap \partial \mathscr{B}(p,a)$. Here $a \in (0, a_0)$ and a_0 is a positive constant depending only on G. (Three possible configurations are shown in Figure 2.9; in the third, $G := \mathscr{B}(p, a_0) \setminus \{p\}$ and F is empty.)

A function with values $W(x, a)$ will be called a *comparison function of the second kind* iff
(a) for each $a \in (0, a_0)$,

$$W(.,a) \in C(\overline{G}_a \setminus F) \cap C^2(G_a) \quad \text{and} \quad W(.,a) \ge 0 \quad \text{on } \overline{G}_a \setminus F, \quad (2.49)$$

$$(\triangle W)(.,a) = 0 \quad \text{in } G_a, \quad (2.50)$$

$$W(.,a) > 0 \quad \text{on } \gamma_a; \quad (2.51)$$

(b) $\inf\{\, W(x,a) \mid x \in \gamma_a \,\} \to \infty$ as $a \to 0$;

(c) there is a function $\lambda : G \to (0,\infty)$ such that $W(x_0,a) \leq \lambda(x_0)$ whenever $x_0 \in G$, $a \leq \frac{1}{2}|x_0 - p|$ and $a < a_0$. $\qquad\square$

Lemma 2.25 *Let G be as in Definition 2.24 and let W be a comparison function of the second kind. Let Ω be a bounded open subset of G such that $p \in \partial\Omega \cap \partial G$. If $u \in C(\overline{\Omega} \setminus \{p\})$, if u is subharmonic in Ω and if*

$$\liminf_{a \to 0} \sup\left\{ \frac{u(x)}{W(x,a)} \,\middle|\, x \in \Omega \cap \partial\mathscr{B}(p,a) \right\} \leq 0, \qquad (2.52)$$

then

$$\sup_\Omega u = \sup_{\partial\Omega \setminus \{p\}} u.$$

Proof The proof resembles that of Lemma 2.22, but to condense it ruthlessly would be a false economy.

(i) We may suppose that $\sup_{\partial\Omega \setminus \{p\}} u < \infty$, otherwise the result is trivial. Let $\tilde{u} := u - \sup_{\partial\Omega \setminus \{p\}} u$. Then $\sup_{\partial\Omega \setminus \{p\}} \tilde{u} = 0$ and $\tilde{u}$ also satisfies (2.52), because $u - \tilde{u}$ is a (finite) constant and by condition (b) in Definition 2.24.

Let both $x_0 \in \Omega$ and $\varepsilon > 0$ be given; we shall prove the lemma by showing that $\tilde{u}(x_0) < \varepsilon$. To this end, write

$$\Omega_a := \Omega \setminus \overline{\mathscr{B}(p,a)} \quad \text{and} \quad \sigma_a := \Omega \cap \partial\mathscr{B}(p,a).$$

(ii) If

$$\liminf_{a \to 0} \sup\{\, \tilde{u}(x) \mid x \in \sigma_a \,\} \leq 0,$$

then no comparison function is needed. We argue as in the proof of Lemma 2.22, using small surfaces $\sigma_{a(n)}$ with $a(n) \to 0$ instead of large hemispheres $\Gamma_{a(n)}$ with $a(n) \to \infty$.

(iii) It remains to consider the following case: there is a number $\alpha > 0$ such that

$$\sup\{\, \tilde{u}(x) \mid x \in \sigma_a \,\} > 0 \quad \text{whenever } a < \alpha,$$

and

$$\liminf_{a \to 0} \sup\left\{ \frac{\tilde{u}(x)}{W(x,a)} \,\middle|\, x \in \sigma_a \right\} = 0. \qquad (2.53)$$

Let

$$s(a) := \sup\left\{ \frac{\tilde{u}(x)}{W(x,a)} \,\middle|\, x \in \sigma_a \right\} \quad \text{for } a < \alpha;$$

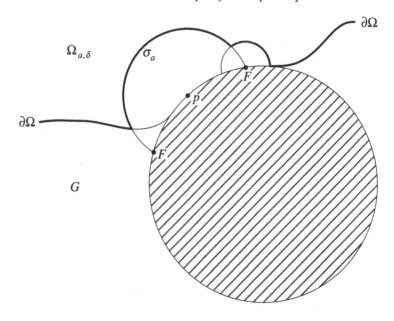

Fig. 2.10.

then $s(a) > 0$. Choose a to be such that, at the given point x_0,

$$s(a) \, W(x_0, a) < \tfrac{1}{2}\varepsilon; \qquad (2.54)$$

this can be done because of (2.53) and hypothesis (c) in Definition 2.24. With a now fixed at this value, define

$$\psi(x) := \tilde{u}(x) - s(a) \, W(x, a) \qquad \text{for } x \in \overline{\Omega}_a \setminus F.$$

If $\partial\Omega$ intersects F, define

$$\Omega_{a,\delta} := \left\{ x \in \Omega \mid |x - p| > a, \ \text{dist}(x, F) > \delta > 0 \right\}$$

(Figure 2.10); choose δ so small that $x_0 \in \Omega_{a,\delta}$ and so small that

$$\text{dist}(x, F) = \delta \ \text{ and } \ x \in \overline{\Omega} \ \Rightarrow \ \tilde{u}(x) < \tfrac{1}{2}\varepsilon;$$

this choice is possible because $\tilde{u}|_{\partial\Omega \cap F} \leq 0$ and $\tilde{u} \in C\left(\overline{\Omega} \setminus \{p\}\right)$. If $\partial\Omega$ does not intersect F (in particular, if F is empty), define $\Omega_{a,\delta} := \Omega_a$. In either case, $W(., a)$ is continuous on $\overline{\Omega}_{a,\delta}$ and $\tilde{u}(x) < \tfrac{1}{2}\varepsilon$ on $\partial\Omega_{a,\delta} \setminus \sigma_a$.

   (iv) Now apply the weak maximum principle, Theorem 2.11, to ψ on $\overline{\Omega}_{a,\delta}$, observing that $\psi \in C(\overline{\Omega}_{a,\delta})$ and that ψ is subharmonic in $\Omega_{a,\delta}$. On $\partial\Omega_{a,\delta} \setminus \sigma_a$ we have $\tilde{u}(x) < \tfrac{1}{2}\varepsilon$ and $W(x, a) \geq 0$, hence $\psi(x) < \tfrac{1}{2}\varepsilon$. On

$\partial\Omega_{a,\delta} \cap \sigma_a$ we have $\psi(x) \leq 0$ by the definition of $s(a)$. The maximum principle ensures that $\psi(x) < \frac{1}{2}\varepsilon$ on $\overline{\Omega}_{a,\delta}$; it follows from (2.54) that

$$\tilde{u}(x_0) = \psi(x_0) + s(a)\, W(x_0, a) < \varepsilon,$$

as desired. □

Like Example 2.23, our first application of Lemma 2.25 will involve only simple comparison functions. But the result is better than that of Example 2.23; it is not restricted to $\mathbb{R}^2$ and it is best possible in a certain sense (Exercise 2.44) when no smoothness is demanded of $\partial\Omega$ at p.

Theorem 2.26 *Let Ω be bounded in $\mathbb{R}^N$, $N \geq 2$, let $p \in \partial\Omega$ and write $\sigma_a := \Omega \cap \partial\mathcal{B}(p, a)$. Let b be so large that $\Omega \subset \mathcal{B}(p, b)$.*
If $u \in C(\overline{\Omega} \setminus \{p\})$, if u is subharmonic in Ω and if

$$\liminf_{a \to 0}\ \sup\left\{ \left.\frac{u(x)}{\log(b/a)}\ \right|\ x \in \sigma_a \right\} \leq 0 \quad \text{when}\ \ N = 2, \qquad (2.55)$$

$$\liminf_{a \to 0}\ \sup\left\{ a^{N-2}u(x)\ \middle|\ x \in \sigma_a \right\} \leq 0 \quad \text{when}\ \ N \geq 3, \qquad (2.56)$$

then

$$\sup\nolimits_{\Omega} u = \sup\nolimits_{\partial\Omega \setminus \{p\}} u.$$

Proof The comparison functions are potentials of point sources (multiples of Newtonian kernels), discussed at some length in Appendix A.

(i) For $N = 2$, we choose $G := \mathcal{B}(p, b) \setminus \{p\}$ for the set in Definition 2.24 and define

$$W(x) := \log\frac{b}{|x - p|} \quad \text{for}\ \ x \in \overline{\mathcal{B}(p, b)} \setminus \{p\}.$$

Then $G_a = \mathcal{B}(p, b) \setminus \overline{\mathcal{B}(p, a)}$ and $\gamma_a = \partial\mathcal{B}(p, a)$; condition (a) of Definition 2.24 is satisfied. Since $W(x) = \log(b/a)$ when $x \in \gamma_a$, condition (b) holds. For (c), we choose $\lambda(x_0) := W(x_0)$. Thus W is a comparison function of the second kind. The growth conditions (2.52) and (2.55) coincide for this function W; therefore Lemma 2.25 implies the present result.

(ii) For $N \geq 3$, we choose $G := \mathbb{R}^N \setminus \{p\}$ for the set in Definition 2.24 and define

$$W(x) := |x - p|^{-N+2} \quad \text{for}\ \ x \in \mathbb{R}^N \setminus \{p\}.$$

One checks without difficulty, very much as in (i), that this function W is a comparison function of the second kind. The growth conditions (2.52) and (2.56) coincide for this W; again Lemma 2.25 implies the present result.   □

2.6 Comparison functions of Siegel type

This section concerns functions $g(.\,;a)$, $g_e(.\,;a)$ and $g_2(.\,;a,b)$ with the property that $ag(.\,;a)$ is a useful comparison function of the first kind, while $a^{-N+1}g_e(.\,;a)$ and $a^{-N+1}g_2(.\,;a,b)$ are corresponding comparison functions of the second kind; $g_e(.\,;a)$ and $g_2(.\,;a,b)$ are defined on different domains. These functions will enable us to extend Example 2.23 to half-spaces in $\mathbb{R}^N$ for all $N \geq 2$; to improve the rate of growth allowed in Example 2.23 from approximately $o(r^\beta)$, where $r := |x|$ and $\beta < 1$, to approximately $o(r^2/x_N)$ as $r \to \infty$; and to improve the rate of growth allowed in Theorem 2.26 from approximately $o(|x - p|^{-N+2})$, for $N \geq 3$ and $x \to p$, to something slightly bigger than $o(|x - p|^{-N+1})$, provided that Ω has the exterior-ball property at p (Definition 2.14).

The functions g, g_e and g_2 will be called *of Siegel type* because g for $N = 2$, displayed here in (2.59), was introduced into Phragmén–Lindelöf theory by D. Siegel (1988). The construction of g for all $N \geq 2$, from the Poisson integral formula for functions harmonic in a ball (§B.5), is the subject of Appendix C. The functions g_e and g_2 result from applications to g of the Kelvin transformation (§B.3).

It will be convenient to use the *signum function*, defined by

$$\operatorname{sgn} t := \begin{cases} -1 & \text{if} \quad t < 0, \\ 0 & \text{if} \quad t = 0, \\ 1 & \text{if} \quad t > 0. \end{cases} \tag{2.57}$$

Theorem 2.27 *Let* $B := \mathscr{B}(0,a)$ *in* $\mathbb{R}^N$, $N \geq 2$, *and let* $E := \{ x \in \partial B \mid x_N = 0 \}$ *denote the equator of* B.

(a) *There exists a function* $g = g(.\,;a)$, *which we call the primary function of Siegel type, such that* $g \in C(\overline{B} \setminus E) \cap C^\infty(B)$ *and*

$$\triangle g = 0 \quad \text{in} \quad B, \tag{2.58a}$$
$$g(x) = a/x_N \quad \text{on} \quad \partial B \setminus E, \tag{2.58b}$$
$$|g(x)| \leq \text{const.}\,|x_N|/a \quad \text{if} \quad r := |x| \leq a/2, \tag{2.58c}$$

where the constant depends only on N. *Also,* $\operatorname{sgn} g(x) = \operatorname{sgn} x_N$ *on* $\overline{B} \setminus E$, *and* $g(x;a)$ *depends only on* x/a.

(b) *For* $N = 2$, *let* $(x, y) \in \mathbb{R}^2$ *and* $z = x + iy \in \mathbb{C}$. *Then, on* $\overline{B} \setminus E \subset \mathbb{R}^2$,

$$g(x, y; a) = \text{Im} \left(\frac{a}{a - z} - \frac{a}{a + z} \right) = \frac{ay}{(a - x)^2 + y^2} + \frac{ay}{(a + x)^2 + y^2}. \quad (2.59)$$

Proof See Appendix C. $\qquad\qquad\qquad\qquad\qquad\qquad\qquad\qquad\qquad\qquad\square$

Corollary 2.28 *The exterior function* $g_e = g_e(.\,; a)$ *of Siegel type is defined by*

$$g_e(x; a) := \left(\frac{a}{r} \right)^{N-2} g \left(\frac{a^2}{r^2} x; a \right), \quad x \in \mathbb{R}^N \setminus (B \cup E), \quad (2.60)$$

where again $r := |x|$ *and* B, E *are as in Theorem 2.27. It follows that* $g_e \in C \left(\mathbb{R}^N \setminus \{B \cup E\} \right) \cap C^\infty(\mathbb{R}^N \setminus \overline{B})$ *and that*

$$\triangle g_e = 0 \quad \text{in } \mathbb{R}^N \setminus \overline{B}, \quad (2.61\text{a})$$

$$g_e(x) = a/x_N \quad \text{on } \partial B \setminus E, \quad (2.61\text{b})$$

$$|g_e(x)| \leq \text{const.}\, a^{N-1} r^{-N} |x_N| \quad \text{if } r \geq 2a, \quad (2.61\text{c})$$

where the constant depends only on N. *Also,* $\text{sgn}\, g_e(x) = \text{sgn}\, x_N$ *on* $\mathbb{R}^N \setminus (B \cup E)$, *and* $g_e(x; a)$ *depends only on* x/a.

Proof A calculation, done fully in Theorem B.15, shows that

$$\triangle g_e(x) = \left(\frac{a}{r} \right)^{N+2} (\triangle g) \left(\frac{a^2}{r^2} x \right) = 0$$

if $r > a$ and hence $|a^2 x/r^2| = a^2/r < a$. The remaining properties of g_e are immediate consequences of the definition (2.60) and the corresponding properties of g. $\qquad\qquad\qquad\qquad\qquad\qquad\qquad\qquad\qquad\square$

Inspection of Corollary 2.28 and Definition 2.24 shows that $a^{-N+1} g_e$, restricted to $D \setminus (B \cup E)$ (where D is our usual half-space), is a comparison function of the second kind, with $G = D$ and $p = 0$ in the notation of Definition 2.24. The restriction to $\overline{D} \setminus (B \cup E)$ is needed in order that $g_e \geq 0$. Therefore $a^{-N+1} g_e$ can be used in Lemma 2.25 for sets Ω that are on one side of a hyperplane containing the point p of $\partial\Omega$ at which u may be discontinuous; this is illustrated in Figure 2.11. In particular, $a^{-N+1} g_e$ can be used for convex sets Ω.

Suppose now that Ω has merely the exterior-ball property (Definition 2.14) at the specified point $p \in \partial\Omega$. Then a suitable comparison function $a^{-N+1} g_2(.\,; a, b)$ is found by inversion relative to a sphere as follows.

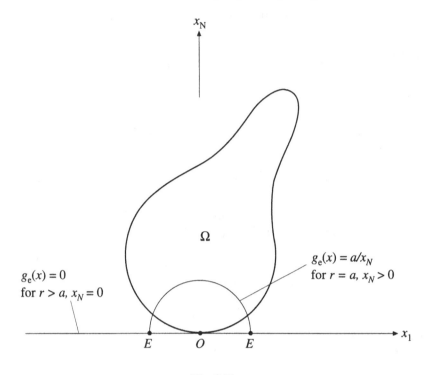

Fig. 2.11.

Choose co-ordinates so that p and an exterior ball B_q at p are given by

$$p = (0,\ldots,0,2b), \quad q = (0,\ldots,0,b), \quad B_q = \mathcal{B}(q,b) \qquad (2.62)$$

for some $b > 0$, as shown in the right half of Figure 2.12. Define

$$V := \{\,\xi \in \mathbb{R}^N \mid \xi_N < 2b\,\}, \quad V_a := V \setminus \overline{\mathcal{B}(p,a)} \quad \text{with } 0 < a < b;$$

we shall use $g_e(p - \xi; a)$ for $\xi \in \overline{V}_a \setminus E$, where E now denotes the equator of $\mathcal{B}(p,a)$. Under the transformation

$$
\left.
\begin{aligned}
\xi &= \frac{4b^2}{r^2}x \quad (r := |x| > 0), \\[2mm]
\text{equivalently} \qquad x &= \frac{4b^2}{\rho^2}\xi \quad (\rho := |\xi| > 0),
\end{aligned}
\right\}
\qquad (2.63)
$$

which is inversion relative to the sphere $\partial \mathcal{B}(0,2b)$,

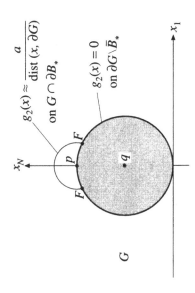

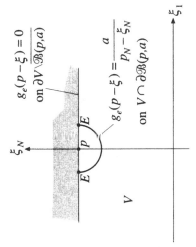

Fig. 2.12.

the punctured half-space $V \setminus \{0\}$ has image $G := \mathbb{R}^N \setminus \overline{B}_q$; (2.64a)

$$\left.\begin{array}{l} \text{the ball } \mathscr{B}(p,a) \text{ has image } B_* := \mathscr{B}(p_*, a_*), \\[2mm] \text{where } \quad p_* := \dfrac{1}{1-(a/2b)^2}\,p, \qquad a_* := \dfrac{1}{1-(a/2b)^2}\,a; \end{array}\right\} \quad (2.64b)$$

the equator E has image $F := \partial G \cap \partial B_*$. (2.64c)

Note that p is a fixed point of the map (2.63) and that, near p, this transformation is approximately reflection in the hyperplane ∂V. Therefore ∂B_* is close to $\partial \mathscr{B}(p,a)$ for small radii a. A more precise statement is that

$$p_N - \xi_N = \frac{4b^2}{r^2}\left(x_N - p_N + \frac{|x-p|^2}{2b}\right). \qquad (2.65)$$

Here the factor $4b^2/r^2$ will be unimportant when g_2 comes to be used in Theorem 2.36 [because $4b^2/r^2 \to 1$ as $x \to p$], but we shall need

$$z(x) := x_N - p_N + \frac{|x-p|^2}{2b}, \qquad (2.66)$$

which is almost $\text{dist}(x, \partial G)$ for points x near ∂G. In fact, a calculation shows that

$$z\big|_{\partial G} = 0, \quad 1 - \frac{1}{2}\frac{z(x)}{b} \le \frac{\text{dist}(x, \partial G)}{z(x)} \le 1 \quad \text{for all } x \in G, \qquad (2.67\text{a,b})$$

and

$$\max_{x \in \partial B_*} z(x) = a\left(1 - \frac{a}{2b}\right)^{-2}. \qquad (2.67\text{c})$$

Corollary 2.29 *Let g_e be as in Corollary 2.28. We adopt the notation (2.62), (2.64a,b,c) and (2.66) for any $b > 0$ and any $a \in (0, b)$. Then the two-ball function $g_2 = g_2(\,.\,; a, b)$ of Siegel type is defined by*

$$g_2(x; a, b) := \left(\frac{2b}{r}\right)^{N-2} g_e\left(p - \frac{4b^2}{r^2}x; a\right), \quad x \in \overline{G} \setminus \left(B_* \cup F \cup \{0\}\right) \qquad (2.68)$$

and by $g_2(0) := \lim_{r \downarrow 0} g_2(x) = 0$. It follows that $g_2 \in C\left(\overline{G} \setminus \left\{B_* \cup F\right\}\right) \cap C^\infty(G \setminus \overline{B}_*)$ and that

$$\triangle g_2 = 0 \quad \text{in } G \setminus \overline{B}_*, \qquad (2.69\text{a})$$

$$g_2(x) = \left(\frac{2b}{r}\right)^{N-4} \frac{a}{z(x)} \quad \text{on } G \cap \partial B_*, \qquad (2.69\text{b})$$

$$\left.\begin{array}{l} \textit{if } x \in \overline{G} \setminus \mathscr{B}(p', 2a'), \quad \textit{where} \quad p' := \dfrac{1}{1 - (a/b)^2}p, \ a' := \dfrac{1}{1 - (a/b)^2}a, \\[3mm] \textit{then } |g_2(x)| \le \text{const.}\, a^{N-1} \left(\dfrac{2b}{r}\right)^{N-2} \left| p - \dfrac{4b^2}{r^2}x \right|^{-N} \left| p_N - \dfrac{4b^2}{r^2}x_N \right|, \end{array}\right\}$$

$$\text{(2.69c)}$$

where the constant depends only on N. Also, $g_2(x) = 0$ on $\partial G \setminus \overline{B}_*$, and $g_2(x) > 0$ in $G \setminus \overline{B}_*$.

Proof We have $\triangle g_2 = 0$ in $G \setminus \overline{B}_*$ by Theorem B.15, already cited in the proof of Corollary 2.28. The remaining properties of g_2 follow from those of g_e by direct calculation. $\qquad\square$

2.7 Some Phragmén–Lindelöf theory for subharmonic functions

We return to the half-space $D := \{ x \in \mathbb{R}^N \mid x_N > 0 \}$, $N \ge 2$.

Theorem 2.30 *If $u \in C(\overline{D})$, if u is subharmonic in D and if*

$$\liminf_{a \to \infty} \max \left\{ \frac{x_N\, u(x)}{a^2} \ \Big| \ x \in \overline{D}, \ |x| = a \right\} = 0, \qquad (2.70)$$

then

$$\sup_D u = \sup_{\partial D} u.$$

Proof Let g continue to denote the primary function of Siegel type (Theorem 2.27), and let $V(.,a) := ag(.\,;a)$ on $\overline{D}_a \setminus E$, where $D_a := D \cap \mathscr{B}(0,a)$ and $E := \partial D \cap \partial \mathscr{B}(0,a)$. Theorem 2.27 shows that this V is a comparison function of the first kind (Definition 2.21); in particular,

$$V(x,a) = \frac{a^2}{x_N} \ge a \qquad \text{for } x \in \Gamma_a := D \cap \partial \mathscr{B}(0,a),$$

and

$$V(x_0, a) \le \text{const.}\, x_{0N} \qquad \text{for } x_0 \in D \ \text{and} \ a \ge 2|x_0|,$$

where the constant depends only on N.

Therefore the present theorem is implied by Lemma 2.22. The supremum over Γ_a in (2.45) of that lemma can now be written as a maximum over $\overline{\Gamma}_a$ because the function with values $x_N u(x)/a^2$ is continuous on $\overline{D}$; the maximum cannot be negative, because of values for $x_N = 0$. $\qquad\square$

Corollary 2.31 *Let G be an unbounded open subset of the half-space D. If* $u \in C(\overline{G})$, *if u is subharmonic in G and if*

$$\liminf_{a \to \infty} \max \left\{ \frac{x_N \, u(x)}{a^2} \;\middle|\; x \in \overline{G}, \;\; |x| = a \right\} \le 0, \qquad (2.71)$$

then

$$\sup_G u = \sup_{\partial G} u.$$

Proof In the proof of Lemma 2.22 we replace D by G, keeping the same comparison function V of the first kind. Thus D_a is replaced by $G_a := G \cap \mathscr{B}(0, a)$ and Γ_a now means $G \cap \partial \mathscr{B}(0, a)$. We define $s(a)$ and choose the radius a exactly as before. If ∂G does not intersect the equator E, we need not remove a neighbourood of E from $\overline{G_a}$. The shape of ∂G is unimportant; what matters is that $\tilde{u}(x) \le 0$ on ∂G by the definition of $\tilde{u}$, and that $\varphi(x) \le 0$ on Γ_a by the definition of $s(a)$.

After this extension of Lemma 2.22, the present corollary results from the choice $V = ag$ made in the proof of Theorem 2.30. The maximum in the growth condition (2.71) may be negative once more, because x_N need not descend to zero when $x \in \overline{G}$ and $|x| = a$.   □

Remark 2.32 (i) *If we add to Corollary 2.31 the hypotheses: G is connected and* $u \in C^1(G)$, *then*

$$u(x) < \sup_{\partial G} u \quad \text{for all } x \in G,$$

unless u is constant on $\overline{G}$. This is an immmediate consequence of the strong maximum principle (Theorem 2.13).

(ii) *Let G be as in Corollary 2.31. If* $u \in C(\overline{G})$, *if u is superharmonic in G and if*

$$\limsup_{a \to \infty} \min \left\{ \frac{x_N \, u(x)}{a^2} \;\middle|\; x \in \overline{G}, \;\; |x| = a \right\} \ge 0, \qquad (2.72)$$

then

$$\inf_G u = \inf_{\partial G} u.$$

This follows from Corollary 2.31 by an argument like that in Remark 2.7.

(iii) For a function u that is continuous on $\overline{G}$ and *harmonic in G* (hence is in $C^\infty(G)$, by Theorem B.6), we wish to conclude that

$$\inf_{\partial G} u \le u(x) \le \sup_{\partial G} u \quad \text{for all } x \in \overline{G}$$

(with strict inequality for $x \in G$ if G is connected and u is not a constant). It may be worthwhile to retain both (2.71) and (2.72) as hypotheses, but the simpler condition

$$\liminf_{a \to \infty} \max \left\{ \frac{x_N |u(x)|}{a^2} \ \middle| \ x \in \overline{G}, \ |x| = a \right\} = 0 \qquad (2.73)$$

is sufficient.   □

When a subset of D is significantly narrower near infinity than is D itself, much larger rates of growth are permissible. Corollary 2.31 is then far from sharp. We illustrate this by two examples; observe that, just as Theorem 2.30 extends to unbounded open subsets of D, so Examples 2.33 and 2.34 extend to unbounded open subsets of S and of H, respectively.

Example 2.33 *Consider the sector*

$$S := \left\{ (r \cos \theta, r \sin \theta) \in \mathbb{R}^2 \ \middle| \ r > 0, \ 0 < \theta < \beta \right\}, \qquad \beta \in (0, 2\pi).$$

If $u \in C(\overline{S})$, if u is subharmonic in S and if

$$\liminf_{R \to \infty} \max \left\{ \frac{\sin(\pi \theta / \beta) \, u(x, y)}{R^{\pi/\beta}} \ \middle| \ (x, y) \in \overline{S}, \ |(x, y)| = R \right\} = 0,$$
$$(2.74)$$

then

$$\sup_S u = \sup_{\partial S} u.$$

Proof This statement is merely a transcription of Theorem 2.30, for $N = 2$, under the conformal map of S onto D. Write $z = x + iy = re^{i\theta}$ for points of $\overline{S}$, and $\zeta = \xi + i\eta = \rho e^{it}$ for points of $\overline{D}$; the appropriate mapping is

$$\left. \begin{array}{c} \zeta = z^{\pi/\beta}, \\ \text{equivalently} \quad \rho = r^{\pi/\beta}, \quad t = \pi \theta / \beta, \end{array} \right\} \quad r \ge 0, \ 0 \le \theta \le \beta. \qquad (2.75)$$

This is a homeomorphism of the closed sector $\overline{S}$ onto the closed half-plane $\overline{D}$; it is also a C^∞ map, with C^∞ inverse, of $\overline{S} \setminus \{0\}$ onto $\overline{D} \setminus \{0\}$.

Let $\hat{u}(\xi, \eta) := u(x(\xi, \eta), y(\xi, \eta))$ under the mapping (2.75). Then $\hat{u} \in C(\overline{D})$ because $u \in C(\overline{S})$. Also, $\hat{u}$ satisfies the growth condition (2.70) because u satisfies (2.74). We now show that $\hat{u}$ is subharmonic in D. Given $\hat{\varphi} \in C_c^\infty(D)$ satisfying $\hat{\varphi} \ge 0$, define $\varphi(x, y) := \hat{\varphi}(\xi(x, y), \eta(x, y))$. Then $\varphi \in C_c^\infty(S)$, $\varphi \ge 0$ and

$$\hat{\varphi}_{\xi\xi} + \hat{\varphi}_{\eta\eta} = \frac{\varphi_{xx} + \varphi_{yy}}{|\, \mathrm{d}\zeta / \mathrm{d}z\, |^2}, \qquad \mathrm{d}\xi \, \mathrm{d}\eta = \left| \frac{\mathrm{d}\zeta}{\mathrm{d}z} \right|^2 \mathrm{d}x \, \mathrm{d}y,$$

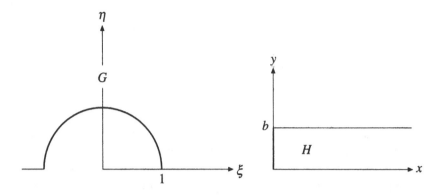

Fig. 2.13.

so that

$$\iint_D (\hat{\varphi}_{\xi\xi} + \hat{\varphi}_{\eta\eta})\, \hat{u}\, d\xi\, d\eta = \iint_S (\varphi_{xx} + \varphi_{yy})\, u\, dx\, dy \geq 0.$$

The result $\sup_D \hat{u} = \sup_{\partial D} \hat{u}$ now implies that $\sup_S u = \sup_{\partial S} u$.   □

If we allow $\beta = 2\pi$ in the definition of S, then the foregoing result remains true even though $\bar{S} = \mathbb{R}^2$ and (2.75) is no longer a homeomorphism of $\bar{S}$ onto $\bar{D}$. Indeed, the weaker condition $\hat{u} \in C(\bar{D})$ can replace $u \in C(\bar{S})$, provided that (2.74) is replaced by

$$\liminf_{R\to\infty} \sup \left\{ \left| \frac{\sin(\theta/2)\, u(x,y)}{R^{1/2}} \right|\ (x,y) \in S,\ |(x,y)| = R \right\} = 0 \quad (2.76)$$

for $\beta = 2\pi$. The condition $\hat{u} \in C(\bar{D})$ is weaker for $\beta = 2\pi$ in that it allows limiting values $u(x,0+)$ as $y \downarrow 0$ and $u(x,0-)$ as $y \uparrow 0$ such that $u(x,0+) \neq u(x,0-)$ for $x > 0$. When contemplating $\sup_{\partial S} u$, we must then regard the upper and lower sides of ∂S as distinct.

Example 2.34 *Consider the half-strip* $H := (0,\infty) \times (0,b)$ *in* $\mathbb{R}^2$. *If* $u \in C(\bar{H})$, *if* u *is subharmonic in* H *and if*

$$\liminf_{c\to\infty} \max \left\{ \left| \frac{\sin(\pi y/b)\, u(c,y)}{\exp(\pi c/b)} \right|\ 0 \leq y \leq b \right\} = 0, \qquad (2.77)$$

then

$$\sup_H u = \sup_{\partial H} u.$$

Proof This result is implied by Corollary 2.31 and the conformal map of H onto $G := D \setminus \mathscr{B}(0,1)$; as in Theorem 2.30, the maximum in the growth condition cannot be negative [because the function in question is zero for $y = 0, b$]. Again let $z = x + iy$ and $\zeta = \xi + i\eta = \rho e^{it}$; the relevant map is now (Figure 2.13)

$$\left. \begin{array}{l} \zeta = \exp\dfrac{\pi z}{b}, \\[2mm] \text{equivalently} \quad \rho = \exp\dfrac{\pi x}{b}, \quad t = \dfrac{\pi y}{b}, \end{array} \right\} \quad x \geq 0,\ 0 \leq y \leq b.$$

The rest is essentially as in Example 2.33. $\qquad\qquad\qquad\qquad\qquad\square$

Finally, we derive two more results for subharmonic functions that may be discontinuous at $p \in \partial\Omega$; as was promised earlier, these theorems allow a rate of growth larger than that in Theorem 2.26, for certain boundaries $\partial\Omega$.

Theorem 2.35 *Let Ω be bounded in $\mathbb{R}^N$, $N \geq 2$, let $p \in \partial\Omega$ and assume that Ω is on one side of a hyperplane A containing p. (Figure 2.11 shows a case with $p = 0$ and $A = \{ x \mid x_N = 0 \}$.) Let $d_A(x) := \mathrm{dist}(x, A)$.*

If $u \in C(\overline{\Omega} \setminus \{p\})$, if u is subharmonic in Ω and if

$$\liminf_{a \to 0} \max \{ a^{N-2} d_A(x)\, u(x) \mid x \in \overline{\Omega} \cap \partial\mathscr{B}(p, a) \} \leq 0, \qquad (2.78)$$

then

$$\sup_{\Omega} u = \sup_{\partial\Omega \setminus \{p\}} u.$$

Proof Choose co-ordinates so that $p = 0$, $A = \{ x \mid x_N = 0 \}$ and Ω lies in the half-space D; then $d_A(x) = x_N$. With g_e denoting the exterior function of Siegel type (Corollary 2.28), define $W(.,a) := a^{-N+1} g_e(.\,; a)$ on $\overline{D} \setminus (B \cup E)$, where $B := \mathscr{B}(0, a)$ and $E := \partial D \cap \partial B$. Then Corollary 2.28 shows W to be a comparison function of the second kind (Definition 2.24); in particular

$$W(x, a) = \frac{a^{-N+2}}{x_N} \geq a^{-N+1} \quad \text{for } x \in \gamma_a := D \cap \partial B,$$

and

$$W(x_0, a) \leq \mathrm{const.}\, |x_0|^{-N} x_{0N} \quad \text{for } x_0 \in D \text{ and } a \leq \tfrac{1}{2}|x_0|,$$

where the constant depends only on N.

Accordingly, the theorem follows from Lemma 2.25. The supremum over $\Omega \cap \partial\mathscr{B}(p,a)$ in (2.52) of that lemma can now be written as a maximum over $\overline{\Omega} \cap \partial\mathscr{B}(p,a)$ because the function with values $a^{N-2}d_A(x)\,u(x)$ is continuous on $\overline{\Omega} \setminus \{p\}$. $\qquad\square$

In the next theorem, the growth condition (2.79) may seem absurd because of the elaborate p_*, a_* and because of the detailed knowledge of u that seems to be assumed. However, as with other growth conditions that we have met, there are simpler statements that imply (2.79). For example, if $u(x) = o(|x-p|^{-N+1})$ as $x \to p$, or if d_B denotes distance to an exterior ball at p and $u(x) = o\left(|x-p|^{-N+2}/d_B(x)\right)$ as $x \to p$, then (2.79) is amply satisfied.

Theorem 2.36 *Let Ω be bounded in $\mathbb{R}^N$, $N \geq 2$, and assume that Ω has the exterior-ball property at $p \in \partial\Omega$ (Definition 2.14). Let $\mathscr{B}(q,b)$, with $b = |p-q|$, be an exterior ball at p such that $2q - p \notin \overline{\Omega}$ (Figure 2.14). Let $d_B(x) := \mathrm{dist}\left(x, \mathscr{B}(q,b)\right)$.*

If $u \in C(\overline{\Omega} \setminus \{p\})$, if u is subharmonic in Ω and if

$$\liminf_{a\to 0} \max\left\{ a^{N-2}d_B(x)\,u(x) \mid x \in \overline{\Omega}\cap\partial\mathscr{B}(p_*,a_*) \right\} \leq 0, \quad (2.79)$$

where

$$p_* := p + \frac{1}{2}\frac{(a/b)^2}{1-(a/2b)^2}(p-q), \quad a_* := \frac{a}{1-(a/2b)^2}, \quad (2.80)$$

then

$$\sup_{\Omega} u = \sup_{\partial\Omega\setminus\{p\}} u.$$

Proof (i) We make two changes in Definition 2.24 and Lemma 2.25. (Readers who distrust such tinkering with previous results may prefer to prove the theorem by means of Exercise 2.45.) First, the ball $\mathscr{B}(p,a)$ is replaced by $\mathscr{B}(p_*,a_*)$, where p_* and a_* are as in (2.80). Second, the condition $a \leq \frac{1}{2}|x_0-p|$, which accompanies the inequality $W(x_0,a) \leq \lambda(x_0)$ in (c) of Definition 2.24, is replaced by

$$\left.\begin{array}{c} a' \leq \frac{1}{2}|x_0 - p'|, \\[2mm] \text{where}\quad p' := p + 2\frac{(a/b)^2}{1-(a/b)^2}(p-q), \\[2mm] a' := \frac{a}{1-(a/b)^2}. \end{array}\right\} \quad (2.81)$$

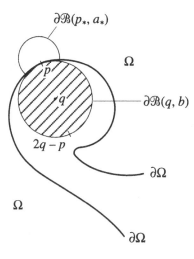

Fig. 2.14.

Here b is fixed and, in the proof of Lemma 2.25, the radius a is always chosen to be so small that certain inequalities hold. Such choices are not thwarted by the new perturbation terms in p_*, a_*, p' and a', so that Lemma 2.25 remains valid [with $\partial \mathscr{B}(p_*, a_*)$ replacing $\partial \mathscr{B}(p, a)$ in (2.52)].

(ii) Choose co-ordinates so that $p = (0, \ldots, 0, 2b)$ and $q = (0, \ldots, 0, b)$, as in Figure 2.12; Let $G := \mathbb{R}^N \setminus \overline{\mathscr{B}(q, b)}$ and $B_* := \mathscr{B}(p_*, a_*)$. Referring to Corollary 2.29, define

$$W_*(x, a) := a^{-N+1} g_2(x; a, b) \qquad \text{for } x \in \overline{G} \setminus \left(B_* \cup F \right).$$

Then W_* is a comparison function of the second kind, according to the modified Definition 2.24; from (2.69b) we obtain

$$W_*(x, a) = \left(\frac{2b}{|x|} \right)^{N-4} \frac{a^{-N+2}}{z(x)} \geq a^{-N+1} \left\{ 1 - O\left(\frac{a}{b} \right) \right\} \text{ for } x \in G \cap \partial B_*,$$

and (2.69c) implies a bound

$$W_*(x_0, a) \leq \lambda(x_0) \qquad \text{for } x_0 \in G \text{ and } a' \leq \tfrac{1}{2} |x_0 - p'|.$$

The growth conditions

$$\liminf_{a \to 0} \ \sup \left\{ \left. \frac{u(x)}{W_*(x, a)} \ \right| \ x \in \Omega \cap \partial B_* \right\} \leq 0 \qquad (2.82)$$

and (2.79) are equivalent because our choice of co-ordinates and (2.67) imply that

$$\frac{|x|}{2b} = 1 + O\left(\frac{a}{b}\right) \quad \text{and} \quad \frac{z(x)}{d_B(x)} = 1 + O\left(\frac{a}{b}\right) \quad \text{for} \quad x \in G \cap \partial B_*,$$

and because the function with values $a^{N-2}d_B(x)u(x)$ is continuous on $\overline{\Omega} \setminus \{p\}$. Therefore the modified Lemma 2.25 implies the present theorem.

$\qquad\qquad\qquad\qquad\qquad\qquad\qquad\qquad\qquad\qquad\qquad\qquad\qquad\qquad$ □

2.8 Exercises

Exercise 2.37 Suppose that $\mathscr{B}(p,a) \subset \Omega \subset \mathscr{B}(q,b)$ in $\mathbb{R}^N$, and that $f := \Omega \to \mathbb{R}$ satisfies $0 \le k \le f(x) \le l$ for all $x \in \Omega$. Prove that, if it exists, the solution $u \in C(\overline{\Omega}) \cap C^2(\Omega)$ of the Dirichlet problem $-\Delta u = f$ in Ω, $u|_{\partial\Omega} = 0$ is bounded by

$$\frac{k}{2N}\left(a^2 - |x - p|^2\right) \le u(x) \le \frac{l}{2N}\left(b^2 - |x - q|^2\right) \quad \text{for all} \quad x \in \overline{\Omega}.$$

If only the foregoing information is given, can this estimate be improved?

Exercise 2.38 Assume that Ω is a bounded region, that u is subharmonic in Ω (Definition 2.20) and v superharmonic in Ω, that $u, v \in C(\overline{\Omega})$ and that $u|_{\partial\Omega} \le v|_{\partial\Omega}$. Prove that either $u(x) < v(x)$ for all $x \in \Omega$, or $u = v$.

Exercise 2.39 Let u be harmonic and continuous in a region Ω; then $u \in C^\infty(\Omega)$, by Theorem B.6. Prove that if $\sup_\Omega |\nabla u|$ is attained in Ω, then ∇u is a constant vector. Show by an example that, if $\inf_\Omega |\nabla u|$ is attained in Ω, then ∇u need not be a constant vector.

Exercise 2.40 Prove that, if $\partial\Omega$ is of class C^2, then Ω has the interior-ball property (Definition 2.14) at every boundary point.

Exercise 2.41 Writing $x = (r\cos\theta, r\sin\theta)$ for points of $\mathbb{R}^2$, consider the region $\Omega := \{(r\cos\theta, r\sin\theta) \in \mathbb{R}^2 \mid r > 0,\ 0 < \theta < \beta\}$ for $\beta \in (0, 2\pi]$, and the function u defined by

$$u(x) := \begin{cases} -r^{\pi/\beta}\sin\dfrac{\pi\theta}{\beta} & \text{if } x \in \overline{\Omega} \setminus \{0\}, \\[2mm] 0 & \text{if } x = 0. \end{cases}$$

Show that (a) $\triangle u = 0$ in Ω; (b) for $\beta \in (0, \pi)$ the boundary-point lemma cannot be applied at the origin, and its conclusion does not hold there; (c) for $\beta \in [\pi, 2\pi]$ the boundary-point lemma does indeed describe the behaviour of u near the origin.

What distinguishes the case $\beta = \pi$ in (c)?

Exercise 2.42 Let Ω be a bounded region with $\partial\Omega$ of class C^1 and with the interior-ball property (Definition 2.14) at every boundary point. The *Neumann problem* for L in Ω (where L is as in Definition 2.3) is to find v such that

$$Lv = f \text{ in } \Omega, \qquad \left.\frac{\partial v}{\partial n}\right|_{\partial\Omega} = g, \qquad v \in C^1(\overline{\Omega}) \cap C^2(\Omega), \qquad (2.83)$$

where f, g are given functions and $\partial v/\partial n$ denotes the outward normal derivative.

Prove that, if they exist, any two solutions of (2.83) differ only by a constant, and that this constant is zero when the coefficient c is not the zero function.

Exercise 2.43 The *Lebesgue spine.* Write $x = (x_1, x_2, z)$ for points of $\mathbb{R}^3$, let $s := (x_1^2 + x_2^2)^{1/2}$ and define

$$\Omega := \left\{ x \in \mathbb{R}^3 \mid 0 < |x| < 1; \ s > \exp(-1/z) \text{ if } z > 0, \ s > 0 \text{ if } z = 0 \right\}.$$

Show that the function v defined by

$$v(x) := \int_0^1 \frac{\zeta \, d\zeta}{\left\{ s^2 + (z - \zeta)^2 \right\}^{1/2}}, \qquad x \in \overline{\Omega} \setminus \{0\},$$

belongs to $C^\infty(\Omega)$ and satisfies $\triangle v = 0$ in Ω; that

$$v(x) = (z + |z|) \log\frac{1}{s} + \varphi(x) \quad \text{with} \quad \varphi \in C(\Omega) \quad \text{and} \quad \varphi(0) = 1;$$

and that v has no extension in $C(\overline{\Omega})$.

Define $g \in C(\partial\Omega)$ by $g(0) := 3$ and $g(x) := v(x)$ for $x \in \partial\Omega \setminus \{0\}$. Prove that the Dirichlet problem of finding u such that

$$\triangle u = 0 \text{ in } \Omega, \qquad u\big|_{\partial\Omega} = g, \qquad u \in C(\overline{\Omega}) \cap C^2(\Omega)$$

has no solution.

[Assume that u exists and apply Theorem 2.26 to $u - v$ and to $-u + v$.]

Exercise 2.44 (i) Let Ω be bounded in $\mathbb{R}^N$, $N \geq 2$, and let $\partial\Omega$ have an isolated point (for example, $\Omega = \mathscr{B}(0,1) \setminus \{0\}$). Use Theorem 2.26 to prove that the Dirichlet problem of finding u such that

$$\triangle u = 0 \text{ in } \Omega, \quad u\big|_{\partial\Omega} = g, \quad u \in C(\overline{\Omega}) \cap C^2(\Omega)$$

has no solution for certain functions $g \in C(\partial\Omega)$.

(ii) Prove that Theorem 2.26 is best possible in the following sense. If the hypotheses (2.55) and (2.56) are changed to

$$\limsup_{a \to 0} \sup \left\{ \frac{u(x)}{\log(b/a)} \,\bigg|\, x \in \sigma_a \right\} < \infty \quad \text{when } N = 2,$$

$$\limsup_{a \to 0} \sup \left\{ a^{N-2} u(x) \mid x \in \sigma_a \right\} < \infty \quad \text{when } N \geq 3,$$

and the other hypotheses remain unchanged, then the conclusion is false.

Exercise 2.45 Prove Theorem 2.36, for the case when $\sup_{\partial\Omega\setminus\{p\}} u < \infty$, by inversion relative to the sphere $\partial\mathscr{B}(2q - p, 2b)$, by use of the corresponding Kelvin transform (§B.3) of the function $\tilde{u} := u - \sup_{\partial\Omega\setminus\{p\}} u$ and by application of Theorem 2.35.

[Use convenient co-ordinates, as in (2.63).]

Exercise 2.46 Let Ω be an unbounded open subset of the half-space $D := \left\{ x \in \mathbb{R}^N \mid x_N > 0 \right\}$, $N \geq 2$, and let Ω have the exterior-ball property (Definition 2.14) at each point of the set $P := \left\{ p^1, \ldots, p^k \right\} \subset \partial\Omega$.

Suppose that $u \in C(\overline{\Omega} \setminus P)$; that u is subharmonic in Ω; that u satisfies both

$$\liminf_{R \to \infty} \max \left\{ \frac{x_N \, u(x)}{R^2} \,\bigg|\, x \in \overline{\Omega}, \; |x| = R \right\} \leq 0$$

(cf. Corollary 2.31) and the growth condition (2.79) on small surfaces $\overline{\Omega} \cap \partial\mathscr{B}(p^m_*, a_*)$ for each $m \in \{1, \ldots, k\}$. (The radius b of exterior balls at the points p^m can be chosen to be independent of m.)

Prove that $\sup_\Omega u = \sup_{\partial\Omega\setminus P} u$.

3

Symmetry for a Non-linear Poisson Equation in a Symmetric Set Ω

3.1 The simplest case

In this chapter we prove theorems about positive solutions u of $\triangle u + f(u) = 0$ in a bounded symmetric set Ω. Theorem 1.2 (which asserted spherical symmetry and monotonicity of u in a ball) appears as Corollary 3.5, with slightly less smoothness required of f now than was demanded before, and again as Corollary 3.9, (b), for a function f that may be discontinuous. Our main concern is with sets Ω and solutions u that are symmetrical only under reflection in a single, fixed hyperplane; results for this situation imply those for balls, as Lemma 1.8 may have indicated already.

Two definitions set the stage as regards f and Ω.

Definition 3.1 Let $W \subset \mathbb{R}$. A function $f : W \to \mathbb{R}$ is *locally Lipschitz continuous* iff, for each compact set $E \subset W$, there is a constant $A(E)$ such that

$$|f(s) - f(t)| \leq A(E)|s - t| \quad \text{whenever } s, t \in E;$$

the number $A(E)$ is called a *Lipschitz constant* of $f|_E$.

The function f is *uniformly Lipschitz continuous* iff a single Lipschitz constant $A(W)$ serves for all $s, t \in W$. $\qquad \square$

For example, if $W = (0, \infty)$ and $f(t) = t^{-1} + |t - 1| - t^2$, then for the compact subset $[\delta, n]$ of $(0, \infty)$ we may use the Lipschitz constant

$$A([\delta, n]) = \delta^{-2} + 1 + 2n.$$

But no single Lipschitz constant will serve for this function and for all s and t in $(0, \infty)$.

Definition 3.2 Recall from Definition 1.7 that $x^{\mu,k}$ denotes the reflection of any point $x \in \mathbb{R}^N$ in the hyperplane $T_\mu(k)$. A set $S \subset \mathbb{R}^N$ will be called *Steiner symmetric relative to* $T_\mu(k)$ iff, whenever $x \in S$, the closed line segment from x to $x^{\mu,k}$ is also in S; that is, iff

$$x \in S \Longrightarrow \left\{ (1-\theta)x + \theta x^{\mu,k} \mid 0 \le \theta \le 1 \right\} \subset S.$$

$\square$

Notation This will be as in Definition 1.7, but with the understanding that $k = e^1 := (1,0,\ldots,0)$ and with no display of k. We write $x'' := (x_2,\ldots,x_N)$. Thus

$$T_\mu := \left\{ \xi \in \mathbb{R}^N \mid \xi_1 = \mu \right\}, \quad x^\mu := (2\mu - x_1, x'')$$

and $g_\mu(x) := g(x^\mu)$ for any function $g : \overline{\Omega} \to \mathbb{R}$ whenever $x^\mu \in \overline{\Omega}$. Given a bounded region Ω that is Steiner symmetric relative to T_0 (Figure 3.1), we also define (for all $\mu \in \mathbb{R}$)

$$Z(\mu) := \left\{ x \in \Omega \mid x_1 > \mu \right\}, \quad Y(\mu) := \left\{ x \in \mathbb{R}^N \mid x^\mu \in Z(\mu) \right\}, \quad (3.1)$$

and call these a *cap* and *reflected cap*, respectively. Note that $Y(\mu) \subset \Omega$ if $\mu \ge 0$. Also, $M := \sup\{ x_1 \mid x \in \Omega \}$, so that $Z(\mu)$ is empty if and only if $\mu \ge M$.

Theorem 3.3 *Let Ω be bounded, connected and Steiner symmetric relative to the hyperplane T_0. Assume that $u := \overline{\Omega} \to \mathbb{R}$ has the following properties.*

(a) $u \in C(\overline{\Omega}) \cap C^2(\Omega), \quad u > 0 \text{ in } \Omega, \quad u\big|_{\partial\Omega} = 0.$

(b) $\triangle u + f(u) = 0 \quad \text{in } \Omega,$ (3.2)

where $f : [0,\infty) \to \mathbb{R}$ is locally Lipschitz continuous.
    Then

$$u(z) < u(z^\mu) \quad \text{if} \quad \mu \in (0, M) \quad \text{and} \quad z \in Z(\mu), \quad (3.3)$$

$$\partial_1 u(x) < 0 \quad \text{if} \quad x_1 > 0 \quad \text{and} \quad x \in \Omega. \quad (3.4)$$

Proof (i) Let $w(.,\mu) := u - u_\mu$ on $\overline{Z(\mu)}$ for $\mu \in [0, M)$. Our main task is to prove (3.3), that

$$z \in Z(\mu) \Longrightarrow w(z,\mu) = u(z) - u(z^\mu) < 0 \quad (*)$$

whenever $\mu \in (0, M)$. We begin by combining the equations

$$\triangle u(z) + f\big(u(z)\big) = 0 \quad \text{for } z \in Z(\mu),$$

$$\triangle u(y) + f\big(u(y)\big) = 0 \quad \text{for } y \in Y(\mu),$$

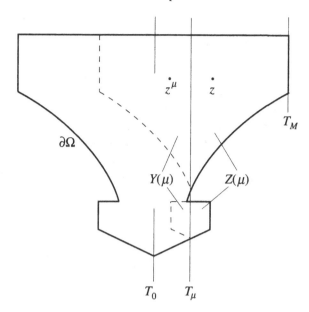

Fig. 3.1.

where $\mu \in [0, M)$. In the second of these, set $y = z^\mu$. Then $u(y) = u(z^\mu) = u_\mu(z)$; also $\partial/\partial y_1 = -\partial/\partial z_1$ and $\partial/\partial y_j = \partial/\partial z_j$ for $j = 2, \ldots, N$; hence

$$\Delta u_\mu(z) + f(u_\mu(z)) = 0 \quad \text{for } z \in Z(\mu).$$

Subtracting this from the equation for $u(z)$, and defining

$$\gamma(.\,,\mu) := \begin{cases} \dfrac{f(u) - f(u_\mu)}{u - u_\mu} & \text{at points } z \text{ where } u(z) \neq u_\mu(z), \\ 0 & \text{at points } z \text{ where } u(z) = u_\mu(z), \end{cases} \tag{3.5}$$

we obtain

$$\Delta w + \gamma(z, \mu)w = 0 \quad \text{for } z \in Z(\mu), \tag{3.6}$$

where $w = w(z, \mu)$. Also, the identity $Z(\mu) = \Omega \cap \{x \in \mathbb{R}^N \mid x_1 > \mu\}$ implies that $\partial Z(\mu) \subset \partial\Omega \cup T_\mu$; therefore

$$w(z, \mu) \leq 0 \quad \text{for } z \in \partial Z(\mu), \tag{3.7}$$

with strict inequality when $z \in \partial Z(\mu) \setminus T_\mu$ and $z^\mu \in \Omega$ [because then $z \in \partial\Omega$, so that $u(z) = 0$, while $u(z^\mu) > 0$], and with equality when $z \in \partial Z(\mu) \cap T_\mu$. Of course, there is a third possibility: that $w(z, \mu) = 0$ because z and z^μ both belong to $\partial\Omega \setminus T_\mu$.

(ii) The next step is to show that (∗) holds for all sufficiently small, positive values $M - \mu$.

(a) First, we apply the maximum principle for thin sets (Theorem 2.19) to the set $Z(\mu)$, the operator $\Delta + \gamma(.,\mu)$ and the function $w(.,\mu)$. In order to have data independent of μ, we use diam $Z(0)$ for the diameter in Theorem 2.19. Since $u \in C(\overline{\Omega})$, there is a number $U > 0$ such that $0 \le u(x) \le U$ on $\overline{\Omega}$; hence there is a Lipschitz constant $A = A([0, U]) > 0$ such that $|f(u) - f(u_\mu)| \le A|u - u_\mu|$; accordingly $|\gamma(z, \mu)| \le A$ for all $z \in Z(\mu)$ and all $\mu \in [0, M)$. Since $w(.,\mu)$ is a C^2-solution of (3.6), it is certainly a distributional solution. Therefore Theorem 2.19 implies existence of a number $\delta > 0$, independent of $\mu \in [0, M)$, such that

$$|Z(\mu)| < \delta \implies w(z, \mu) \le 0 \quad \text{for } z \in \overline{Z(\mu)}. \tag{3.8}$$

(b) Second, we sharpen (3.8) to $w(z, \mu) < 0$ in $Z(\mu)$ by means of the strong maximum principle (Theorem 2.13). Now that we have $\sup_{Z(\mu)} w(.,\mu) = 0$ [the value 0 being attained on T_μ], we may ignore the sign of $\gamma(z, \mu)$, by Remark 2.16; we apply Theorem 2.13 to each component of $Z(\mu)$ and to the function $w(.,\mu)$. Assume that $w(z_0, \mu) = 0$ at some point $z_0 \in Z(\mu)$; let $Z_0(\mu)$ be the component of $Z(\mu)$ that contains z_0. Then, by Theorem 2.13 and the continuity of u on $\overline{\Omega}$, we have $w(z, \mu) = 0$ at all $z \in \overline{Z_0(\mu)}$. This is a contradiction because for $\mu > 0$ there exist points $p \in \partial Z_0(\mu) \setminus T_\mu$ such that $p^\mu \in \Omega$ [see a hint to Exercise 3.13]; then $w(p, \mu) < 0$, as was noted after (3.7). Consequently,

$$|Z(\mu)| < \delta \implies w(z, \mu) < 0 \quad \text{for } z \in Z(\mu). \tag{3.9}$$

(iii) Let (m, M) be the largest open interval of μ in which (∗) holds; assume (for contradiction) that $m > 0$.

(a) Fix $z \in Z(m)$ and let $\mu \downarrow m$. Since $u(z) < u(z^\mu)$ for $\mu \in (m, z_1)$, and since $u(z^\mu)$ varies continuously for fixed z and varying μ, we have $w(z, m) \le 0$. This inequality can be sharpened to

$$w(z, m) < 0 \quad \text{for } z \in Z(m), \tag{3.10}$$

by means of (3.6), (3.7) and the strong maximum principle; the argument is like that in (ii)(b).

(b) Let δ be the small positive number, independent of $\mu \in [0, M)$, inferred from Theorem 2.19 and introduced before (3.8). Let $F \subset Z(m)$ be a compact set satisfying $|Z(m) \setminus F| < \frac{1}{2}\delta$ (Figure 3.2); the existence of

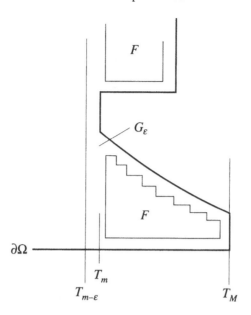

Fig. 3.2.

such an F follows from the representation of $Z(m)$ as a countable union of dyadic cubes (Exercise 3.14). Then

$$w(z, m) \leq -c \quad \text{for all } z \in F \text{ and for some } c > 0, \tag{3.11}$$

by (3.10), by the continuity of $w(., m)$ and by the compactness of F.

(c) Now consider $w_\varepsilon := w(., m - \varepsilon)$ for $0 < \varepsilon \leq \varepsilon_0$, and let $G_\varepsilon := Z(m - \varepsilon) \setminus F$. The number ε_0 is to be so small that $|G_\varepsilon| < \delta$, that $m - \varepsilon > 0$ and that $w_\varepsilon(z) \leq -\frac{1}{2}c$ for $z \in F$. This last is possible by the uniform continuity of u on $\overline{\Omega}$. We apply the maximum principle for thin sets (Theorem 2.19) to the set G_ε, the operator $\Delta + \gamma(., m - \varepsilon)$ and the function w_ε; the data $\operatorname{diam} Z(0)$, $|\gamma(z, \mu)| \leq A$ and δ used in (ii)(a) can be used again here. By (3.7) and because $w_\varepsilon \leq -\frac{1}{2}c$ on F we have $w_\varepsilon \leq 0$ on ∂G_ε; therefore Theorem 2.19 implies that $w_\varepsilon \leq 0$ on $\overline{G_\varepsilon}$. Consequently $w_\varepsilon \leq 0$ in $Z(m - \varepsilon)$; then, by the strong maximum principle once more [applied as in (ii)(b) to each component of $Z(m - \varepsilon)$], $w_\varepsilon(z) < 0$ for $z \in Z(m - \varepsilon)$. This contradicts the maximality of the interval (m, M) with $m > 0$, and thus completes the proof that $(*)$ holds for all $\mu \in (0, M)$.

(iv) That $\partial_1 u < 0$ in $Z(0)$ will now be inferred from the boundary-point lemma for balls (Lemma 2.12). Given $p \in \Omega$ with $p_1 > 0$, we set $\mu = p_1$

and observe that there is a ball $B := \mathcal{B}(p + \rho e^1, \rho)$ in $Z(\mu)$ [because Ω is open, hence contains a ball $\mathcal{B}(p, 2\rho)$]. Now $w(., \mu)$ is a C^2-solution relative to $\triangle + \gamma(., \mu)$ and B; also, $w(z, \mu) < 0$ for $z \in B$ while $w(p, \mu) = 0$. It follows from Lemma 2.12 and Remark 2.16 that $-(\partial_1 w)(p, \mu) > 0$. This proves that $(\partial_1 u)(p) < 0$ because the identity

$$\partial_1 w(x, \mu) = \frac{\partial}{\partial x_1}\left\{ u(x) - u(2\mu - x_1, x'') \right\} = (\partial_1 u)(x) + (\partial_1 u)(2\mu - x_1, x'')$$
(3.12)

shows that $(\partial_1 w)(p, \mu) = 2(\partial_1 u)(p)$ when $p_1 = \mu$. $\square$

Corollary 3.4 *Under the hypotheses of Theorem 3.3, u is an even function of x_1:*

$$u(-x_1, x'') = u(x_1, x'') \quad \text{for all } x \in \overline{\Omega}.$$
(3.13)

Proof Fix $z \in Z(0)$ and let $\mu \downarrow 0$. Since $u(z) < u(z^\mu)$ for $\mu \in (0, z_1)$, and since $u(z^\mu)$ varies continuously for fixed z and varying μ, we obtain $u(z) \leq u(z^0)$ for $z \in Z(0)$. Then, by continuity of u on $\overline{\Omega}$,

$$u(x_1, x'') \leq u(-x_1, x'') \quad \text{for } x_1 \geq 0 \text{ and } x \in \overline{\Omega}.$$

But, by applying to points of $\overline{Y(0)}$ the various arguments that we have used for points of $\overline{Z(0)}$ (in other words, by moving T_μ from $\mu = -M$ to $\mu = 0$ rather than from $\mu = M$ to $\mu = 0$), we obtain

$$u(-x_1, x'') \leq u(x_1, x'') \quad \text{for } x_1 \geq 0 \text{ and } x \in \overline{\Omega}.$$

Together, these two inequalities imply (3.13). $\square$

Corollary 3.5 *If in Theorem 3.3 the set Ω is a ball, say $\Omega = \mathcal{B}(0, a) \subset \mathbb{R}^N$, then u is spherically symmetric (depends only on $r := |x|$) and $du/dr < 0$ for $0 < r < a$.*

Proof (i) Let k be a given unit vector ($k \in \mathbb{R}^N$ and $|k| = 1$); let $R(k)$ be an orthogonal $N \times N$ matrix such that the transformation $\tilde{x} = R(k)x$ yields $\tilde{x}_1 = x \cdot k$; let $\tilde{u}(\tilde{x}) = u(x)$ under this transformation. Then $\tilde{u}$ satisfies the hypotheses of Theorem 3.3, hence is an even function of $\tilde{x}_1$ by Corollary 3.4. In the notation of Definition 1.7 we have $u(x^{0,k}) = u(x)$ for all $x \in \mathbb{R}^N$, provided that we set $u(x) := 0$ outside $\mathcal{B}(0, a)$. This holds for every unit vector k; therefore Lemma 1.8 establishes the spherical symmetry of u.

(ii) That $du/dr < 0$ for $0 < r < a$ follows from (3.4); indeed, now that u is known to depend only on r, it suffices to have (3.4) on a single ray from the origin.   □

3.2 A discontinuous non-linearity f

There are two reasons for relaxing the condition in Theorem 3.3 that f be Lipschitz continuous. First, functions f with simple discontinuities arise naturally in problems of Newtonian potentials, steady vortex flows and magnetohydrodynamics (§4.4 and Exercises 3.16 and 4.21 to 4.24). Second, although Theorem 3.3 and its corollaries serve well as introduction to a strategy of proof, they do not show how robust and flexible the method is. The elliptic equation (3.6) satisfied by w was perhaps the most important ingredient of the proof of Theorem 3.3; in this section we shall see that Lemmas 3.7 and 3.8, which are mere shadows of equation (3.6), are quite enough to prove the results of the previous section for a less pleasant function f.

Notation We continue to use the notation described after Definition 3.2. In particular, it is implicit that the unit vector $k = e^1$.

Theorem 3.6 *Let Ω be bounded, connected and Steiner symmetric relative to the hyperplane T_0. Assume that $u : \overline{\Omega} \to \mathbb{R}$ has the following properties.*
  (a) $u \in C(\overline{\Omega}) \cap C^1(\Omega)$, $u > 0$ in Ω, $u|_{\partial\Omega} = 0$.
  (b) *For all $\varphi \in C_c^\infty(\Omega)$,*

$$\int_\Omega \left\{ -\nabla\varphi \cdot \nabla u + \varphi f(u) \right\} = 0, \qquad (3.14)$$

where f has a decomposition $f = f_1 + f_2$ such that $f_1 : [0,\infty) \to \mathbb{R}$ is locally Lipschitz continuous, while $f_2 : [0,\infty) \to \mathbb{R}$ is non-decreasing and is identically 0 on $[0,\kappa]$ for some $\kappa > 0$.
  Then the previous conclusions hold:

$$u(z) < u(z^\mu) \quad \text{if} \quad \mu \in (0, M) \text{ and } z \in Z(\mu), \qquad (3.15)$$

$$\partial_1 u(x) < 0 \quad \text{if} \quad x_1 > 0 \text{ and } x \in \Omega. \qquad (3.16)$$

Remarks 1. *Equivalent hypotheses on f.* Consider the conditions: $f = g_1 + g_2$, where $g_1 : [0,\infty) \to \mathbb{R}$ is locally Lipschitz continuous, while $g_2 : [0,\infty) \to \mathbb{R}$ is non-decreasing and is uniformly Lipschitz continuous on $[0,\kappa]$ for some $\kappa > 0$. These conditions are *not* more general than those in the theorem: let

$$g_3(t) := \begin{cases} g_2(t) & \text{if } 0 \le t \le \kappa, \\ g_2(\kappa) & \text{if } t > \kappa, \end{cases}$$

and define $f_1 := g_1 + g_3$, $f_2 := g_2 - g_3$.

2. *Upward jumps and downward jumps of f.* The function f in Theorem 3.6 may 'jump upwards' (as its argument increases) in that the hypotheses allow simple discontinuities

$$\left. \begin{aligned} f(c+) - f(c-) > 0, \quad &\text{where } f(c+) := \lim_{t \downarrow c} f(t), \\ &f(c-) := \lim_{t \uparrow c} f(t), \end{aligned} \right\} \tag{3.17}$$

and where $c \ge \kappa$ [with $f(\kappa) = f(\kappa-)$]. If f jumps downwards, then the result need not hold (Exercise 3.19). Indeed, if f is merely Hölder continuous [see remarks preceding Definition A.10] at a point of an interval in which it decreases, then (3.15) and (3.16) may be false (Exercise 3.18).

3. *Notation and terminology.* Let $d(x) := \text{dist}(x, \partial\Omega)$ for points $x \in \overline{\Omega}$. The condition $u|_{\partial\Omega} = 0$ and the uniform continuity of u on $\overline{\Omega}$ imply that

$$\exists h > 0 \quad \text{such that} \quad d(x) < h \Longrightarrow u(x) < \kappa; \tag{3.18}$$

we define

$$Z_h(\mu) := Z(\mu) \cap \{ x \in \Omega \mid d(x) < h \}. \tag{3.19}$$

The phrase *in the generalized sense* will be a slight extension of the terminology in Definition 2.10; its meaning is displayed in Lemma 3.7. As before,

$$w(.,\mu) := u - u_\mu \quad \text{on } \overline{Z(\mu)} \quad \text{for} \quad \mu \in [0, M).$$

4. *Method.* Since equation (3.6), which was basic to the proof of Theorem 3.3, is no longer available, we must make do with the substitutes in the following two lemmas.

Lemma 3.7 *If $\mu \in [0, M)$, then*

$$\triangle w + \gamma(z, \mu)w \ge 0 \quad \text{in } Z_h(\mu) \text{ in the generalized sense}, \tag{3.20a}$$

where $w = w(z, \mu)$ and γ is defined as before but with f_1 replacing f. That is,

$$\left. \begin{aligned} \int_{Z_h(\mu)} \left\{ -\nabla\varphi \cdot \nabla w + \varphi\gamma(z,\mu)w \right\} \, dz \ge 0 \\ \text{whenever} \quad \varphi \in C_c^\infty\big(Z_h(\mu)\big) \quad \text{and} \quad \varphi \ge 0, \end{aligned} \right\} \tag{3.20b}$$

$$\gamma(.,\mu) := \begin{cases} \dfrac{f_1(u) - f_1(u_\mu)}{u - u_\mu} & \text{at points } z \text{ where } u(z) \neq u_\mu(z), \\ 0 & \text{at points } z \text{ where } u(z) = u_\mu(z). \end{cases}$$

$$\text{(3.20c)}$$

Proof (i) In order to derive an equation for w that will be useful both for this lemma and for the next, we abbreviate $Z(\mu)$, $Y(\mu)$ to Z, Y and proceed from

$$\int_Z \{-\nabla\varphi \cdot \nabla u + \varphi f(u)\} \, dz = 0 \qquad \text{for all } \varphi \in C_c^\infty(Z),$$

$$\int_Y \{-\nabla\psi \cdot \nabla u + \psi f(u)\} \, dy = 0 \qquad \text{for all } \psi \in C_c^\infty(Y).$$

Given $\varphi \in C_c^\infty(Z)$, we choose $\psi = \varphi_\mu$ [so that $\operatorname{supp}\psi \subset Y$] and set $y = z^\mu$. Then $\partial/\partial y_1 = -\partial/\partial z_1$ and $\partial/\partial y_j = \partial/\partial z_j$ for $j = 2,\ldots,N$; also,

$$\psi(y) = \varphi_\mu(y) = \varphi(y^\mu) = \varphi(z) \quad \text{and} \quad u(y) = u(z^\mu) = u_\mu(z),$$

whence

$$\int_Z \{-\nabla\varphi \cdot \nabla u_\mu + \varphi f(u_\mu)\} \, dz = 0 \quad \text{for all } \varphi \in C_c^\infty(Z).$$

Subtract this equation from the equation for $u(z)$; then

$$\left. \begin{aligned} &\int_Z \{-\nabla\varphi \cdot \nabla w + \varphi[f_1(u) - f_1(u_\mu)]\} \, dz \\ &= -\int_Z \varphi[f_2(u) - f_2(u_\mu)] \, dz \qquad \text{for all } \varphi \in C_c^\infty(Z). \end{aligned} \right\} \quad \text{(3.21)}$$

(ii) Now restrict attention to $\varphi \in C_c^\infty(Z_h)$ with $\varphi \geq 0$, where $Z_h = Z_h(\mu)$, and integrate only over Z_h in (3.21). The definition (3.20c) of γ shows that, on the left-hand side of (3.21),

$$f_1(u) - f_1(u_\mu) = \gamma(.,\mu)w.$$

On the right-hand side, $f_2(u) = 0$ because $u(z) < \kappa$ in Z_h, by (3.19) and (3.18), while $f_2(u_\mu) \geq 0$ because $f_2 \geq 0$ on $[0,\infty)$. Thus

$$-\int_{Z_h} \varphi[f_2(u) - f_2(u_\mu)] \, dz \geq 0 \quad \text{whenever } \varphi \in C_c^\infty(Z_h) \text{ and } \varphi \geq 0,$$

from which (3.20b) follows. □

Lemma 3.8 *If* $\mu \in [0,\infty)$ *and* $w(z,\mu) \leq 0$ *for all* $z \in Z(\mu)$, *then*

$$\triangle w - Aw \geq 0 \qquad \text{in } Z(\mu) \text{ in the generalized sense;} \quad \text{(3.22)}$$

here $w = w(z, \mu)$ *and* A *is a Lipschitz constant for* f_1 *on the interval* $[0, \sup_\Omega u]$. *Thus* $A \geq 0$ *and* A *is independent of* μ.

Proof In (3.21) we have

$$f_1(u) - f_1(u_\mu) \leq A|u - u_\mu| = -Aw,$$

by the definition of A and because $w(., \mu) \leq 0$ in $Z(\mu)$, and

$$f_2(u) - f_2(u_\mu) \leq 0,$$

because $u \leq u_\mu$ in $Z(\mu)$ [by hypothesis] and f_2 is non-decreasing. Accordingly, for all $\varphi \in C_c^\infty(Z)$ with $\varphi \geq 0$,

$$
\begin{aligned}
-\int_Z \nabla\varphi \cdot \nabla w \, dz &= -\int_Z \varphi \big\{ f_1(u) - f_1(u_\mu) + f_2(u) - f_2(u_\mu) \big\} \, dz \\
&\geq \int_Z \varphi A w \, dz, \quad\quad\quad\quad (3.23)
\end{aligned}
$$

and this is (3.22).   □

Proof of Theorem 3.6 (i) As before, our main task is to prove that

$$z \in Z(\mu) \implies w(z, \mu) = u(z) - u(z^\mu) < 0 \quad\quad (*)$$

whenever $\mu \in (0, M)$. Again we begin with small values of $M - \mu$, and proceed very much as in the proof of Theorem 3.3, step (ii).

(a) In order to apply the maximum principle for thin sets (Theorem 2.19) to the set $Z_h(\mu)$, the operator $\triangle + \gamma(z, \mu)$ and the function $w(., \mu)$, we choose data A and δ independent of $\mu \in [0, M)$, as in the earlier proof. Of course, f_1 replaces f; the constant A is as in Lemma 3.8, so that $|\gamma(z, \mu)| \leq A$; and diam $Z(0)$ serves as the diameter in Theorem 2.19.

If $M - h \leq \mu < M$, then $Z(\mu) = Z_h(\mu)$; therefore Lemma 3.7 applies to $Z(\mu)$. We have boundary values $w(z, \mu) \leq 0$ for $z \in \partial Z(\mu)$; consequently, by Theorem 2.19,

$$M - h \leq \mu < M \text{ and } |Z(\mu)| < \delta \implies w(z, \mu) \leq 0 \text{ for } z \in \overline{Z(\mu)}. \quad (3.24)$$

(b) With μ as in (3.24), let $Z_0(\mu)$ be an arbitrary component of $Z(\mu)$. Exactly as before, there are points $p \in \partial Z_0(\mu) \setminus T_\mu$ such that $p^\mu \in \Omega$, so that $w(p, \mu) < 0$. In view of (3.24), Lemma 3.8 now applies. Then the strong maximum principle (Theorem 2.13), this time for a generalized subsolution relative to $\triangle - A$ and $Z_0(\mu)$, shows that

$$M - h \leq \mu < M \text{ and } |Z(\mu)| < \delta \implies w(z, \mu) < 0 \text{ for } z \in Z(\mu). \quad (3.25)$$

(ii) Let (m, M) be the largest open interval of μ in which (∗) holds; assume (for contradiction) that $m > 0$.

(a) Fixing $z \in Z(m)$ and letting $\mu \downarrow m$, we obtain [precisely as in the proof of Theorem 3.3, step (iii)(a)] that $w(z, m) \leq 0$. Then Lemma 3.8 applies for $\mu = m$ and shows that $w(., m)$ is a generalized subsolution relative to $\triangle - A$ and to each component of $Z(m)$. As before, each component of $Z(m)$ has boundary points p at which $w(p, m) < 0$; the strong maximum principle (Theorem 2.13) shows that

$$w(z, m) < 0 \quad \text{for} \quad z \in Z(m). \tag{3.26}$$

(b) Next, we prove that

$$\partial_1 u < 0 \quad \text{on} \quad \overline{Z(m)} \setminus \partial\Omega. \tag{3.27}$$

Given $p \in \overline{Z(m)} \setminus \partial\Omega$, we set $\mu = p_1$; there is a ball $\mathscr{B}(p + \vartheta e^1, \vartheta)$ in $Z(\mu)$ for some $\vartheta > 0$ and $w(z, \mu) < 0$ in this ball because $\mu = p_1 \geq m$. The boundary-point lemma for balls (Lemma 2.12), applied as in the proof of Theorem 3.3, step (iv) [but now for a generalized subsolution relative to $\triangle - A$], shows that $(\partial_1 u)(p) = \frac{1}{2}(\partial_1 w)(p, \mu) < 0$.

(c) Letting $w_\varepsilon := w(., m - \varepsilon)$ for $0 < \varepsilon \leq \varepsilon_0$, as before, we wish to obtain a contradiction by showing that $w_\varepsilon < 0$ in $Z(m - \varepsilon)$ if ε_0 is sufficiently small. We continue to imitate the proof of Theorem 3.3, but a further construction is required because Lemma 3.7 applies only to sets near $\partial\Omega$, while the hypothesis $w(., \mu) \leq 0$ of Lemma 3.8 remains to be proved for $\mu < m$. The new step is first to observe that (3.27) extends to certain subsets of $Z(m - \rho)$ for some $\rho > 0$, and then to introduce a cylindrical set

$$E_\varepsilon := (m - \varepsilon, m + \tau] \times S, \quad \tau > 0, \quad S \subset \mathbb{R}^{N-1}, \quad E_\varepsilon \subset Z(m - \varepsilon), \tag{3.28}$$

as well as a compact subset F of $Z(m)$; see Figure 3.3. We can prove that $w_\varepsilon(z) < 0$ for $z \in E_\varepsilon$ by integrating $\partial_1 u$ from $z^{m-\varepsilon}$ to z, provided that $\{m\} \times S$ is a compact subset of $T_m \cap \Omega$ and that τ and ε_0 are sufficiently small.

Let $G_\varepsilon := Z(m - \varepsilon) \setminus (E_\varepsilon \cup F)$. We shall prove in (iii) that E_ε and F can be so chosen that, for $0 < \varepsilon \leq \varepsilon_0$ and for some $\varepsilon_0 > 0$,

$$w_\varepsilon < 0 \quad \text{on} \quad E_\varepsilon, \qquad w_\varepsilon < 0 \quad \text{on} \quad F, \tag{I, II}$$

$$G_\varepsilon \subset Z_h(m - \varepsilon), \qquad |G_\varepsilon| < \delta. \tag{III, IV}$$

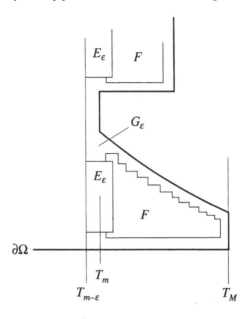

Fig. 3.3.

Here δ is the positive constant, independent of $\mu \in [0, M)$, inferred from Theorem 2.19 and introduced in step (i)(a). We are now on a familiar path to the desired contradiction. Conditions (III) and (IV) allow application to w_ε and G_ε first of Lemma 3.7 and then of Theorem 2.19; for the latter, we have $w_\varepsilon \leq 0$ on ∂G_ε by (I), (II) and by the usual conditions on $\partial Z(m - \varepsilon)$. Therefore $w_\varepsilon \leq 0$ on $\overline{G}_\varepsilon$, whence $w_\varepsilon \leq 0$ on $Z(m - \varepsilon)$, with strict inequality on E_ε and F.

By Lemma 3.8, w_ε is now a generalized subsolution relative to $\triangle - A$ and to each component of $Z(m - \varepsilon)$; provided that $m - \varepsilon > 0$, each component of $Z(m - \varepsilon)$ has boundary points p at which $w_\varepsilon(p) < 0$; consequently, the strong maximum principle (Theorem 2.13) shows that $w_\varepsilon < 0$ in $Z(m - \varepsilon)$. This contradicts the maximality of the interval (m, M) with $m > 0$.

(iii) Here is a detailed description of E_ε and F that leads to conditions (I) to (IV).

(a) Let $S_m := \{m\} \times S$, where S is a closed subset of $\mathbb{R}^{N-1}$ such that $S_m \subset (T_m \cap \Omega)$ and such that, in the hyperplane T_m, all points between S_m and $\partial \Omega$ are sufficiently near $\partial \Omega$; say

$$x \in (T_m \cap \Omega) \setminus S_m \implies \text{dist}(x, T_m \cap \partial \Omega) < \tfrac{1}{2}h. \qquad (3.29)$$

The existence of such a set S_m follows from the representation of $T_m \cap \Omega$ as a countable union of dyadic squares, that is, of $(N - 1)$-dimensional dyadic cubes (Exercise 3.14).

By (3.27) and the continuity of $\partial_1 u$ on the compact set S_m, there is a number $c = c(S) > 0$ such that $\partial_1 u \leq -c$ on S_m. Moreover, $\partial_1 u$ is uniformly continuous on each compact subset of Ω. Therefore, there exists a cylindrical set $[m - \rho, m + \sigma] \times S$ within Ω, with $\rho > 0$ and $\sigma > 0$, such that

$$\partial_1 u < 0 \quad \text{on } [m - \rho, m + \sigma] \times S. \tag{3.30}$$

In view of (3.27), the number σ is restricted only by the need to have this set within Ω.

Now, for every $\tau > 0$,

$$m - \tfrac{1}{2}\tau \leq \mu < m \quad \text{and} \quad \mu < z_1 \leq m + \tau \implies 2\mu - z_1 \geq m - 2\tau.$$

Hence in the definition (3.28) of E_ε it suffices for condition (I) that $0 < \tau \leq \sigma$ and $2\tau \leq \rho$, and then that $0 < \varepsilon_0 \leq \tfrac{1}{2}\tau$; for it follows that, when $z \in E_\varepsilon$, the line segment from $z^{m-\varepsilon}$ to z is in the set $[m - \rho, m + \sigma] \times S$, so that $w_\varepsilon < 0$ on E_ε, by (3.30).

(b) Let $E' := E_\varepsilon \cap Z(m)$; the sets E' and F are to be independent of ε. The complement of E' in $Z(m) \setminus Z(m + \tau)$ is, or can be made, small in two senses. First, (3.29) and the fact (made precise in Exercise 3.13, (iii)) that $T_\mu \cap \Omega$ cannot widen as μ increases through positive values, imply that

$$z \in \{ Z(m) \setminus Z(m + \tau) \} \setminus E' \implies d(z) < \tfrac{1}{2}h. \tag{3.31}$$

Second, if τ is sufficiently small, then

$$\left| \{ Z(m) \setminus Z(m + \tau) \} \setminus E' \right| < \tfrac{1}{4}\delta. \tag{3.32}$$

The number τ is now chosen so that the condition (I) and (3.32) hold; S and τ are fixed henceforth, but ε_0 may be reduced further.

(c) Condition (II) will follow, as in the earlier proof, from the compactness of F and the uniform continuity of $u(z^\mu)$ for fixed z and varying μ, provided that ε_0 is sufficiently small. We now choose F so that (III) and (IV) hold for sufficiently small ε_0.

Let F_n denote, for $n \in \mathbb{N}_0$, the union of all (closed) dyadic cubes of edge-length 2^{-n} (Exercise 3.14) that are subsets of the open set $Z(m)$. Then $\text{dist}(x, \partial Z(m)) \leq N^{1/2} 2^{-n}$ whenever $x \in Z(m) \setminus F_n$; also, $|Z(m) \setminus F_n| \to 0$

as $n \to \infty$. Therefore we can choose k to be so large that

$$x \in Z(m) \setminus F_k \Longrightarrow \text{dist}\big(x, \partial Z(m)\big) < \min\big\{ h, \tau \big\}, \tag{3.33}$$

and

$$\big|Z(m) \setminus F_k\big| < \tfrac{1}{4}\delta. \tag{3.34}$$

Define $F := F_k \setminus \text{int } E'$. [The definition $F := F_k$ would also serve and would not change G_ε, but the interiors of E' and F would intersect.] We argue as follows to establish (III) and (IV) for sufficiently small ε_0.

(d) Consider (III). For $z \in G_\varepsilon \setminus Z(m)$, contemplate the point (m, z''). If $(m, z'') \in \Omega$, then, by (3.29),

$$d(z) \le \big|z - (m, z'')\big| + \text{dist}\big((m, z''), \partial\Omega\big) < \varepsilon + \tfrac{1}{2}h.$$

If $(m, z'') \notin \Omega$, then there is a point of $\partial\Omega$ on the closed line segment from z to (m, z''). In either case, $d(z) < h$ if $\varepsilon_0 < \tfrac{1}{2}h$.

For $z \in G_\varepsilon \cap \big\{ Z(m) \setminus Z(m + \tau) \big\}$, (3.31) states that $d(z) < \tfrac{1}{2}h$.

For $z \in G_\varepsilon \cap Z(m + \tau)$, we have $z \in Z(m) \setminus F_k$ and $z_1 > m + \tau$; by (3.33), a point of $\partial Z(m)$ that is nearest z must be in $\partial\Omega$ rather than in T_m. Therefore $d(z) < h$, by (3.33).

(e) To prove (IV), we observe that $\big|G_\varepsilon \setminus Z(m)\big| < \tfrac{1}{2}\delta$ if ε_0 is sufficiently small, and that

$$\big|G_\varepsilon \cap Z(m)\big| \le \big|\big\{ Z(m) \setminus Z(m + \tau) \big\} \setminus E'\big| + \big|Z(m + \tau) \setminus F_k\big| < \tfrac{1}{2}\delta$$

by (3.32) and (3.34).

(iv) That $\partial_1 u < 0$ in $Z(0)$ is proved as before [proof of Theorem 3.3, step (iv)], except that now $w(., \mu)$ is a generalized subsolution relative to $\Delta - A$ and a suitable ball. $\qquad\square$

Corollary 3.9 (a) *Under the hypotheses of Theorem 3.6, u is an even function of x_1:*

$$u(-x_1, x'') = u(x_1, x'') \quad \text{for all } \ x \in \overline{\Omega}.$$

(b) *If in Theorem 3.6 the set Ω is a ball, say $\Omega = \mathscr{B}(0, a) \subset \mathbb{R}^N$, then u is spherically symmetric (depends only on $r := |x|$) and $du / dr < 0$ for $0 < r < a$.*

Proof The proofs of Corollaries 3.4 and 3.5 serve here also. $\qquad\square$

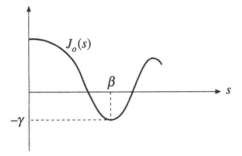

Fig. 3.4.

3.3 Exercises

Exercise 3.10 In Corollary 3.5 the conclusion $du/dr < 0$ for $0 < r < a$ cannot be extended to $r = a$, even when the hypothesis $u \in C^1(\overline{\Omega})$ is added. Demonstrate this by means of the set $\Omega := \mathscr{B}(0,1) \subset \mathbb{R}^2$ and functions f and u as follows.

$$f(t) := \beta^2(t - \gamma) \quad \text{for all} \quad t \geq 0, \qquad u(x) := J_0(\beta r) + \gamma \quad \text{for} \quad r := |x| \leq 1,$$

where the Bessel function J_0 is described by Exercise 1.19 and Figure 3.4, β denotes the smallest positive zero of J_0' and $\gamma := -J_0(\beta)$.

Exercise 3.11 Let $\Omega := \mathscr{B}(0,1) \subset \mathbb{R}^N$ for $N = 1$ or 3. Find analogues of the functions f and u in Exercise 3.10.
   [Fragments of Exercise 1.21 are relevant for $N = 3$.]

Exercise 3.12 Adopting the notation introduced after Definition 3.2, give an example of a set $\Omega \subset \mathbb{R}^2$ that is as in Theorem 3.3, has $\partial\Omega$ of class C^∞ and is such that the cap $Z(\mu)$ has infinitely many components for some value $\mu \in (0, M)$.

Exercise 3.13 Let Ω be bounded and Steiner symmetric relative to $T_0 = T_0(e^1)$; here we use Definition 3.2 and the notation introduced after it. Prove the following.

   (i) $\overline{\Omega}$ is Steiner symmetric relative to T_0.
   (ii) For an arbitrary component $Z_0(\mu)$ of the cap $Z(\mu)$, where $0 < \mu < M$, there exist boundary points $p \in \partial Z_0(\mu) \setminus T_\mu$ such that $p^\mu \in \Omega$.

[Given any point $z_0 \in Z_0(\mu)$, define $p = z_0 + \eta e^1$, where η is the smallest positive number such that $p \in \partial\Omega$.]

(iii) If $x, y \in \Omega$ and $y_1 > x_1 \geq 0$, $y'' = x''$, then dist$(y, \partial\Omega) \leq$ dist$(x, \partial\Omega)$. [Assume the contrary and consider the largest balls in Ω with centres x and y.]

(iv) The boundary $\partial\Omega$ need not be of class C.

Exercise 3.14 By a *dyadic cube in* $\mathbb{R}^N$ we mean a set of form

$$Q_n(p) := \left\{ x \in \mathbb{R}^N \mid p_j 2^{-n} \leq x_j \leq (p_j + 1)2^{-n} \text{ for } j = 1, \ldots, N \right\},$$

where $p = (p_1, \ldots, p_N)$, each $p_j \in \mathbb{Z}$ and $n \in \mathbb{N}_0$. [Thus a dyadic cube is closed, has edges of length 2^{-n} for some $n \in \mathbb{N}_0$, and has faces on each of which a co-ordinate is constant and equal to an integral multiple of 2^{-n}.]

Given a bounded open subset Ω of $\mathbb{R}^N$, define F_n to be the union of all dyadic cubes of edge-length 2^{-n} that are subsets of Ω. Prove the following.

(i) If $x \in \Omega$ and dist$(x, \partial\Omega) > N^{1/2}2^{-n}$ (where $n \in \mathbb{N}_0$), then $x \in F_n$.

(ii) $F_n \subset F_{n+1}$ and $\Omega = \bigcup_{n=0}^{\infty} F_n$.

(iii) Ω is a countable union of dyadic cubes.

(iv) $|\Omega \setminus F_n| \to 0$ as $n \to \infty$.

Exercise 3.15 We seek all functions $u \in C(\overline{\Omega}) \cap C^1(\Omega)$, other than the zero function, that are generalized solutions (Definition A.7) of

$$u'' + \lambda f_H(u - 1) = 0 \quad \text{in} \quad \Omega := (-1, 1) \subset \mathbb{R}, \qquad u\big|_{\partial\Omega} = 0,$$

where λ is a positive constant and f_H is the Heaviside function (see Chapter 0, (v)). Prove the following.

(a) For $\lambda > 4$ there are exactly two non-trivial, generalized solutions; they involve the points

$$a_1, a_2 := \tfrac{1}{2}\left(1 \mp \sqrt{1 - 4/\lambda}\right) \qquad \text{respectively,}$$

and are, for $j = 1, 2$,

$$u_j(x) = \begin{cases} \dfrac{1+x}{1-a_j} & \text{if } -1 \leq x \leq -a_j, \\[2mm] 1 + \tfrac{1}{2}\lambda(a_j^2 - x^2) & \text{if } -a_j \leq x \leq a_j, \\[2mm] \dfrac{1-x}{1-a_j} & \text{if } a_j \leq x \leq 1. \end{cases}$$

(b) For $\lambda = 4$ there is exactly one non-trivial, generalized solution.

(c) For $0 < \lambda < 4$ there is no non-trivial, generalized solution.

To obtain bonus marks, prove (a) to (c) without appeal to Theorem 3.6 or Corollary 3.9.

[Theorem B.6, applied to $u(x)$ wherever $u(x) < 1$ and to $u(x) + \frac{1}{2}\lambda x^2$ wherever $u(x) > 1$, implies *a priori* that $u \in C^\infty(G)$, where $G := \{x \in (-1,1) \mid u(x) \neq 1\}$.]

Exercise 3.16 The solution of this exercise becomes, under the transformation in Exercise A.25, the Stokes stream function of 'Hill's vortex in a ball'. Exercise 4.23 will explain this phrase and will account for the strange notation adopted here. However, the reader needs no knowledge of all this in order to do the exercise.

Let $B := \mathcal{B}(0, b) \subset \mathbb{R}^5$. We seek all functions $w \in C(\overline{B}) \cap C^1(B)$, other than the zero function, that are generalized solutions (Definition A.7) of

$$\triangle w + \lambda f_H \left(w - \tfrac{1}{2}U\right) = 0 \quad \text{in} \quad B, \qquad w\big|_{\partial B} = 0,$$

where λ and U are positive constants and f_H is the Heaviside function (see Chapter 0, (v)). It will be convenient to define

$$\lambda_0 := \min_{0<\alpha<1} \frac{15U}{2b^2\alpha^2(1-\alpha^3)} = \frac{75}{6}\left(\frac{2}{5}\right)^{-2/3}\frac{U}{b^2}.$$

Use Corollary 3.9 to prove the following.

(a) For $\lambda > \lambda_0$ there are exactly two non-trivial, generalized solutions; they involve the two solutions α_1 and α_2 of the equation

$$\alpha^2(1 - \alpha^3) = \frac{15U}{2\lambda b^2}, \qquad 0 < \alpha < 1 \;;$$

with the notation $r := |x|$, $a := \alpha_1 b$ when $\alpha = \alpha_1$, or $a := \alpha_2 b$ when $\alpha = \alpha_2$, the two solutions are

$$w(x) = \begin{cases} \dfrac{1}{2}\dfrac{U}{1-\alpha^3}\left(\dfrac{5}{2} - \alpha^3 - \dfrac{3}{2}\dfrac{r^2}{a^2}\right) & \text{if } r \leq a, \\[3mm] -\dfrac{1}{2}\dfrac{U}{1-\alpha^3}\left(\alpha^3 - \dfrac{a^3}{r^3}\right) & \text{if } a \leq r \leq b. \end{cases}$$

(b) For $\lambda = \lambda_0$ there is exactly one non-trivial, generalized solution.

(c) For $0 < \lambda < \lambda_0$ there is no non-trivial, generalized solution.

[A remark like the hint for Exercise 3.15 applies here.]

Exercise 3.17 Suppose that in Exercises 3.15 and 3.16 we do not prescribe λ, but rather prescribe the 'energy norms'

$$\|u\| \;:=\; \left\{ \int_{-1}^{1} u'^2 \, dx \right\}^{1/2} = \eta > 0 \qquad \text{(Exercise 3.15)},$$

$$\|w\| \;:=\; \left\{ \int_{B} |\nabla w|^2 \, dx \right\}^{1/2} = \eta > 0 \qquad \text{(Exercise 3.16)},$$

and calculate λ *a posteriori*. Prove that each problem has exactly one non-trivial, generalized solution for prescribed $\eta > 0$.

Exercise 3.18 Consider, for arbitrary $N \in \mathbb{N}$,

$$\left. \begin{array}{rl} \Delta u + f(u) = 0 & \text{in } B := \mathscr{B}\left(0, \sqrt{6}\right) \subset \mathbb{R}^N, \\ u\big|_{\partial B} = 0, & \end{array} \right\} \qquad (3.35)$$

where, for a constant $p > 2$,

$$f(t) := \begin{cases} 4p(p-1)(1-t)^{1-(2/p)} + p(N+2p-2)(1-t)^{1-(1/p)} & \text{if} \\ & 0 \le t \le 1, \\ -4p(p-1)(t-1)^{1-(2/p)} + 2p(N+2p-2)(t-1)^{1-(1/p)} & \text{if } t \ge 1. \end{cases} \qquad (3.36)$$

If p is large, does f fall far short of the local Lipschitz continuity demanded in Theorem 3.3?

Verify that there is a solution $u(.,c) \in C^2(\overline{B})$ as follows for each $c \in \overline{\mathscr{B}(0,1)}$, so that the conclusions of Theorem 3.3 and its corollaries are false in this case.

$$u(x,c) = \begin{cases} 1 + (1-\rho^2)^p & \text{if } \rho := |x-c| \le 1, \\ 1 & \text{if } \rho \ge 1 \text{ and } r := |x| \le 2, \\ 1 - \left(\tfrac{1}{2}r^2 - 2\right)^p & \text{if } 2 \le r \le \sqrt{6}. \end{cases}$$

Write further solutions of (3.35) with (3.36).

Exercise 3.19 Consider the problem (3.35) with the change that now

$$f(t) := \begin{cases} 8 + 2(N+2)(1-t)^{1/2} & \text{if } 0 \le t < 1, \\ 0 & \text{if } t = 1, \\ -8 + 4(N+2)(t-1)^{1/2} & \text{if } t > 1. \end{cases}$$

Why does this function fail to satisfy the hypotheses on f in Theorem 3.6?

Exhibit a generalized solution $u(.,c) \in C(\overline{B}) \cap C^1(B)$, for each $c \in \overline{\mathscr{B}(0,1)}$, that shows the conclusions of Theorem 3.6 and Corollary 3.9 to be false in this case.

[Generalized solutions are introduced in Definition A.7.]

4

Symmetry for the Non-linear Poisson Equation in $\mathbb{R}^N$

4.1 Statement of the main result

This chapter continues exploration of symmetry for the equation $\triangle u + f(u) = 0$; the function f is very much like that in Theorem 3.6 and may jump upwards as its argument increases. However, the equation now governs u on the whole of $\mathbb{R}^N$; relative to the situation when Ω is a ball, we lose both the boundary condition $u = 0$ on $\partial\Omega$ and the knowledge that, if u is spherically symmetric, then the centre of symmetry of u must be the centre of the ball Ω. It turns out that this loss of information is tolerable if the behaviour of $u(x)$ as $|x| \to \infty$ is admissible in the sense of Definition 4.1. Such admissible asymptotic behaviour supplies both a unique point as a possible centre of symmetry and a means of establishing, outside some large ball, the monotonicity condition (*) which was our main objective when proving Theorems 3.3 and 3.6.

By the end of Lemma 4.6, appropriate behaviour outside a large ball will have been demonstrated; two methods are then available for a proof of monotonicity and of spherical symmetry on all of $\mathbb{R}^N$. The first method, which we adopt in the main text of the chapter, makes no appeal to the maximum principle for thin sets. The second method, which is the subject of Exercise 4.27 and 4.28, imitates the proofs of Theorems 3.3 and 3.6; this is possible, under an additional condition on discontinuities of f, because the monotonicity condition has already been established outside a bounded set.

Corollary 4.10 returns to a more modest symmetry: if $f(u(x))$ is replaced by $f(u(x), x'')$ in the non-linear Poisson equation, where $f(., x'')$ resembles the previous f for all $x'' \in \mathbb{R}^{N-1}$, and if u has admissible asymptotic behaviour, then u enjoys the earlier monotonicity and symmetry relative to a hyperplane $\{\, x \mid x_1 = \text{const.} \,\}$ that can be located *a priori*.

Definition 4.1 Let $u \in C^1(\mathbb{R}^N)$. We shall say that u has *admissible asymptotic behaviour* iff one of the following conditions holds outside some ball, say for $r := |x| \geq R_u$. The numbers κ, m, δ and γ are all [strictly] positive constants, with $\delta \in (0,1]$, and $c = (c_1, \ldots, c_N)$ is a constant vector.

(A) $\qquad u(x) = -\kappa r^m + (c \cdot x) r^{m-2} + h(x),$

where $|\nabla h(x)| = O\left(r^{m-2-\delta}\right)$ and, if $m - 1 - \delta < 0$, then $h(x) \to 0$ as $r \to \infty$.

(B) $\qquad u(x) = \kappa \log \dfrac{\gamma}{r} + (c \cdot x) r^{-2} + h(x),$

where $|\nabla h(x)| = O(r^{-2-\delta})$ and $h(x) \to 0$ as $r \to \infty$.

(C) $\qquad u(x) = \kappa r^{-m} + (c \cdot x) r^{-m-2} + h(x),$

where $|\nabla h(x)| = O(r^{-m-2-\delta})$ and $h(x) \to 0$ as $r \to \infty$. $\qquad \square$

The essential features of Definition 4.1 are these. First, $u(x)$ *decreases* as $r \to \infty$; that $u(x) \to -\infty$ in cases (A) and (B), while $u(x) \to 0$ in case (C), is less important. Second, if $c \neq 0$, then $c \cdot x$ does not depend only on r. Therefore, if u should turn out to be spherically symmetric about some point $q \in \mathbb{R}^N$, then $u(x + q)$ will contain no term $(c \cdot x) r^\alpha$ (where $\alpha \in \mathbb{R}$ is a constant). This enables us to find *a priori* the only point q that can be a centre of symmetry of the function u.

Theorem 4.2 *Suppose that a function $u : \mathbb{R}^N \to \mathbb{R}$ has the following properties.*

(a) *$u \in C^1(\mathbb{R}^N)$; it has admissible asymptotic behaviour; and $u > 0$ on $\mathbb{R}^N$ in case* (C).

(b) *For all $\varphi \in C_c^\infty(\mathbb{R}^N)$,*

$$\int_{\mathbb{R}^N} \{-\nabla \varphi \cdot \nabla u + \varphi f(u)\} = 0, \qquad (4.1)$$

where f has a decomposition $f = f_1 + f_2$ such that $f_1 : \mathbb{R} \to \mathbb{R}$ is locally Lipschitz continuous, while $f_2 : \mathbb{R} \to \mathbb{R}$ is non-decreasing. In case (C)*, f_1 and f_2 need be defined and need have these properties only on $[0, \infty)$.*

Let

$$
q := \begin{cases} \dfrac{1}{\kappa m}c & \text{for cases (A) and (C),} \\[2mm] \dfrac{1}{\kappa}c & \text{for case (B),} \end{cases} \qquad\qquad v(x) := u(x+q). \quad (4.2)
$$

Then v *is spherically symmetric (depends only on* r*) and* $dv/dr < 0$ *for* $r > 0$.

The proof of Theorem 4.2 will take the form of five lemmas and a proposition. A first step is to verify that, under the transformation (4.2), the function v has every good property of u and has better asymptotic behaviour in that the second-order terms in the expansion of u are absent from the expansion of v.

Lemma 4.3 *The function* v *in* (4.2) *satisfies conditions* (a) *and* (b) *of Theorem 4.2, but with improved asymptotic behaviour in that one of the following conditions holds for* $r \geq R_u + 2|q| =: R_v$. *As before,* κ, m, δ *and* γ *are positive constants, with* $\delta \in (0, 1]$.

(A) $v(x) = -\kappa r^m + g(x)$, *where* $|\nabla g(x)| = O(r^{m-2-\delta})$,

and, if $m - 1 - \delta < 0$, *then* $g(x) \to 0$ *as* $r \to \infty$.

(B) $v(x) = \kappa \log \dfrac{\gamma}{r} + g(x)$, *where* $|\nabla g(x)| = O\left(r^{-2-\delta}\right)$,

and $g(x) \to 0$ *as* $r \to \infty$.

(C) $v(x) = \kappa r^{-m} + g(x)$, *where* $|\nabla g(x)| = O(r^{-m-2-\delta})$,

and $g(x) \to 0$ *as* $r \to \infty$.

Proof (i) It is clear from (4.2) that $v \in C^1(\mathbb{R}^N)$ and that $v > 0$ on $\mathbb{R}^N$ in case (C).

(ii) Here is a merely formal calculation showing the removal of terms $(c \cdot x)r^\alpha$ in Definition 4.1. For case (A),

$$
\begin{aligned}
-\kappa |x+q|^m &+ c \cdot (x+q)|x+q|^{m-2} \\
&= -\kappa (r^2 + 2q \cdot x + |q|^2)^{m/2} + c \cdot (x+q)|x+q|^{m-2} \\
&= -\kappa r^m \left\{ 1 + m(q \cdot x)r^{-2} + \cdots \right\} + (c \cdot x)r^{m-2} + \cdots,
\end{aligned}
$$

so that terms of order r^{m-1} are cancelled by the choice $q = c/\kappa m$.

For case (B),

$$
\kappa \log \frac{\gamma}{|x+q|} = \kappa \left\{ \log \frac{\gamma}{r} - \frac{1}{2}\log[1 + 2(q \cdot x)r^{-2} + |q|^2 r^{-2}] \right\}
$$

$$= \kappa \left\{ \log \frac{\gamma}{r} - (q \cdot x)r^{-2} + \cdots \right\},$$

so that terms of order r^{-1} are cancelled by the choice $q = c/\kappa$.
For case (C),

$$\kappa|x + q|^{-m} = \kappa r^{-m} \left\{ 1 - m(q \cdot x)r^{-2} + \cdots \right\},$$

so that terms of order r^{-m-1} are cancelled by the choice $q = c/\kappa m$.

(iii) To show that v satisfies (4.1), we suppose that any ψ in $C_c^\infty(\mathbb{R}^N)$ is given. Choose $\varphi(y) = \psi(y - q)$ in (4.1); then

$$\int_{\mathbb{R}^N} \left\{ -(\nabla \psi)(y - q) \cdot (\nabla u)(y) + \psi(y - q) f(u(y)) \right\} \, dy = 0,$$

and the substitution $y = x + q$ yields (4.1) for ψ and v.

(iv) For a rigorous justification of the claims made for ∇g and g, we define functions F, G and H by

$$|x + q|^a =: r^a + a(q \cdot x)r^{a-2} + F(x, a),$$

$$\log \frac{1}{|x + q|} =: \log \frac{1}{r} - (q \cdot x)r^{-2} + G(x),$$

$$c \cdot (x + q)|x + q|^{a-2} =: (c \cdot x)r^{a-2} + H(x, a),$$

where $a \in \mathbb{R}$ is an arbitrary constant. Then, with h still denoting the remainder function in Definition 4.1, g is given by one of the formulae

(A)	$g(x) = -\kappa F(x, m) + H(x, m) + h(x + q),$	
(B)	$g(x) = \kappa G(x) + H(x, 0) + h(x + q),$	(4.3)
(C)	$g(x) = \kappa F(x, -m) + H(x, -m) + h(x + q).$	

Accordingly, to prove the claims made for ∇g it is sufficient to show that, for each $k \in \{1, \ldots, N\}$,

$$r \geq 2|q| \quad \Rightarrow \quad \left| \frac{\partial}{\partial x_k} F(x, a) \right| \leq \text{const.} \, r^{a-3},$$

$$\left| \frac{\partial}{\partial x_k} G(x) \right| \leq \text{const.} \, r^{-3},$$

$$\left| \frac{\partial}{\partial x_k} H(x, a) \right| \leq \text{const.} \, r^{a-3},$$

(4.4)

where the constants may depend on a, q, N and k, but not on x. To this end, let $\varphi(t) := |x + tq|^a$, $0 \le t \le 1$; then the definition of F becomes

$$
\begin{aligned}
F(x,a) &= \varphi(1) - \varphi(0) - \varphi'(0) = \int_0^1 (1-t)\, \varphi''(t)\, dt \\
&= \int_0^1 (1-t) \left\{ \left(\sum_{j=1}^N q_j \partial_j \right)^2 |x + tq|^a \right\} dt, \quad\quad (4.5a)
\end{aligned}
$$

where x and q are fixed and ∂_j means, as always, differentiation with respect to the jth argument of the operand. Similarly,

$$
G(x) = \int_0^1 (1-t) \left\{ \left(\sum_{j=1}^N q_j \partial_j \right)^2 \log \frac{1}{|x+tq|} \right\} dt, \quad\quad (4.5b)
$$

$$
H(x,a) = \int_0^1 \left(\sum_{j=1}^N q_j \partial_j \right) \left\{ c \cdot (x+tq) |x + tq|^{a-2} \right\} dt. \quad\quad (4.5c)
$$

It is sufficient to justify differentiation with respect to x_k under the integral sign for $2|q| \le r \le M$ and each M. We may suppose that $|q| > 0$ (otherwise, there is nothing to prove), and then the bounds

$$
0 < |q| \le \tfrac{1}{2} r \le |x + tq| \le \tfrac{3}{2} r \le \tfrac{3}{2} M \quad\quad (4.6)
$$

imply uniform continuity of each differentiated integrand.

The desired result (4.4) is now implied by the following estimates, which are slight extensions of Lemma A.4 and are proved in the same way. For any $y \in \mathbb{R}^N \setminus \{0\}$ and any multi-index β (Definition A.3),

$$
\left| \partial^\beta |y|^a \right| \le \text{const.}\ |y|^{a-|\beta|}, \quad\quad (4.7a)
$$

$$
\left| \partial^\beta \log \frac{1}{|y|} \right| \le \text{const.}\ |y|^{-|\beta|} \quad\quad \text{if } |\beta| \ge 1, \quad\quad (4.7b)
$$

$$
\left| \partial^\beta (c \cdot y) |y|^{a-2} \right| \le \text{const.}\ |y|^{a-1-|\beta|}; \quad\quad (4.7c)
$$

the Leibniz rule (Exercise A.23) can be used for derivation of (4.7c), and the constants may depend on everything except y. To prove (4.4), we first set $y = x + tq$ and then use (4.6).

(v) Regarding the claims made for g itself: in view of (4.3) it is enough to prove that, if $a < 1 + \delta$, then $F(x,a)$, $G(x)$ and $H(x,a)$ all tend to zero as $r \to \infty$. This is the case; since $\delta \le 1$, we have $a < 2$, and (4.5a) to (4.7c) show that $F(x, a) = O(r^{a-2})$, $G(x) = O(r^{-2})$ and $H(x, a) = O(r^{a-2})$. □

4.2 Four lemmas about reflection of v

Definition 4.4 Let k be a given unit vector ($k \in \mathbb{R}^N$ and $|k| = 1$), and let

$$\tilde{x}_i := \sum_{j=1}^{N} R_{ij}(k) \, x_j \qquad (i = 1, \ldots, N),$$

where $R(k)$ is an orthogonal $N \times N$ matrix such that $\tilde{x}_1 = x \cdot k$. By *variables aligned with k* we mean

(a) the co-ordinates $\tilde{x}_1$ and $\tilde{x}'' := (\tilde{x}_2, \ldots, \tilde{x}_N)$;

(b) the description $\tilde{T}_\mu := \{ \xi \in \mathbb{R}^N \mid \tilde{\xi}_1 = \mu \}$ of the hyperplane $T_\mu(k)$ (Definition 1.7);

(c) the description $\tilde{x}^\mu := (2\mu - \tilde{x}_1, \tilde{x}'')$ of the reflected point $x^{\mu,k}$;

(d) the version $\tilde{\varphi}$, such that $\tilde{\varphi}(\tilde{x}_1, \ldots, \tilde{x}_N) = \tilde{\varphi}(R(k)x) = \varphi(x)$, of any given function φ from $\mathbb{R}^N$ into $\mathbb{R}$;

(e) the open half-spaces $\tilde{Y}(\mu) := \{ x \in \mathbb{R}^N \mid \tilde{x}_1 < \mu \}$ and $\tilde{Z}(\mu) := \{ x \in \mathbb{R}^N \mid \tilde{x}_1 > \mu \}$ that are separated by the hyperplane $\tilde{T}_\mu$. □

Until the contrary is stated, k will be arbitrary but *fixed*; we shall use variables aligned with k, but shall *omit the tilde* as long as there is no danger of confusion.

Let v be the function in Lemma 4.3 (more precisely, the version of that function aligned with k); we wish to show that $v(-x_1, x'') = v(x_1, x'')$ for all $x \in \mathbb{R}^N$, and that $\partial_1 v(x) < 0$ whenever $x_1 > 0$. If these conditions hold (Figure 4.1), then

$$\mu > 0 \quad \text{and} \quad y \in Y(\mu) \implies v(y^\mu) < v(y),$$

and we begin by showing that this inequality holds whenever $|y|$ is sufficiently large.

The four key results of this section will be labelled by Roman numerals.

Lemma 4.5 *There is a number $R(\mu)$, depending only on v and μ, such that*

$$\mu > 0, \quad y \in Y(\mu) \quad \text{and} \quad |y| > R(\mu) \implies v(y^\mu) < v(y). \qquad \text{(I)}$$

The function R is non-increasing on $(0, 1]$ and is constant on $[1, \infty)$.

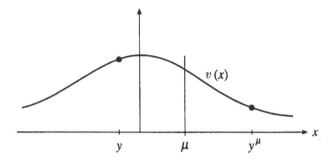

Fig. 4.1.

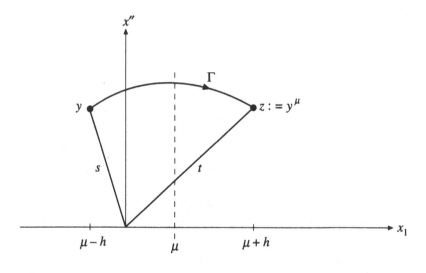

Fig. 4.2.

Proof (i) We consider only points outside the ball $\mathscr{B}(0, R_v)$, where R_v is as in Lemma 4.3, and adopt the notation (Figure 4.2)

$$z := y^\mu, \qquad h := \tfrac{1}{2}(z_1 - y_1) > 0,$$

$$s := |y| = \left\{ (\mu - h)^2 + |y''|^2 \right\}^{1/2}, \qquad t := |z| = \left\{ (\mu + h)^2 + |y''|^2 \right\}^{1/2},$$

so that

$$t^2 - s^2 = 4\mu h > 0, \qquad t > s \geq R_v. \tag{4.8}$$

The basis of the proof is the size of the gap, in the asymptotic description of v, between the leading term and the remainder function g.

We shall demonstrate this for two of four cases: case (A) of Lemma 4.3 with $m \geq 2$, and case (C) of that lemma. Case (A) with $0 < m < 2$, and case (B), are left as exercises for the reader. The letters $C_1, C_2, \ldots$ will denote constants independent of y and μ, but possibly depending on v (and hence on $N, m, \ldots$).

(ii) Case (A) with $m \geq 2$ and $N \geq 2$. By Lemma 4.3,

$$v(y) - v(z) = \kappa(t^m - s^m) + g(y) - g(z), \qquad m \geq 2,$$

$$|\nabla g(x)| \leq C_1 r^{m-2-\delta} \qquad \text{for} \quad r \geq R_v.$$

We estimate separately for $4\mu h \leq s^2$ and for $4\mu h > s^2$.

(a) $4\mu h \leq s^2$. By convexity of the function $\tau \mapsto \tau^{m/2}$, $0 < \tau < \infty$, and by (4.8),

$$t^m - s^m = (s^2 + 4\mu h)^{m/2} - (s^2)^{m/2} \geq \frac{m}{2}(s^2)^{m/2-1} 4\mu h = 2m s^{m-2} \mu h.$$

To estimate $g(y) - g(z)$, define P_{yz} to be the two-dimensional plane containing the points $0, y$ and z (or any such plane if $y'' = 0$), and define Γ to be the circular arc in P_{yz} from y to z, centred at $(\mu, 0)$. Then points $x \in \Gamma$ have $r = |x| \in [s, t]$, and Γ has length πh at most, so that

$$|g(y) - g(z)| = \left| \int_\Gamma \nabla g(x) \cdot dx \right| \leq \begin{cases} C_1 \pi t^{m-2-\delta} h & \text{if } m - 2 - \delta \geq 0, \\ C_1 \pi s^{m-2-\delta} h & \text{if } m - 2 - \delta < 0, \end{cases}$$

and $t^2 = s^2 + 4\mu h \leq 2s^2$. Hence for this case [(A) with $m \geq 2$, $N \geq 2$ and $4\mu h \leq s^2$],

$$v(y) - v(z) \geq s^{m-2-\delta} h(2\kappa m \mu s^\delta - C_2) > 0 \quad \text{if } s^\delta > \frac{C_2}{2\kappa m \mu}.$$

(b) $4\mu h > s^2$. This condition implies that $t^2 > 2s^2$ and hence that

$$t^m - s^m = t^m \left\{ 1 - \left(\frac{s}{t}\right)^m \right\} > t^m \left\{ 1 - \left(\frac{1}{2}\right)^{m/2} \right\} \geq \frac{1}{2} t^m.$$

To bound $g(y) - g(z)$, we observe that $m - 1 - \delta \geq 0$ (because $m \geq 2$ and $\delta \leq 1$). Reducing δ slightly if necessary, we may suppose that $m - 1 - \delta > 0$ whenever $m \geq 2$; then g can be estimated as follows by integrating radially outwards from $\partial \mathscr{B}(0, R_v)$. Let x_0 be any point outside $\mathscr{B}(0, R_v)$; we write $r_0 := |x_0|$ and

$$g(x_0) = g\left(\frac{R_v}{r_0} x_0\right) + \int_{R_v/r_0}^1 \frac{dg(\tau x_0)}{d\tau} d\tau,$$

where

$$\left| \frac{dg(\tau x_0)}{d\tau} \right| = |x_0 \cdot (\nabla g)(\tau x_0)| \le C_1 r_0 (\tau r_0)^{m-2-\delta},$$

so that

$$|g(x_0)| \le \max_{r=R_v} |g(x)| + C_1 r_0^{m-1-\delta} \int_0^1 \tau^{m-2-\delta} \, d\tau$$

$$\le C_3 r_0^{m-1-\delta} \qquad (r_0 \ge R_v),$$

if C_3 is sufficiently large. Consequently,

$$|g(y) - g(z)| \le 2C_3 t^{m-1-\delta},$$

and, for this case [(A) with $m \ge 2$, $N \ge 2$ and $4\mu h > s^2$],

$$v(y) - v(z) > t^{m-1-\delta} \left(\frac{\kappa}{2} t^{1+\delta} - 2C_3 \right)$$

$$> t^{m-1-\delta} \left\{ \frac{\kappa}{2} (2^{1/2} s)^{1+\delta} - 2C_3 \right\}$$

$$> 0 \quad \text{if } s^{1+\delta} > \frac{2^{3/2-\delta/2} C_3}{\kappa} =: C_4.$$

(c) Collecting results, we define

$$R(\mu) := \max \left\{ R_v, \left(\frac{C_2}{2\kappa m \min\{\mu, 1\}} \right)^{1/\delta}, C_4^{1/(1+\delta)} \right\}, \qquad (4.9)$$

and the lemma is proved for case (A) with $m \ge 2$ and $N \ge 2$.

(iii) Case (C) with $N \ge 2$. By Lemma 4.3,

$$v(y) - v(z) = \kappa(s^{-m} - t^{-m}) + g(y) - g(z), \qquad m > 0,$$

$$|\nabla g(x)| \le C_5 r^{-m-2-\delta} \quad \text{for } r \ge R_v.$$

Again we estimate separately for $4\mu h \le s^2$ and for $4\mu h > s^2$.

(a) $4\mu h \le s^2$. By strict convexity of the function $\tau \mapsto \tau^{-m/2}$, $0 < \tau < \infty$,

$$s^{-m} - t^{-m} = (s^2)^{-m/2} - (s^2 + 4\mu h)^{-m/2} > \frac{m}{2}(s^2 + 4\mu h)^{-m/2-1} 4\mu h$$

$$\ge \frac{m}{2}(2s^2)^{-m/2-1} 4\mu h = C_6 \, s^{-m-2} \mu h.$$

Integration along Γ gives, essentially as before,

$$|g(y) - g(z)| \le C_5 \pi s^{-m-2-\delta} h.$$

Hence for this case [(C) with $N \geq 2$ and $4\mu h \leq s^2$],

$$v(y) - v(z) > s^{-m-2-\delta} h \left(C_6 \kappa \mu s^\delta - C_5 \pi \right) > 0 \quad \text{if } s^\delta > \frac{C_5 \pi}{C_6 \kappa \mu}.$$

(b) $4\mu h > s^2$. Since now $t^2 > 2s^2$,

$$s^{-m} - t^{-m} = s^{-m} \left\{ 1 - \left(\frac{s}{t}\right)^m \right\} > s^{-m} \left\{ 1 - \left(\frac{1}{2}\right)^{m/2} \right\} = C_7 s^{-m}.$$

Next, for any point x_0 outside $\mathscr{B}(0, R_v)$ we obtain more easily than before

$$|g(x_0)| = \left| \int_1^\infty \frac{dg(\tau x_0)}{d\tau} \, d\tau \right| \leq \frac{C_5 r_0^{-m-1-\delta}}{m+1+\delta},$$

so that

$$|g(y) - g(z)| \leq \frac{2C_5}{m+1+\delta} s^{-m-1-\delta}.$$

Then, for this case [(C) with $N \geq 2$ and $4\mu h > s^2$],

$$v(y) - v(z) > s^{-m-1-\delta} \left(C_7 \kappa s^{1+\delta} - \frac{2C_5}{m+1+\delta} \right) > 0$$

$$\text{if } s^{1+\delta} > \frac{2C_5}{C_7 \kappa (m+1+\delta)} =: C_8.$$

(c) Collecting results, we define

$$R(\mu) := \max \left\{ R_v, \left(\frac{C_5 \pi}{C_6 \kappa \min\{\mu, 1\}} \right)^{1/\delta}, C_8^{1/(1+\delta)} \right\}, \qquad (4.10)$$

and the lemma is proved for case (C) with $N \geq 2$.

(iv) *Modifications for $N = 1$.* When the domain of v is $\mathbb{R}$, we must modify certain estimates of $|g(y) - g(z)|$ because no path outside the ball $(-R_v, R_v)$ connects y and z if $y \leq -R_v$. In other words, we have no substitute for the arc Γ when $N = 1$ and $y \leq -R_v$. However, if $y \geq R_v$ we can replace integration of ∇g along Γ by integration along $\mathbb{R}$ from y to z. Also, integrals of ∇g along radial lines from $(R_v/r_0) x_0$ to x_0, or from x_0 to infinity, can be used as before. (For $N = 1$ and $x_0 \leq -R_v$, the latter are from x_0 to $-\infty$.) Since the arc Γ was used only for $4\mu h \leq s^2$, we need change only estimates that are subject to both

$$4\mu h \leq s^2, \quad \text{equivalently}, \quad t^2 \leq 2s^2,$$

and

$$y = \mu - h = -s \leq -R_v, \quad \text{equivalently}, \quad h = s + \mu \geq R_v + \mu.$$

Then, for case (C) with $N = 1$, by inequalities in (iii) that remain valid and by $h > s$,

$$s^{-m} - t^{-m} > C_6 \, s^{-m-2} \mu h > C_6 \, s^{-m-1} \mu,$$

$$\left| g(y) - g(z) \right| \leq \frac{2C_5}{m+1+\delta} s^{-m-1-\delta},$$

whence

$$v(y) - v(z) \; > \; s^{-m-1-\delta} \left(C_6 \kappa \mu s^\delta - \frac{2C_5}{m+1+\delta} \right) > 0$$

$$\text{if } \; s^\delta > \frac{2C_5}{C_6 \kappa \mu (m+1+\delta)}.$$

For case (A) with $m \geq 2$ and $N = 1$, by inequalities in (ii) that remain valid and by $h > s$, $t \leq 2^{1/2}s$,

$$t^m - s^m \geq 2ms^{m-2} \mu h > 2ms^{m-1} \mu,$$

$$\left| g(y) - g(z) \right| \; \leq \; 2C_3 \, t^{m-1-\delta} \leq 2C_3 \left(2^{1/2}s \right)^{m-1-\delta} =: C_9 \, s^{m-1-\delta},$$

whence

$$v(y) - v(z) > s^{m-1-\delta} \left(2\kappa m \mu s^\delta - C_9 \right) > 0 \quad \text{if } \; s^\delta > \frac{C_9}{2\kappa m \mu}.$$

$\square$

Lemma 4.6 *There is a number $\mu_* > 1$ such that*

$$\mu \geq \mu_* \quad \text{and} \quad y \in Y(\mu) \implies v(y^\mu) < v(y). \tag{II}$$

Proof (i) Let $R(\mu)$ be as in Lemma 4.5, so that $R(\mu) = R(1)$ for $\mu \geq 1$, and define

$$R_1 := \max\{ 1, R(1) \}, \qquad c_1 := \min_{r \leq R_1} v(x).$$

[Here $r := |x|$, as before.]

(ii) For cases (A) and (B) of Lemma 4.3, choose μ_* to be such that

$$r \geq \mu_* \implies v(x) \leq c_1 - 1;$$

this is possible because $v(x) \to -\infty$ as $r \to \infty$. Then $\mu_* > R_1$; Figure 4.3 illustrates the situation.

If $\mu \geq \mu_*$, $y \in Y(\mu)$ and $|y| > R_1$, then also $|y| > R(\mu)$ [because $\mu_* > 1$ and $R_1 \geq R(1)$]; Lemma 4.5 ensures that $v(y^\mu) < v(y)$. If $\mu \geq \mu_*$,

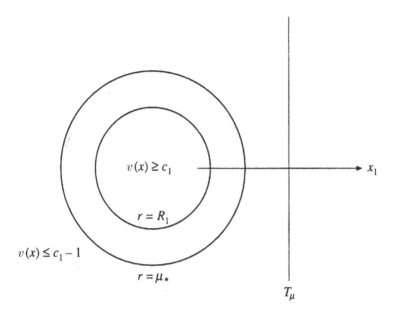

Fig. 4.3.

$y \in Y(\mu)$ and $|y| \le R_1$, then $v(y) \ge c_1$ [by definition of c_1] and $|y^\mu| > \mu_*$ [because $y_1 \le R_1 < \mu_*$] so that

$$v(y^\mu) \le c_1 - 1 < v(y).$$

(iii) For case (C) of Lemma 4.3, choose μ_* to be such that

$$r \ge \mu_* \implies v(x) \le \tfrac{1}{2}c_1;$$

this is possible because now $c_1 > 0$ and $v(x) \to 0$ as $r \to \infty$. The rest is as in (ii), with $\tfrac{1}{2}c_1$ replacing $c_1 - 1$.   □

Although all four results of this section will be needed to prove Theorem 4.2, it is the next lemma that is the heart of the matter. Observe that the generalized Poisson equation (4.1) is used only in Lemma 4.7, and that the whole theorem rests on our ability to sharpen $\le$ to $<$ in Lemma 4.7.

Lemma 4.7 *Assume that for some* $\mu > 0$ *we have* (a) $v(y^\mu) \le v(y)$ *for all* $y \in Y(\mu)$, (b) $v(y_0^\mu) \ne v(y_0)$ *for some* $y_0 \in Y(\mu)$.

Then

$$v(y^\mu) < v(y) \quad \textit{for all} \ \ y \in Y(\mu), \tag{III.a}$$

and

$$\partial_1 v(x) < 0 \quad \textit{for all} \ \ x \in T_\mu. \tag{III.b}$$

Here (III.a) *implies* (III.b).

Proof Let $Y := Y(\mu)$ and $Z := Z(\mu)$ for the given μ, and let $w := v_\mu - v$ [which means, according to Definition 1.7, that $w(x) = v(x^\mu) - v(x)$ for all $x \in \mathbb{R}^N$]. By hypothesis, $w(y) \le 0$ for $y \in Y$; we shall prove strict inequality by means of the maximum principle.

(i) A generalized Poisson equation for w.

By (4.1) and Lemma 4.3, we have

$$\int_{\mathbb{R}^N} \left\{ -\nabla\psi \cdot \nabla v + \psi f(v) \right\} = 0 \quad \text{for all} \ \ \psi \in C_c^\infty(\mathbb{R}^N). \tag{4.11}$$

Given $\varphi \in C_c^\infty(Y)$, choose $\psi = \varphi_\mu$ in this equation; then $\operatorname{supp}\varphi_\mu \subset Z$ and

$$\int_Z \left\{ -\nabla\varphi_\mu(z) \cdot \nabla v(z) + \varphi_\mu(z) f\big(v(z)\big) \right\} \, \mathrm{d}z = 0.$$

Set $z = y^\mu$. Then $y = z^\mu \in Y$, $\partial/\partial z_1 = -\partial/\partial y_1$ and $\partial/\partial z_j = \partial/\partial y_j$ for $j = 2, \ldots, N$; also, $\varphi_\mu(z) = \varphi(z^\mu) = \varphi(y)$ and $v(z) = v(y^\mu) = v_\mu(y)$. Accordingly,

$$\int_Y \left\{ -\nabla\varphi(y) \cdot \nabla v_\mu(y) + \varphi(y) f\big(v_\mu(y)\big) \right\} \, \mathrm{d}y = 0. \tag{4.12}$$

Now choose $\psi = \varphi$ in (4.11), and subtract the result from (4.12); then

$$\int_Y \left\{ -\nabla\varphi \cdot \nabla w + \varphi[f(v_\mu) - f(v)] \right\} = 0 \quad \text{for all} \ \ \varphi \in C_c^\infty(Y). \tag{4.13}$$

(ii) Application of the strong maximum principle.

In order to control f, we consider balls in Y rather than the whole of Y. Let

$$D := \mathscr{B}(c, \rho), \quad c \in Y, \quad \rho := \min\{ 1, \operatorname{dist}(c, T_\mu) \},$$

so that $D \subset Y$. To estimate $f\big(v_\mu(y)\big) - f\big(v(y)\big)$ at points y of D, we recall that $f = f_1 + f_2$ with f_1 locally Lipschitz continuous and f_2 non-decreasing. By the continuity of v, there is a number $M = M(c, v, \mu)$

such that $|v(y)| \leq M$ and $|v(y^\mu)| \leq M$ whenever $y \in D$; hence there is a Lipschitz constant $A = A(M) > 0$ such that, in D,

$$f_1(v_\mu) - f_1(v) \leq A|v_\mu - v| = -Aw \qquad \text{[since } w \leq 0].$$

Also,

$$f_2(v_\mu) - f_2(v) \leq 0,$$

because $v_\mu \leq v$ and f_2 is non-decreasing. By addition of these bounds and by (4.13),

$$0 = \int_D \left\{ -\nabla\varphi \cdot \nabla w + \varphi[f(v_\mu) - f(v)] \right\} \leq \int_D \left\{ -\nabla\varphi \cdot \nabla w - \varphi Aw \right\}$$

whenever $\varphi \in C_c^\infty(D)$ and $\varphi \geq 0$; in other words, w is a generalized subsolution relative to $\triangle - A$ and D.

Now suppose that $w(c) = 0 = \sup_D w$ at some $c \in Y$. Then the strong maximum principle, Theorem 2.13, states that $w\big|_D = 0$, and this shows that the zero set of w, say $X := \{ y \in Y \mid w(y) = 0 \}$, is open in Y. Since w is continuous, X is also closed in Y. Then, since Y is connected, either $X = Y$ or $X = \emptyset$; our hypothesis that $w(c) = 0$ at some $c \in Y$ implies that $X = Y$, which contradicts hypothesis (b) of the lemma. Consequently, X must be empty, and this proves (III.a).

(iii) Application of the boundary-point lemma for balls.

Let $p \in T_\mu$ be given. Define

$$B := \mathscr{B}\big((p_1 - \rho, p''), \rho\big) \quad \text{for any } \rho \in (0, 1].$$

We know from (ii) that w is a generalized subsolution relative to $\triangle - A$ and B, and that $w < 0$ in B. In addition, $w(p) = 0$ because $p^\mu = p$. Then the boundary-point lemma for balls, Lemma 2.12, states that $(\partial_1 w)(p) > 0$. Now,

$$(\partial_1 w)(p) = -2(\partial_1 v)(p)$$

because

$$\partial_1 w(x) = \frac{\partial}{\partial x_1} \left\{ v(2\mu - x_1, x'') - v(x_1, x'') \right\},$$

so that $(\partial_1 v)(p) < 0$. This proves (III.b). $\qquad\qquad\qquad\qquad\qquad$ $\square$

Lemma 4.8 *The set*

$$\big\{ \mu > 0 \mid \text{(III.a)} \;\; holds : y \in Y(\mu) \Rightarrow v(y^\mu) < v(y) \big\} \;\; is\ open\ in\ \mathbb{R}. \quad \text{(IV)}$$

Proof Assume that the lemma is false. Then, for some value $\mu = 2\alpha > 0$ at which (III.a) holds, there is a sequence (μ_n) in $(\alpha, 3\alpha)$ such that (III.a) fails at each μ_n and $\mu_n \to 2\alpha$ as $n \to \infty$. This failure of (III.a) must be for $|y| \le R(\alpha)$, because (I) ensures that

$$\mu \ge \alpha, \ y \in Y(\mu) \quad \text{and} \quad |y| > R(\alpha) \implies v(y^\mu) < v(y),$$

where $R(\mu) \le R(\alpha)$. Thus there are sequences (μ_n) and (y_n) [here y_n is not a co-ordinate, but a point of $\mathbb{R}^N$] such that

$$\mu_n \to 2\alpha, \ y_n \in Y(\mu_n), \ |y_n| \le R(\alpha) \quad \text{and} \quad v\left(y_n^{\mu_n}\right) \ge v(y_n).$$

By the relative compactness of (y_n), and resort to a subsequence if need be, we may suppose that y_n tends to some point q in $\mathscr{B}\left(0, R(\alpha)\right)$ as $n \to \infty$. Then $q \in \overline{Y(2\alpha)}$ and $v\left(q^{2\alpha}\right) \ge v(q)$. Since (III.a) holds at $\mu = 2\alpha$, we must have $q \in T_{2\alpha}$. But (III.b) holds at $\mu = 2\alpha$ [because (III.a) does]; hence $(\partial_1 v)(q) < 0$, and this contradicts $v\left(y_n^{\mu_n}\right) \ge v(y_n)$ when n is sufficiently large. □

4.3 Proof of Theorem 4.2 and a corollary

Proposition 4.9 *The results* (I) *to* (IV) *imply that v is spherically symmetric and that* $dv/dr < 0$ *for* $r > 0$; *that is, they imply Theorem 4.2.*

Proof (i) Let (m, ∞) be the largest open interval of μ in which (III.a) holds. This m exists, by (II) and (IV).

Assume that $m > 0$. We show that the hypotheses preceding (III.a) hold for $\mu = m$. First,

$$y \in Y(m) \Rightarrow v(y^m) \le v(y),$$

by (III.a) for $\mu > m$ and by continuity: fix $y \in Y(m)$ and let $\mu \downarrow m$. Second, by (I),

$$y \in Y(m) \quad \text{and} \quad |y| > R(m) \implies v(y^m) < v(y).$$

Therefore (III.a) holds at $\mu = m$. But then (IV) states that (III.a) holds in $(m - \varepsilon, \infty)$ for some $\varepsilon > 0$, so that (m, ∞) is not maximal for (III.a). This contradiction shows that $m = 0$.

(ii) We now have

$$v(x_1, x'') \le v(-x_1, x'') \quad \text{whenever} \quad x_1 \ge 0,$$

by (III.a) for $\mu > 0$ and by continuity: fix $-x_1 < 0$ and let $\mu \downarrow 0$. [For $x_1 = 0$ we have equality.]

Keeping the co-ordinates $x_1 = \tilde{x}_1 = x \cdot k$ and $x'' = \tilde{x}''$, which are aligned with k, we repeat the argument for the unit vector $-k$; this yields

$$v(x_1, x'') \geq v(-x_1, x'') \quad \text{whenever} \quad x_1 \geq 0.$$

Thus $\tilde{v}$ is an even function of $\tilde{x}_1$ for arbitrary unit vector k, hence v is spherically symmetric, by Lemma 1.8.

That $dv/dr < 0$ for $r > 0$ is shown by (III.b). $\qquad\square$

The method also yields symmetry relative to a single co-ordinate, in place of spherical symmetry, for a wider class of functions f. In the following corollary there is only *one* system of co-ordinates (no alignment with an arbitrary unit vector). We still write

$$x'' := (x_2, \ldots, x_N).$$

The remark about $f_1(., x'')$ means that, for each compact set $E \subset \mathbb{R}$, there is a constant $A(E)$ *independent of* x'' such that $\left| f_1(s, x'') - f_1(t, x'') \right| \leq A(E)|s - t|$ whenever $s, t \in E$ and $x'' \in \mathbb{R}^{N-1}$.

Corollary 4.10 *Suppose that* $u : \mathbb{R}^N \to \mathbb{R}$ *satisfies condition* (a) *of Theorem 4.2 and that, in condition* (b), $f\big(u(x)\big)$ *is replaced by* $f\big(u(x), x''\big)$, *where* f *now has a decomposition* $f = f_1 + f_2$ *such that* $f_1(., x'') : \mathbb{R} \to \mathbb{R}$ *is locally Lipschitz continuous uniformly over* $x'' \in \mathbb{R}^{N-1}$, *while* $f_2(., x'') : \mathbb{R} \to \mathbb{R}$ *is non-decreasing for each* $x'' \in \mathbb{R}^{N-1}$. *In case* (C), *these conditions need hold only on* $[0, \infty)$.

Again define q *and* v *by* (4.2). *Then* $v(-x_1, x'') = v(x_1, x'')$ *for all* $x \in \mathbb{R}^N$, *and* $\partial_1 v(x) < 0$ *whenever* $x_1 > 0$.

Proof The only changes are these.

(i) The function v satisfies not (4.1) but, for all $\psi \in C_c^\infty(\mathbb{R}^N)$,

$$\int_{\mathbb{R}^N} \left\{ -\nabla\psi(x) \cdot \nabla v(x) + \psi(x)\, F\big(v(x), x''\big) \right\} \, dx = 0, \qquad (4.14)$$

where

$$F(t, x'') := f(t, x'' + q'') \qquad \text{for all} \;\; t \in \mathbb{R},$$

or for all $t \in [0, \infty)$ in case (C). This follows from step (iii) of Lemma 4.3. Then the decomposition $f = f_1 + f_2$ implies a decomposition $F = F_1 + F_2$ with the same properties.

(ii) The results (I) to (IV) are derived not for an arbitrary unit vector k, but only for $k = (1, 0, \ldots, 0)$ [relative to the original co-ordinate system]. Proposition 4.9 is restricted similarly. $\qquad\square$

4.4 Application to some Newtonian potentials

Notation As in (A.18) of Appendix A, the *Newtonian kernel* for $\mathbb{R}^N$ is

$$
K(x) = K(x; N) := \begin{cases} -\tfrac{1}{2}|x| & \text{in } \mathbb{R}, \\[2mm] \dfrac{1}{2\pi} \log \dfrac{1}{|x|} & \text{in } \mathbb{R}^2 \setminus \{0\}, \\[3mm] \kappa_N \dfrac{1}{|x|^{N-2}} & \text{in } \mathbb{R}^N \setminus \{0\}, \quad N \geq 3, \end{cases}
$$

where

$$
\kappa_N = \frac{1}{(N-2)|\partial \mathscr{B}_N(0,1)|} \quad (N \neq 2), \qquad |\partial \mathscr{B}_N(0,1)| = \frac{N\pi^{N/2}}{(N/2)!}.
$$

The formula for the surface area of the unit sphere is a result of Exercise 1.14.

The *characteristic function* χ_A of any subset A of a universal set U is defined by

$$
\chi_A(x) := \begin{cases} 1 & \text{if } x \in A, \\ 0 & \text{if } x \in U \setminus A. \end{cases}
$$

We define the *Heaviside function* to be $f_H := \chi_{(0,\infty)}$, so that

$$
f_H(t) = \begin{cases} 1 & \text{if } t > 0, \\ 0 & \text{if } t \leq 0. \end{cases}
$$

(The value $f_H(0) = 0$ is a little unorthodox.)

Recall Theorem 1.1: *if G is a bounded open subset of $\mathbb{R}^N$ ($N \geq 1$), λ is a positive constant,*

$$
u(x) := \lambda \int_G K(x - \xi)\, d\xi \qquad \text{for all } x \in \mathbb{R}^N, \tag{4.15}
$$

and

$$
u\big|_{\partial G} = \text{constant} = \beta, \tag{4.16}
$$

then G is a ball. The truth of this will emerge from the more general Theorem 4.13, but it is worthwhile to anticipate here the reasons that (4.15) and (4.16) allow application of Theorem 4.2. First, results in Appendix A show that $u \in C^1(\mathbb{R}^N)$ and that the asymptotic behaviour of u (as $r \to \infty$) is admissible. Second, we shall infer from the maximum principle that $u(x) > \beta$ if and only if $x \in G$; therefore, in place of the classical result (Appendix A) that

$$
\triangle u = \begin{cases} -\lambda & \text{in } G, \\ 0 & \text{in } \mathbb{R}^N \setminus \overline{G}, \end{cases}
$$

we shall find that u is a generalized solution of the equation

$$\triangle u + \lambda f_H(u - \beta) = 0 \quad \text{in } \mathbb{R}^N. \tag{4.17}$$

In other words, u satisfies (4.1) with $f = \lambda f_H(. - \beta)$, and this f has a suitable decomposition: choose $f_1 = 0$.

We consider, in the first instance, the Newtonian potential of a given density function F on a given set G; let

$$\left. \begin{array}{l} w(x) := \int_G K(x - \xi) \, F(\xi) \, d\xi \quad \text{for all } x \in \mathbb{R}^N, \\[2mm] \text{where} \quad G \text{ is open and bounded in } \mathbb{R}^N, \quad F \in L_p(G) \text{ with } p > N. \end{array} \right\} \tag{4.18}$$

Lemma 4.11 *The potential w in (4.18) enjoys the following properties.*
(i) $w \in C^1(\mathbb{R}^N)$.
(ii) *Extend F to be zero outside G; then*

$$\int_{\mathbb{R}^N} \left\{ -\nabla \varphi \cdot \nabla w + \varphi F \right\} = 0 \quad \text{for all } \varphi \in C_c^\infty (\mathbb{R}^N). \tag{4.19}$$

(iii) *If $\int_G F > 0$, then the asymptotic behaviour of w is admissible in the sense of Definition 4.1; for $N = 1$ we have case (A) with $m = 1$, for $N = 2$ we have case (B), and for $N \geq 3$ we have case (C) with $m = N - 2$.*

Proof (i), (ii) Theorem A.11 establishes somewhat more than the C^1 property of w and also shows that w is a generalized solution of $-\triangle w = F$ in $\mathbb{R}^N$; in other words, shows that (4.19) holds.

(iii) Let R_w be so large that $G \subset \mathscr{B}(0, \frac{1}{2}R_w)$, and let $r := |x| \geq R_w$. For $N = 1$, we observe from (A.56), or else directly from the definition (4.18), that

$$w(x) = -\kappa r + cxr^{-1} \quad (N = 1, \ r \geq R_w),$$

where

$$\kappa = \frac{1}{2} \int_G F(\xi) \, d\xi, \qquad c = \frac{1}{2} \int_G \xi F(\xi) \, d\xi.$$

For $N = 2$, Theorem A.12 states that

$$w(x) = \kappa \log \frac{1}{r} + (c \cdot x) r^{-2} + h(x) \quad (N = 2, \ r \geq R_w),$$

where

$$\kappa = \frac{1}{2\pi} \int_G F(\xi) \, d\xi, \qquad c_j = \frac{1}{2\pi} \int_G \xi_j F(\xi) \, d\xi \quad (j = 1, 2),$$

$$|h(x)| \leq \text{const.}\, r^{-2}, \qquad |\nabla h(x)| \leq \text{const.}\, r^{-3}.$$

For $N \geq 3$, Theorem A.12 states that

$$w(x) = \kappa r^{-N+2} + (c \cdot x) r^{-N} + h(x) \qquad (N \geq 3, \; r \geq R_w),$$

where

$$\kappa = \kappa_N \int_G F(\xi) \, d\xi, \quad c_j = (N-2)\kappa_N \int_G \xi_j F(\xi) \, d\xi \qquad (j = 1, \ldots, N),$$

$$|h(x)| \leq \text{const.}\, r^{-N}, \qquad |\nabla h(x)| \leq \text{const.}\, r^{-N-1}.$$

$\square$

Lemma 4.12 *Suppose that in* (4.18) *we have* $F(\xi) > 0$ *almost everywhere in* G, *and that* $w|_{\partial G} = \text{constant} = \beta$. *Then* $w(x) > \beta$ *if and only if* $x \in G$; *equivalently,*

$$\chi_G(x) = f_H(w(x) - \beta) \quad \text{for all} \;\; x \in \mathbb{R}^N. \tag{4.20}$$

Proof It is enough to prove that $x \in G \Rightarrow w(x) > \beta$ and that $x \in \mathbb{R}^N \setminus \overline{G} \Rightarrow w(x) \leq \beta$.

(i) Consider points of G. By Lemma 4.11, $w \in C^1(\overline{G})$ and

$$-\int_G \nabla \varphi \cdot \nabla w \; = \; -\int_G \varphi F \qquad \text{for all} \;\; \varphi \in C_c^\infty(G) \tag{4.21a}$$

$$\leq \; 0 \qquad \text{if also} \;\; \varphi \geq 0. \tag{4.21b}$$

Thus w is a generalized supersolution relative to Δ and G. By the weak maximum principle, Theorem 2.11 with $c = 0$, we have $w \geq \beta$ in G. Now apply the strong maximum principle, Theorem 2.13, to any component [maximal connected subset] G_0 of G. If $w(x_0) = \beta$ at some point $x_0 \in G_0$, then $w = \beta$ in G_0, hence $\nabla w = 0$ in G_0. This contradicts (4.21a) if we choose φ to be such that $\text{supp}\, \varphi \subset G_0$, $\varphi \geq 0$ everywhere and $\varphi > 0$ somewhere in G_0.

(ii) Given any point $y \in \mathbb{R}^N \setminus \overline{G}$, we prove that $w(y) \leq \beta$. Let

$$\Omega := \mathscr{B}(0, R) \setminus \overline{G},$$

with R so large that $y \in \Omega$ and $w < \beta$ on $\partial \mathscr{B}(0, R)$. This last is possible because, if $N = 1$ or 2, then $w(x) \to -\infty$ as $r \to \infty$; if $N \geq 3$, then $\beta = w|_{\partial G} > 0$ and $w(x) \to 0$ as $r \to \infty$. Now in (4.19) we have $F = 0$ in Ω; hence $w \in C^1(\overline{\Omega})$ and

$$-\int_\Omega \nabla \varphi \cdot \nabla w = 0 \quad \text{for all} \;\; \varphi \in C_c^\infty(\Omega),$$

so that w is a generalized solution relative to $\triangle$ and Ω. [In fact, subsolution would suffice.] By the weak maximum principle, Theorem 2.11 with $c = 0$,

$$\max_{\bar{\Omega}} w = \max_{\partial \Omega} w = \beta.$$

$\square$

Accordingly, the density function of the potential u in (4.15) and (4.16) is $F = \lambda \chi_G = \lambda f_H(u(.)) - \beta)$ on $\mathbb{R}^N$, and, as explained earlier, Theorem 1.1 follows from Theorem 4.2. However, we need not restrict attention to density functions that are constant in G. In the next theorem we consider certain density functions $F = g \circ u$ that are positive and may vary with the potential u itself; if u is constant on the boundary ∂G, then G is a ball once more. Whether such functions u exist, and whether, if they do, they are of interest, will be considered briefly after the theorem has been proved.

Theorem 4.13 *Let G be a bounded open subset of $\mathbb{R}^N$ ($N \geq 1$), and suppose that a measurable function $u : \mathbb{R}^N \to \mathbb{R}$ satisfies*

$$u(x) = \int_G K(x - \xi)\, g(u(\xi))\, d\xi \quad \text{for all } x \in \mathbb{R}^N, \quad (4.22)$$

$$u\big|_{\partial G} = \text{constant} = \beta, \quad (4.23)$$

where $g : \mathbb{R} \to \mathbb{R}$ is as follows.

(a) $0 < g(t) \leq M$ for all $t \in \mathbb{R}$ and for some constant M.

(b) g is continuous except at finitely many points.

(c) The restriction of g to $[\beta, \infty)$ has a decomposition $g\big|_{[\beta,\infty)} = g_1 + g_2$ such that $g_1(\beta) = 0$ and g_1 is locally Lipschitz continuous, while $g_2(\beta) > 0$ and g_2 is non-decreasing.

Then G is a ball, there is a point $q \in \mathbb{R}^N$ such that the definition $v(x) := u(x + q)$ makes v spherically symmetric, and $dv/dr < 0$ for $r > 0$.

Proof We verify that u satisfies the hypotheses of Theorem 4.2.

(i) $u \in C^1(\mathbb{R}^N)$. This will follow from Lemma 4.11 if $g \circ u \in L_p(G)$ with $p > N$. In fact, $g \circ u \in L_\infty(G)$ if $g \circ u$ is measurable, because hypothesis (a) ensures that $g \circ u$ is bounded. To prove that $g \circ u$ is measurable, we let F denote the finite set of points at which g is discontinuous and let $h := g\big|_{\mathbb{R} \setminus F}$. If W is open in $\mathbb{R}$, then $g^{-1}(W) = E \cup h^{-1}(W)$ for some set $E \subset F$, and $h^{-1}(W)$ is open in $\mathbb{R}$ because it can be written $\bigcup_m (h_m)^{-1}(W)$, where each restriction h_m of h is continuous. Thus $g^{-1}(W)$ is a Borel set in $\mathbb{R}$, which makes g Borel-measurable; by a standard result (Kingman & Taylor 1966, p.108; Rudin 1970, p.31), $g \circ u$ is then measurable.

(ii) That u has admissible asymptotic behaviour also follows from Lemma 4.11. Case(C) of such behaviour occurs if and only if $N \geq 3$; then the positivity of K and of g imply that $u > 0$ on $\mathbb{R}^N$.

(iii) The generalized Poisson equation (4.1) is established as follows. By (4.19),

$$\int_{\mathbb{R}^N} \left\{ -\nabla\varphi \cdot \nabla u + \varphi F \right\} = 0 \quad \text{for all} \quad \varphi \in C_c^\infty(\mathbb{R}^N),$$

where

$$
\begin{aligned}
F(x) &= g(u(x)) \, \chi_G(x) \quad &&\text{for all } x \in \mathbb{R}^N \\
&= g(u(x)) \, f_H(u(x) - \beta) \quad &&\text{[by Lemma 4.12]} \\
&= f(u(x))
\end{aligned}
$$

if we set $f(t) := g(t) f_H(t - \beta)$ for all $t \in \mathbb{R}$. Also, let

$$f_j(t) := g_j(t) f_H(t - \beta) \quad \text{for} \quad j = 1, 2 \quad \text{and all} \quad t \in R;$$

then $f_1 : \mathbb{R} \to \mathbb{R}$ is locally Lipschitz continuous, $f_2 : \mathbb{R} \to \mathbb{R}$ is non-decreasing, and $f = f_1 + f_2$.

(iv) Accordingly, Theorem 4.2 applies to u; there is a point $q \in \mathbb{R}^N$ such that

$$\left. \begin{aligned} u(x + q) &= v(x) =: V(r), \quad \text{say} \quad (r := |x|), \\ \text{and} \qquad V'(r) &< 0 \quad \text{for } r > 0. \end{aligned} \right\} \tag{4.24}$$

Since $V(0) = \max u > \beta$, there is exactly one number b (for the function V in question) such that $V(b) = \beta$, and since $x + q \in G$ if and only if $V(r) = u(x + q) > \beta$ (by Lemma 4.12), we have $G = \mathscr{B}(q, b)$. ☐

Apart from the case in which g is a constant, are there functions u that satisfy both (4.22) and (4.23)? The answer is that such functions exist in profusion; we shall see a few of them in Example 4.15 and in Exercises 4.25 and 4.26. However, equations (4.22) and (4.23) do not form the best starting point for an existence proof; it is more usual and profitable to proceed from (4.1), constructing solutions u in a Sobolev space by means of some variational principle or other. This allows functions g that are less restricted than those in Theorem 4.13. Further properties of u then follow from the theory of Sobolev spaces and from regularity theory for solutions of elliptic equations. We do not follow this course here because some of these steps are outside the range of the present book.

The proof of Theorem 4.13 shows that it is essentially a particular case of Theorem 4.2. Whether functions u as in Theorem 4.2 are of interest or of importance is, of course, in the eye of the beholder. For $N = 2$ and $N = 5$ there are applications to steady vortex flows of an ideal fluid and to the problem in magnetohydrostatics of the confinement of a plasma in equilibrium; some of these are indicated by Exercises 4.18 to 4.24.

Example 4.14 In Theorem 4.13, let

$$g(t) = \text{const.} = \lambda > 0 \quad \text{for all } t \in \mathbb{R}.$$

(This is a return to Theorem 1.1 and Exercise 1.12.) If $N = 1$, (4.22) implies that $\beta < 0$; if $N \geq 3$, that $\beta > 0$. In these cases there is a unique radius $b = b(\lambda, \beta)$ such that $G = \mathscr{B}(q, b)$ for some $q \in \mathbb{R}^N$. In fact, (A.23) shows that

$$b = \left\{ N(N - 2)\beta/\lambda \right\}^{1/2} \quad (N \neq 2).$$

If $N = 2$, then (A.23) shows that the radius b of the ball G is determined by

$$b^2 \log \frac{1}{b^2} = \frac{4\beta}{\lambda} \quad (N = 2) \tag{4.25}$$

Consider the graph of $b^2 \log(1/b^2)$ for $b > 0$. If $\beta > 0$, then (4.25) has no solution when $4\beta/\lambda > 1/e$, one solution when $4\beta/\lambda = 1/e$, and two solutions when $4\beta/\lambda < 1/e$. If $\beta \leq 0$, then (4.25) has exactly one positive solution. $\quad\square$

Note that, for a *given solution* u in Theorem 4.13, there is always a unique radius b such that $G = \mathscr{B}(q, b)$, as was mentioned in the proof. However, for *given data g and β*, there may well be more than one solution V as in (4.24), and hence more than one radius b; this is the case in Example 4.14 if $N = 2$ and $0 < 4\beta/\lambda < 1/e$.

Example 4.15 In Theorem 4.13, let

$$\left. \begin{array}{l} N = 2, \quad g(t) = \lambda + (\mu - \lambda)f_H(t - \alpha) \quad \text{for all } t \in \mathbb{R}, \\[2mm] \text{where} \qquad 0 < \lambda < \mu, \quad \beta < 0, \quad \beta < \alpha, \end{array} \right\} \tag{4.26}$$

and λ, μ, α are constants. The function f in Theorem 4.2 is then as in Figure 4.4. We propose to list all possible solutions V of (4.22) and (4.23) for the N and g in (4.26); here V is as in (4.24).

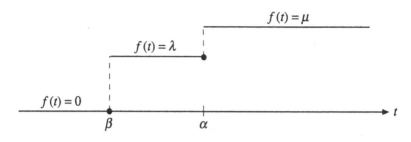

Fig. 4.4.

(i) If $V_1(0) \le \alpha$ for a given solution V_1, then $V_1(r) < \alpha$ for all $r > 0$ and we call V_1 a *one-step solution* of the present problem. It is a particular one of the solutions in Example 4.14 (with $N = 2$ and $\beta < 0$), and, by (A.23),

$$V_1(b) = \frac{\lambda b^2}{2} \log \frac{1}{b} = \beta, \tag{4.27}$$

$$V_1(0) = \frac{\lambda b^2}{2} \left(\log \frac{1}{b} + \frac{1}{2} \right) \le \alpha. \tag{4.28}$$

Define σ to be the inverse of the function $s \mapsto s \log s$, $s \ge 1$ (Figure 4.5). Then (4.27) becomes $b^2 = \sigma(-4\beta/\lambda)$; the condition (4.28) for existence of a one-step solution becomes

$$\beta + \frac{\lambda}{4} \sigma \left(-\frac{4\beta}{\lambda} \right) \le \alpha. \tag{4.29}$$

(ii) If $V(0) > \alpha$ for a given solution V, then there is exactly one radius a for that solution such that $V(a) = \alpha$, and exactly one radius b for that solution such that $V(b) = \beta$; we note that $0 < a < b$ and call this V a *two-step solution*. It is easy to write V in terms of λ, μ a and b: one adds the potentials of density λ in $\mathscr{B}(0,b)$ and of density $\mu - \lambda$ in $\mathscr{B}(0,a)$, each of these being given by (A.23). However, our aim is to find V when λ, μ, α and β are given. Knowing that

$$V(r) = \begin{cases} -\dfrac{\mu r^2}{4} + \dfrac{(\mu - \lambda)a^2}{2} \left(\log \dfrac{1}{a} + \dfrac{1}{2} \right) + \dfrac{\lambda b^2}{2} \left(\log \dfrac{1}{b} + \dfrac{1}{2} \right), & r \le a, \\[3mm] \dfrac{(\mu - \lambda)a^2}{2} \log \dfrac{1}{r} - \dfrac{\lambda r^2}{4} + \dfrac{\lambda b^2}{2} \left(\log \dfrac{1}{b} + \dfrac{1}{2} \right), & a \le r \le b, \\[3mm] \left\{ \dfrac{(\mu - \lambda)a^2}{2} + \dfrac{\lambda b^2}{2} \right\} \log \dfrac{1}{r}, & r \ge b, \end{cases}$$

$$\tag{4.30}$$

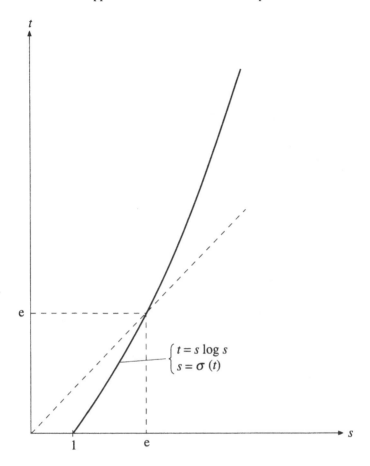

Fig. 4.5.

we seek a and b such that

$$\frac{(\mu - \lambda)a^2}{2} \log \frac{1}{a} - \frac{\lambda a^2}{4} + \frac{\lambda b^2}{2} \left(\log \frac{1}{b} + \frac{1}{2} \right) = \alpha, \qquad (4.31a)$$

$$\left\{ \frac{(\mu - \lambda)a^2}{2} + \frac{\lambda b^2}{2} \right\} \log \frac{1}{b} = \beta. \qquad (4.31b)$$

This may seem an unattractive task, but it is of a kind that must be faced in non-linear problems. The corresponding calculations for $N = 1$ (Exercise 4.25) and for $N \geq 3$ (Exercise 4.26) turn out to be easier.

(iii) With the notation

$$\left.\begin{array}{ll} \xi := \left(\dfrac{\mu}{\lambda} - 1\right) a^2, & \eta := b^2, \\[2mm] A := -\dfrac{4\alpha}{\lambda}, \quad B := -\dfrac{4\beta}{\lambda} > 0, & k := \dfrac{\mu}{\lambda} - 1 > 0, \end{array}\right\} \qquad (4.32)$$

we now seek $\xi > 0$ and $\eta > 0$ such that

$$\xi \left(\log \xi + \log \frac{1}{k} + \frac{1}{k} \right) + \eta(\log \eta - 1) = A, \qquad (4.33a)$$

$$(\xi + \eta) \log \eta = B. \qquad (4.33b)$$

The non-negative solution ξ of (4.33b) is (Figure 4.6)

$$\xi(\eta) := \frac{B}{\log \eta} - \eta, \qquad 1 < \eta \le \eta_2, \qquad (4.34)$$

where η_1 and η_2 are the unique solutions of

$$\eta_1 \log \eta_1 = \frac{B}{k+1}, \qquad \eta_2 \log \eta_2 = B, \qquad (4.35)$$

whence $1 < \eta_1 < \eta_2$. We are interested only in solutions for which $0 < a < b$, hence $0 < \xi < k\eta$; therefore $\xi(\eta)$ is acceptable only on (η_1, η_2), but it is convenient to have the function defined on $(1, \eta_2]$. It follows from (4.34) that

$$\xi(\eta) > 0 \quad \text{on} \ (1, \eta_2), \qquad \xi'(\eta) < 0 \ \text{and} \ \xi''(\eta) > 0 \quad \text{on} \ (1, \eta_2]. \quad (4.36)$$

In view of (4.33a), the next step is to define

$$h(\eta) := \xi(\eta) \left\{ \log \xi(\eta) + \log \frac{1}{k} + \frac{1}{k} \right\} + \eta(\log \eta - 1), \qquad 1 < \eta \le \eta_2, \tag{4.37}$$

and to contemplate all solutions of $h(\eta) = A$. Here $h(\eta) = h(\eta; k, B)$ and we have $k > 0$, $B > 0$ and $A < B$; the parameter A may have either sign or be zero.

Remark 1 For each $k > 0$ and $B > 0$,

$$h'(\eta_1) < 0 \qquad \text{and} \qquad \lim_{\eta \uparrow \eta_2} h'(\eta) = \infty. \qquad (4.38)$$

Proof The definition (4.37) implies that

$$h'(\eta) = \xi'(\eta) \left\{ \log \xi(\eta) + \log \frac{1}{k} + \frac{1}{k} + 1 \right\} + \log \eta. \qquad (4.39)$$

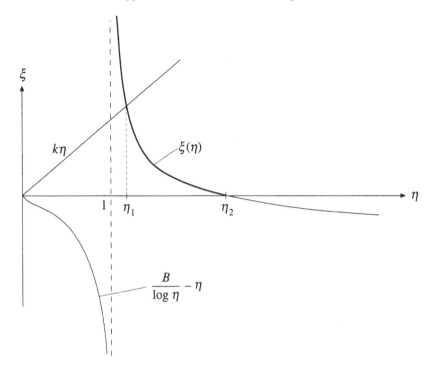

Fig. 4.6.

For $\eta = \eta_1$, we have $\xi(\eta_1) = k\eta_1$ and [by (4.34) and (4.35)]

$$\xi'(\eta_1) = -\frac{B}{(\log \eta_1)^2 \eta_1} - 1 = -\frac{k+1}{\log \eta_1} - 1,$$

whence

$$h'(\eta_1) = -(k+1) - \left(\frac{k+1}{\log \eta_1} + 1\right)\left(\frac{1}{k} + 1\right) < 0.$$

As $\eta \uparrow \eta_2$, $\log \xi(\eta) \to -\infty$, then $h'(\eta) \to \infty$ because

$$\xi'(\eta) \to \xi'(\eta_2) = -\frac{B}{(\log \eta_2)^2 \eta_2} - 1 = -\frac{1}{\log \eta_2} - 1 < 0.$$

Remark 2 There is exactly one point $\eta_0 \in (\eta_1, \eta_2)$ such that $h'(\eta_0) = 0$. At this stationary point η_0, the restriction of h to $[\eta_1, \eta_2]$ has a strict local and global minimum.

Proof Remark 1 shows that η_1 and η_2 are not stationary points of h, and that there is at least one stationary point, say η_0, in (η_1, η_2). Thus

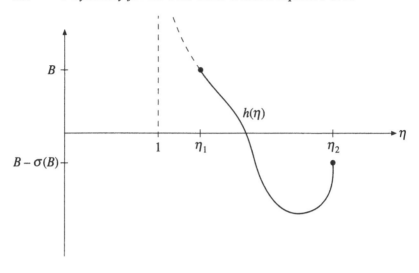

Fig. 4.7.

$h'(\eta_0) = 0$; since $\xi'(\eta_0) < 0$ by (4.36), and $\log \eta_0 > 0$, we infer from (4.39) that

$$\log \xi(\eta_0) + \log \frac{1}{k} + \frac{1}{k} + 1 > 0.$$

Also, $\xi''(\eta_0) > 0$ by (4.36), so that

$$h''(\eta_0) = \xi''(\eta_0) \left\{ \log \xi(\eta_0) + \log \frac{1}{k} + \frac{1}{k} + 1 \right\} + \frac{\xi'(\eta_0)^2}{\xi(\eta_0)} + \frac{1}{\eta_0} > 0.$$

Thus at any stationary point in (η_1, η_2) the function h has a strict local minimum. Hence there can be at most one such stationary point, and we have seen that there is at least one.

Remark 3 The definitions of $\xi(\eta), \eta_1, \eta_2$ and $h(\eta)$ imply that

$$
\begin{aligned}
h(\eta_1) &= (k+1)\eta_1 \log \eta_1 = B, \\
h(\eta_2) &= \eta_2(\log \eta_2 - 1) = B - \sigma(B),
\end{aligned}
$$

where σ is the function defined after (4.28).

(iv) We can now collect ingredients for a list of solutions.

(a) Remarks 1 to 3 show that h is essentially as in Figure 4.7. The figure is misleading in that $B - \sigma(B)$ may have either sign or be zero (in fact, it is negative, zero or positive according as $B < e$, $B = e$ or $B > e$), and in that negative values $h''(\eta)$ occur only for large values of k, but the figure is sufficient for counting solutions of the equation $h(\eta) = A$.

(b) A two-step solution V is determined completely by the solution $\eta \in (\eta_1, \eta_2)$ of $h(\eta) = A$, because ξ is then given by (4.34), the radii a and b are given by (4.32), and V is given by (4.30).

(c) The necessary and sufficient condition for existence of a one-step solution V_1 is [by (4.29) and (4.32)] $A \le B - \sigma(B)$.

(v) *Conclusions*   The results in (iv) imply the following list of all possible solutions. Recall that $B > 0$ and $A < B$ by hypothesis, and that $h(\eta) = h(\eta; k, B)$.

(a) For $B - \sigma(B) < A < B$, there is one solution V; it is a two-step solution.

(b) For $A = B - \sigma(B)$, there are two solutions: one has $\eta \in (\eta_1, \eta_2)$ and is a two-step solution, the other has $\eta = \eta_2$, $V_1(0) = \alpha$ and is a one-step solution.

(c) For $\min_\eta h(\eta) < A < B - \sigma(B)$, there are two distinct two-step solutions and a one-step solution with $V_1(0) < \alpha$.

(d) For $A = \min_\eta h(\eta)$, there is a two-step solution and a one-step solution with $V_1(0) < \alpha$.

(e) For $A < \min_\eta h(\eta)$, there is a one-step solution with $V_1(0) < \alpha$.   □

4.5 Exercises

Exercise 4.16 Prove Lemma 4.5 for case (A) of Lemma 4.3 with $0 < m < 2$.

[You may find it helpful to separate estimates of g for $4\mu h > s^2$, $m - 1 - \delta > 0$ from those for $4\mu h > s^2$, $m - 1 - \delta < 0$. If $m - 1 - \delta = 0$, we may reduce δ slightly.]

Exercise 4.17 Prove Lemma 4.5 for case (B) of Lemma 4.3.

Exercise 4.18 This exercise and the next are concerned only with the occurrence in hydrodynamics of the equation $\triangle \psi + f(\psi) = 0$.

Let $v : \mathbb{R}^3 \to \mathbb{R}^3$ denote the steady (that is, time-independent) velocity, and $p : \mathbb{R}^3 \to \mathbb{R}$ the steady pressure, of an inviscid fluid having constant density $\rho > 0$. Because of the identity

$$(v \cdot \nabla)v = (\nabla \times v) \times v + \nabla \left(\tfrac{1}{2} |v|^2 \right),$$

the equations of motion, under an extraneous force field $-\nabla\Phi$, are

$$\left.\begin{aligned} \operatorname{div} v := \nabla \cdot v = 0, \quad \operatorname{curl} v := \nabla \times v =: \zeta, \\ \zeta \times v = -\nabla\left(\frac{p}{\rho} + \Phi + \frac{1}{2}|v|^2\right); \end{aligned}\right\} \quad (4.40)$$

one calls ζ the *vorticity* and $p/\rho + \Phi + \frac{1}{2}|v|^2$ a *Bernoulli* function (see Batchelor 1967, pp.74 and 160; Lamb 1932, pp. 4–6).

(i) Flow independent of x_3, with stream function $\psi(x_1, x_2)$. Show that, for any given functions $W \in C^1(\mathbb{R})$ and $H \in C^1(\mathbb{R})$, the velocity field and Bernoulli function

$$v = (-\partial_2\psi, \partial_1\psi, W(\psi)) \qquad \text{and} \qquad \frac{p}{\rho} + \Phi + \frac{1}{2}|v|^2 = H(\psi)$$

satisfy equations (4.40) if $\psi \in C^2(\mathbb{R}^2)$ and

$$\Delta\psi(x) = H'(\psi(x)) - W(\psi(x))W'(\psi(x)) \qquad (4.41)$$

at all points $x = (x_1, x_2) \in \mathbb{R}^2$.

(ii) Cylindrically symmetric flow, with Stokes stream function $\Psi(x_1, s)$. Write $x = (x_1, s\cos\vartheta, s\sin\vartheta)$, $s \geq 0$, for points of $\mathbb{R}^3$, and let

$$e^1 := (1, 0, 0), \ e^s := (0, \cos\vartheta, \sin\vartheta), \ e^\vartheta := (0, -\sin\vartheta, \cos\vartheta).$$

Show that, for any given functions $\Gamma \in C^1(\mathbb{R})$ and $H \in C^1(\mathbb{R})$, the velocity field and Bernoulli function

$$v = \frac{1}{s}\left(-\frac{\partial\Psi}{\partial s}e^1 + \frac{\partial\Psi}{\partial x_1}e^s + \Gamma(\Psi)e^\vartheta\right) \quad (s > 0)$$

and

$$\frac{p}{\rho} + \Phi + \frac{1}{2}|v|^2 = H(\Psi)$$

satisfy equations (4.40) for $s > 0$ if $\Psi \in C^2(\mathbb{R} \times (0, \infty))$ and

$$\left(\frac{\partial^2}{\partial x_1^2} + \frac{\partial^2}{\partial s^2} - \frac{1}{s}\frac{\partial}{\partial s}\right)\Psi = s^2 H'(\Psi) - \Gamma(\Psi)\Gamma'(\Psi) \qquad (4.42)$$

at all points $(x_1, s) \in \mathbb{R} \times (0, \infty)$.

Exercise 4.19 Transformation of the Hicks equation (4.42). Show that under the transformation $\Psi(x_1, s) = s^2 X(x_1, s)$, in which X is Greek capital chi, (4.42) becomes

$$\Delta_5 X = \left(\frac{\partial^2}{\partial x_1^2} + \frac{\partial^2}{\partial s^2} + \frac{3}{s}\frac{\partial}{\partial s}\right)X = H'(s^2 X) - \frac{1}{s^2}\Gamma(s^2 X)\Gamma'(s^2 X), \quad (4.43)$$

where $s > 0$ and $\triangle_5 X = \left\{ (\partial/\partial y_1)^2 + \cdots + (\partial/\partial y_5)^2 \right\} X$ if we write $x_1 = y_1$ and $s = (y_2^2 + \cdots + y_5^2)^{1/2}$ (cf. Exercises 1.18 and A.25).

In the context of generalized solutions, H and Γ need not be in $C^1(\mathbb{R})$. Let

$$H(t) = -\lambda t f_H(t) + \text{const.} \qquad \text{and} \qquad \Gamma(t) = \gamma t f_H(t) \qquad (4.44)$$

for all $t \in \mathbb{R}$; here λ, γ are constants $(\lambda > 0, \gamma \geq 0)$ and f_H is the Heaviside function (Chapter 0, (v)). Also, write

$$X(x_1, s) = \chi(x_1, s) - \tfrac{1}{2} U, \qquad U = \text{const.} > 0.$$

This corresponds to $\Psi = \psi - \tfrac{1}{2} U s^2$, where ψ will be required to 'vanish at infinity' while $-\tfrac{1}{2} U s^2$ is the Stokes stream function of velocity $(U, 0, 0)$.

Show that (4.43) becomes

$$\triangle_5 \chi = -\left\{ \lambda + \gamma^2 \left(\chi - \tfrac{1}{2} U \right) \right\} f_H(\chi - \tfrac{1}{2} U) \qquad (4.45)$$

at all points (x_1, s) such that $s > 0$ and $\chi(x_1, s) \neq \tfrac{1}{2} U$.

Exercise 4.20 Let ψ be a C^2-solution of (4.41) on $\mathbb{R}^2$, let H' and W' be locally Lipschitz continuous on $\mathbb{R}$, and suppose that, as $r := |(x_1, x_2)| \to \infty$, either

$$\nabla \psi(x) = -(\omega x_1, \omega x_2) + (V_2, -V_1) + O(r^{-1}), \left.\begin{array}{c} \\ \\ \end{array}\right\}$$
$$\text{where} \quad \omega = \text{const.} > 0, \qquad V_1 = \text{const.}, \qquad V_2 = \text{const.},$$

(this corresponds, for the fluid, to 'solid-body' rotation about the origin plus a uniform velocity plus something smaller), or

$$\nabla \psi(x) = -\frac{\kappa}{r^2}(x_1, x_2) - (c_1, -c_2)\frac{x_1^2 - x_2^2}{r^4} - (c_2, c_1)\frac{2x_1 x_2}{r^4} + O(r^{-3}), \left.\begin{array}{c} \\ \\ \end{array}\right\}$$
$$\text{where} \quad \kappa = \text{const.} > 0, \qquad c_1 = \text{const.}, \qquad c_2 = \text{const.},$$

(this could be due to negative vorticity in a bounded set).

(i) Show that the level sets of ψ are the point q where ψ has its maximum, and circles with centre q. Find q in terms of the data for each of the two cases.

(ii) The *streamlines* in $\mathbb{R}^3$ are the solution curves of $dx/dt = v(x)$, where now $x \in \mathbb{R}^3$ and v is as in Exercise 4.18, (i). Show that a streamline is either the line $A := \left\{ (q_1, q_2, x_3) \mid -\infty < x_3 < \infty \right\}$, or a circle in a plane of constant x_3 and with centre q, or a helix contained in a cylindrical surface with axis A.

Exercise 4.21 Consider the elliptic disk $D := \{ (x, y) \in \mathbb{R}^2 \mid \alpha x^2 + \beta y^2 < 1 \}$, where α and β are positive constants, and consider the function

$$\psi_0(x, y) := \frac{1 - \alpha x^2 - \beta y^2}{2(\alpha + \beta)}, \quad (x, y) \in \overline{D}.$$

Note that $\triangle \psi_0 = -1$ in D, $\psi_0 = 0$ on ∂D. (Thus ψ_0 is the stream function of flow in D with vorticity -1 and with velocity tangential to ∂D.)

We seek an extension ψ of ψ_0 as follows: $\psi \in C^1(\mathbb{R}^2) \cap C^2(\mathbb{R}^2 \setminus \partial D)$, $\psi = \psi_0$ on $\overline{D}$, $\triangle \psi = 0$ in $\mathbb{R}^2 \setminus \overline{D}$, $\psi < 0$ in $\mathbb{R}^2 \setminus \overline{D}$ and the asymptotic behaviour of ψ, as $|(x, y)| \to \infty$, is admissible (Definition 4.1).

Prove that such a function ψ exists if and only if $\alpha = \beta$.

If D is replaced by an arbitrary bounded open set $\Omega \subset \mathbb{R}^2$, with smooth boundary $\partial \Omega$, is there an analogous result?

Exercise 4.22 Let E be the Hilbert space formed by completion of the set $C_c^\infty(\mathbb{R}^5)$ in the norm $\| \cdot \|$ defined by

$$\|u\| := \sqrt{\langle u, u \rangle}, \quad \text{where} \quad \langle u, v \rangle := \int_{\mathbb{R}^5} \nabla u \cdot \nabla v.$$

The space E is embedded in $L_{10/3}(\mathbb{R}^5)$.

Referring to Exercise 4.19, we discard the cylindrical symmetry there; we call $w : \mathbb{R}^5 \to \mathbb{R}$ a *finite-energy solution* of (4.45) iff $w \in E \setminus \{0\}$ and

$$\int_{\mathbb{R}^5} \left[-\nabla \varphi \cdot \nabla w + \varphi \left\{ \lambda + \gamma^2 \left(w - \tfrac{1}{2} U \right) \right\} f_H(w - \tfrac{1}{2} U) \right] = 0$$

for all $\varphi \in C_c^\infty(\mathbb{R}^5)$. From these two properties of w it follows (by methods beyond us here) that the equivalence class $w \in E$ has a representative $w \in C^1(\mathbb{R}^5)$; that w has admissible asymptotic behaviour (Definition 4.1) of type (C), with $m = 3$ and $\delta = 1$; that $w > 0$ on $\mathbb{R}^5$; and that $w \in C^2(\mathbb{R}^5 \setminus S)$, where $S := \{ y \in \mathbb{R}^5 \mid w(y) = \tfrac{1}{2} U \}$.

Given these facts, deduce that every finite-energy solution w of (4.45) is as follows, for some point $q \in \mathbb{R}^5$.

(i) If $\lambda > 0$ and $\gamma = 0$, define $a := (15U/2\lambda)^{1/2}$; then

$$w(y + q) = \chi_0(r) := \begin{cases} \dfrac{1}{4} U \left(5 - 3\dfrac{r^2}{a^2} \right) & \text{if } r \le a, \\[2mm] \dfrac{1}{2} U \dfrac{a^3}{r^3} & \text{if } r \ge a, \end{cases} \tag{4.46}$$

where $r := |y|$ (cf. Example 4.14 with $N = 5$ and $\beta = \tfrac{1}{2} U$).

(ii) If $\lambda > 0$ and $\gamma > 0$, a unique radius a is determined by

$$0 < \gamma a < \beta_1 = 4.49\ldots \qquad \text{and} \qquad \frac{\gamma a J_{5/2}(\gamma a)}{J_{3/2}(\gamma a)} = \frac{3}{2}\frac{U\gamma^2}{\lambda},$$

where the J_ν are Bessel functions (Exercises 1.19, 1.21) and β_1 is the smallest positive zero of $J_{3/2}$; then

$$w(y+q) = \chi_1(r) := \begin{cases} \dfrac{3}{2}U\left\{ A\dfrac{J_{3/2}(\gamma r)}{(\gamma r)^{3/2}} - B + \dfrac{1}{3}\right\} & \text{if } r \le a, \\[2mm] \dfrac{1}{2}U\dfrac{a^3}{r^3} & \text{if } r \ge a, \end{cases} \tag{4.47}$$

where

$$r := |y|, \quad A := \frac{(\gamma a)^{1/2}}{J_{5/2}(\gamma a)}, \quad B := \frac{J_{3/2}(\gamma a)}{\gamma a J_{5/2}(\gamma a)}.$$

Exercise 4.23 *The uniqueness of Hill's vortex and of Hicks's vortex.* Let F be the set of functions $\varphi : \mathbb{R} \times [0,\infty) \to \mathbb{R}$ defined by

$$\varphi(y_1, s) = s^2 \tilde{u}(y_1, s) = s^2 u(y), \qquad s = (y_2^2 + \cdots + y_5^2)^{1/2},$$

where u is a cylindrically symmetric function in the space E (Exercise 4.22). Show that, if φ, u and ψ, v are such pairs of functions, then

$$\iint\limits_{\mathbb{R}\times(0,\infty)} \frac{1}{s^2}\left\{ \frac{\partial\varphi}{\partial y_1}\frac{\partial\psi}{\partial y_1} + \frac{\partial\varphi}{\partial s}\frac{\partial\psi}{\partial s}\right\} s\, dy_1\, ds = \frac{1}{2\pi^2}\int\limits_{\mathbb{R}^5} \nabla u(y) \cdot \nabla v(y)\, dy.$$

It follows that F can be made a Hilbert space; also, that a flow in $\mathbb{R}^3$ with Stokes stream function $\psi \in F$ (and with no velocity in the direction e^ϑ) has finite kinetic energy. [Take $\varphi = \psi$.]

Now consider all generalized solutions $\Psi = \psi - \frac{1}{2}Us^2$, with $\psi \in F$, of (4.42) with H and Γ as in (4.44). Prove that, for some constant $c \in \mathbb{R}$ and with $r := (x_1^2 + s^2)^{1/2}$,

$$\psi(x_1 + c, s) = \begin{cases} s^2\chi_0(r) & \text{if } \lambda > 0 \quad \text{and} \quad \gamma = 0, \\ s^2\chi_1(r) & \text{if } \lambda > 0 \quad \text{and} \quad \gamma > 0, \end{cases}$$

where χ_0 is as in (4.46) and χ_1 as in (4.47).

These functions were discovered, without reference to $\mathbb{R}^5$, by M.J.M. Hill (1894, for $\gamma = 0$) and W.M. Hicks (1899, for $\gamma > 0$). The streamlines of Hill's vortex are shown in Figure 4.8; for those of Hicks's vortex, see Moffatt (1969).

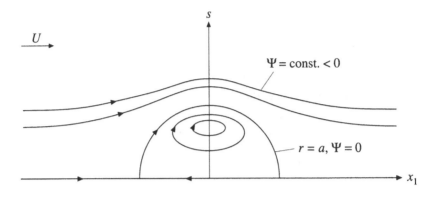

Fig. 4.8.

Exercise 4.24 The equations of *magnetohydrostatics* relate the magnetic induction $B : \mathbb{R}^3 \to \mathbb{R}^3$ to the current density $j : \mathbb{R}^3 \to \mathbb{R}^3$ and pressure $p : \mathbb{R}^3 \to \mathbb{R}$ of an electrically conducting fluid in equilibrium. Under an idealization that is frequently made (Ferraro & Plumpton 1966, p.35; Thompson 1962, pp.47 and 53), these equations are

$$\left. \begin{array}{cc} \operatorname{div} B = 0, & \operatorname{curl} B = \mu j, \\ j \times B = \nabla(p + \rho \Phi), \end{array} \right\}$$

where the constants $\mu > 0$ and $\rho > 0$ are respectively the magnetic permeability and density of the fluid, and $-\nabla\Phi$ is an extraneous force field.

Translate (as far as it interests you to do so) the results of Exercises 4.18 to 4.23 into the language of magnetohydrostatics.

Exercise 4.25 In Theorem 4.13, let

$$\left. \begin{array}{l} N = 1, \quad g(t) = \lambda + (\mu - \lambda)f_H(t - \alpha) \quad \text{for all} \ \ t \in \mathbb{R}, \\ \qquad\qquad 0 < \lambda < \mu, \ \ \beta < \alpha < 0. \end{array} \right\}$$
where

List all possible solutions V for these data; here V is as in (4.24).

[The strategy is as in Example 4.15; the tactics differ. For a two-step solution, defined to be one satisfying $V(0) > \alpha$, a possible method is to let $V(a) = \alpha$, $V(b) = \beta$ and $\theta := \lambda/\mu$; to seek $\xi := a/b \in (0, 1)$ such that

$$\varphi(\xi, \theta) := \frac{(1 - \frac{1}{2}\theta)\xi^2 + \frac{1}{2}\theta}{(1 - \theta)\xi + \theta} = \frac{\alpha}{\beta} \in (0, 1);$$

and to prove that $\varphi_\xi(., \theta)$ has exactly one zero $\xi_0 \in (0, 1)$ for fixed θ.]

Exercise 4.26 In Theorem 4.13, let

$$N \geq 3, \quad g(t) = \lambda + (\mu - \lambda)f_H(t - \alpha) \quad \text{for all } t \in \mathbb{R},$$
$$\text{where} \qquad\qquad 0 < \lambda < \mu, \quad 0 < \beta < \alpha.$$

List all possible solutions V for these data; again V is as in (4.24).

[For a two-step solution, again defined to be one satisfying $V(0) > \alpha$, a possible method is to let $V(a) = \alpha$, $V(b) = \beta$ and $\theta := \lambda/\mu$; to seek $\xi := a/b \in (0,1)$ such that

$$\psi(\xi, \theta) := \frac{\left(1 - \frac{1}{2}N\theta\right)\xi^2 + \frac{1}{2}N\theta}{(1-\theta)\xi^N + \theta} = \frac{\alpha}{\beta} > 1;$$

to prove that, if $\theta \in (0, 2/N)$ and is fixed, then $\psi_\xi(., \theta)$ has exactly one zero $\xi_0 \in (0,1)$; and to prove that, if $\theta \in [2/N, 1)$, then $\psi_\xi(\xi, \theta) < 0$ for $0 < \xi \leq 1$.]

Exercise 4.27 This exercise concerns an alternative to Lemmas 4.7 and 4.8 and to Proposition 4.9 in the case when $f_2 = 0$, so that f is locally Lipschitz continuous.

Let (m, ∞) be the largest open interval of μ such that

$$y \in Y(\mu) \;\Rightarrow\; w(y, \mu) := v(y^\mu) - v(y) < 0; \qquad\qquad (*)$$

assume (for contradiction) that $m > 0$. With $R(\mu)$ as in Lemma 4.5, define

$$R_0 := R\left(\tfrac{1}{2}m\right) + 1,$$

$$X(\mu) := \left\{ x \in \mathbb{R}^N \;\middle|\; -R_0 < x_1 < \mu, \quad |x''| < R_0 \right\},$$

where $\mu \in [\frac{1}{2}m, m]$. Consider $w(., m - \varepsilon)$ in $X(m - \varepsilon)$ for $0 < \varepsilon \leq \varepsilon_0$ with ε_0 sufficiently small; obtain the desired contradiction, and complete the proof of Theorem 4.2 for $f_2 = 0$, by imitating the proof of Theorem 3.3.

Exercise 4.28 Here we extend the method in Exercise 4.27 to certain discontinuous functions f.

(a) Let f be as in Theorem 4.2 with the additional condition that $f_2(t) = 0$ whenever $t \leq \kappa$, for some $\kappa \in \mathbb{R}$ when u satisfies (A) or (B) of Definition 4.1, or for some $\kappa > 0$ when u satisfies (C).
[Cf. Theorem 3.6 and Remark 1 following it.]

(b) Let R_2 be so large that $v(x) < \kappa$ whenever $r > R_2$, and choose

$$R_0 := \max\left\{ R\left(\tfrac{1}{2}m\right) + 1, \; R_2 + 1 \right\}$$

in order to construct a set $X(\mu)$ like that in Exercise 4.27. For functions f as in (a), complete the proof of Theorem 4.2 beyond Lemma 4.6 by combining the method of Exercise 4.27 with that used to prove Theorem 3.6.

5

Monotonicity of Positive Solutions in a Bounded Set Ω

5.1 Prospectus

This chapter concerns positive solutions of the equations

$$\triangle u + f(u) = 0 \tag{5.1}$$

and

$$\triangle u + b \cdot \nabla u + f(u) = 0 \tag{5.2}$$

in a bounded set Ω that need not be symmetric; in (5.2), the coefficient b is a constant vector. Although (5.1) is a particular case of (5.2), the two are treated separately because (5.2) will be examined under stronger restrictions on f and on the subset of Ω to be considered. These stronger restrictions are not necessary, but without them one must either work long and hard or use advanced techniques; in fact, equations far more general than (5.2) can be treated by advanced versions of the method. Our purpose here is merely to show, at an elementary level, that the techniques of this chapter are not restricted to the non-linear Poisson equation (5.1).

Our first task (§5.2) is to explore the nature of caps and reflected caps for a set Ω that lacks Steiner symmetry. For suitable sets Ω and unit vectors k, we shall introduce *alpha caps* $Z(\alpha(k), k)$ that generalize the caps $Z(0, e^1)$ of the Steiner symmetric sets Ω in Chapter 3. We shall also define *beta caps* $Z(\beta(k), k)$; these form a smaller class of possibly smaller caps and have an additional property that simplifies certain proofs.

The first monotonicity theorem in §5.3 is a variant of Theorem 3.6 for alpha caps and for equation (5.1). A further result for (5.1) concerns the hyperplane $T_{\alpha(k)}(k)$ that bounds an alpha cap: if there is a point $x_0 \in T_{\alpha(k)}(k) \cap \Omega$ at which the derivative of u normal to the hyperplane

141

vanishes [that is, $(k \cdot \nabla u)(x_0) = 0$], then both u and Ω must be symmetric relative to $T_{\alpha(k)}(k)$.

For positive solutions of (5.2), similar, but slightly more elaborate, results will appear.

The reader who dislikes the geometry of caps, and who is interested only in equation (5.1), need read §5.2 only as far as Theorem 5.3.

5.2 On the geometry of caps and reflected caps

Definition 5.1 Let Ω be a bounded region. For every unit vector $k \in \mathbb{R}^N$ and every $\mu \in \mathbb{R}$, the hyperplane $T_\mu(k)$ and the reflection $x^{\mu,k}$ of x are as in Definition 1.7.

(a) The sets

$$Z(\mu,k) := \left\{ x \in \Omega \mid x \cdot k > \mu \right\}$$

and

$$Y(\mu,k) := \left\{ x \in \mathbb{R}^N \mid x^{\mu,k} \in Z(\mu,k) \right\}$$

will be called a *cap* and *reflected cap*, respectively. The *right-hand boundary of the cap* $Z(\mu,k)$ is

$$\Gamma_\mu(k) := \partial Z(\mu,k) \setminus T_\mu(k).$$

We also define

$$M(k) := \sup\left\{ x \cdot k \mid x \in \Omega \right\} = \sup\left\{ \mu \mid Z(\mu,k) \text{ is not empty} \right\}.$$

(b) Iff, for a given unit vector k, there is a number $\sigma < M(k)$ such that $Y(\mu,k) \subset \Omega$ whenever $\mu \in (\sigma, M(k))$, then Ω will be called *admissible for the direction* k. In that case, we define

$$\alpha(k) := \inf\left\{ \sigma \mid \mu \in (\sigma, M(k)) \Rightarrow Y(\mu,k) \subset \Omega \right\}$$

and call $Z(\alpha(k), k)$ the *alpha cap for the direction* k.   □

Figure 5.1 shows examples of caps, reflected caps and alpha caps with $k = e^1 := (1, 0, \dots, 0)$ and with no display of k. Note that if, for the moment, we define $\gamma := \inf\left\{ \mu \mid Y(\mu) \subset \Omega \right\}$, then γ can be very different from α.

The following lemma and theorem are very simple, but have important consequences.

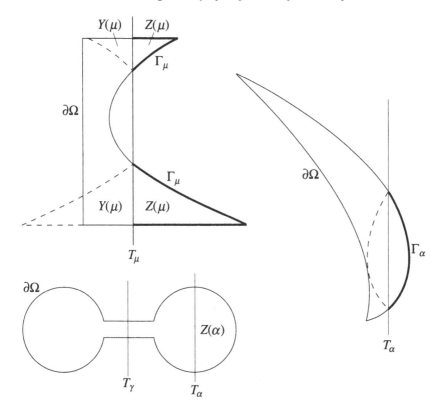

Fig. 5.1.

Lemma 5.2 *Let Ω be admissible for a given direction k. Let $\mu \in [\alpha, M)$ and $z \in Z(\mu)$, the label k being omitted from the symbols in Definition 5.1. Denote the closed line segment from z to z^μ by*

$$\ell := \left\{ (1 - \theta)z + \theta z^\mu \mid 0 \le \theta \le 1 \right\}.$$

Then $\ell \subset \Omega$.

Proof Since Ω is open, there is a point $z_* := z + 2\varepsilon k$ in Ω for some $\varepsilon > 0$. Let $x \in \ell$. Then x is the reflection of z_* in T_σ [that is, $x = (z_*)^\sigma$] if

$$\sigma = \tfrac{1}{2}\left(x \cdot k + z_* \cdot k\right) = \tfrac{1}{2}x \cdot k + \tfrac{1}{2}z \cdot k + \varepsilon,$$

where

$$2\mu - z \cdot k \le x \cdot k \le z \cdot k \qquad \text{because } x \in \ell,$$

so that

$$\mu + \varepsilon \le \sigma \le z \cdot k + \varepsilon.$$

Thus $\alpha < \sigma < z_* \cdot k$. Since $z_* \in \Omega$ and $z_* \cdot k > \sigma$, we have $z_* \in Z(\sigma)$ and hence $x = (z_*)^\sigma \in Y(\sigma)$. Since $\sigma > \alpha$, the definition of α ensures that $Y(\sigma) \subset \Omega$. Therefore $x \in \Omega$, as desired. $\square$

The significance of the next little theorem is that the inclusion $Y(\mu, k) \subset \Omega$ extends from $\mu > \alpha(k)$ to $\mu = \alpha(k)$.

Theorem 5.3 *If Ω is admissible for a given direction k, then for that direction the reflection in $T_{\alpha(k)}$ of the alpha cap is in Ω. In other words, $Y\big(\alpha(k), k\big) \subset \Omega$.*

Proof Omitting the label k again, we wish to show that, if $z \in Z(\alpha)$, then $z^\alpha \in \Omega$. This follows from Lemma 5.2 with $\mu = \alpha$. $\square$

The remainder of this §5.2 leads to the result that a bounded region Ω with $\partial\Omega$ of class C^1 is admissible for every direction k. At the same time we establish properties of the outward unit normal n (to the boundary of such a set) that become useful in the context of Theorem 5.12, which concerns equation (5.2). This material may be of interest in its own right, but *it is not needed for Theorems 5.10 and 5.11*, which concern equation (5.1).

Definition 5.4 Let $\partial\Omega$ be of class C^ℓ for some $\ell \in \mathbb{N}_0$ and let $p \in \partial\Omega$. As in Chapter 0, (viii), the set Ω is defined globally in terms of co-ordinates x_j, while y_j are 'local' co-ordinates such that $\partial\Omega$ has a representation

$$y_N = h(y'), \quad \text{where } y' := (y_1, \ldots, y_{N-1}), \tag{5.3}$$

near the boundary point p.

(a) The co-ordinate transformation will now be written

$$y = Y_p(x) := A(p)(x - p), \tag{5.4}$$

where $A(p)$ is an orthogonal $N \times N$ matrix depending on p.

(b) Let $Q'(0, \rho) := \big\{ y' \in \mathbb{R}^{N-1} \;\big|\; -\rho < y_j < \rho, j = 1, \ldots, N - 1 \big\}$ denote a cube about the origin in $\mathbb{R}^{N-1}$, with edges of length $2\rho > 0$. We define (Figure 5.2)

$$\mathscr{V}_{\rho,\sigma}(p) := \big\{ y \in \mathbb{R}^N \;\big|\; y' \in Q'(0, \rho), \quad -\sigma < y_N - h(y') < \sigma \big\},$$

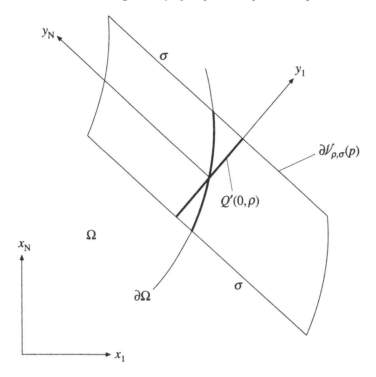

Fig. 5.2.

and its x-image

$$\mathcal{U}_{\rho,\sigma}(p) := Y_p^{-1}\big(\mathcal{V}_{\rho,\sigma}(p)\big),$$

and call these sets *rhomboid neighbourhoods*, of the points 0 and p respectively. The y_N-axis is to point into Ω, and the positive lengths ρ and σ are to be so small that

$$x \in \partial\Omega \cap \mathcal{U}_{\rho,\sigma}(p) \quad \text{iff} \quad y = Y_p(x) \in \mathcal{V}_{\mu,\upsilon}(p) \quad \text{and} \quad y_N = h(y'), \qquad (5.5a)$$

$$x \in \Omega \cap \mathcal{U}_{\rho,\sigma}(p) \quad \text{iff} \quad y = Y_p(x) \in \mathcal{V}_{\rho,\sigma}(p) \quad \text{and} \quad y_N > h(y'). \qquad (5.5b)$$

(The word rhomboid is used *faute de mieux* and because $\mathcal{V}_{\rho,\sigma}(p)$ is approximately rhomboid when $N = 2$, $\ell \geq 1$ and ρ is sufficiently small.) □

Exercise 5.5 Let $\partial\Omega$ be of class C^ℓ for some $\ell \geq 1$, let $p \in \partial\Omega$ and let the outward unit normal

$$v(y') = A(p)n(x), \quad \text{where} \quad y = A(p)(x - p),$$

be as in equation (0.2). Abbreviating $A(p)$ to A and writing $\nabla_x :=$ $(\partial/\partial x_1, \ldots, \partial/\partial x_N)$, define new local co-ordinates $z_1, \ldots, z_N$ by

$$z = By = BA(x - p),$$

where B is an orthogonal $N \times N$ matrix such that

$$n(p) \cdot \nabla_x z_N = v(0) \cdot \nabla_y z_N = \sum_{j=1}^{N} v_j(0) B_{Nj} \neq 0. \tag{5.6}$$

Prove the following.

(a) There is a set W, open in $\mathbb{R}^N$ and containing p, such that $\partial\Omega \cap W$ has a representation

$$z_N = g(z'), \quad z' \in H, \quad g \in C^{\ell}(\overline{H}), \tag{5.7}$$

where H is open in $\mathbb{R}^{N-1}$ and convex.

(b) There exist new rhomboid neighbourhoods, say $V_{a,b}(p)$ and its x-image $U_{a,b}(p)$, characterized by $z' \in Q'(0, a)$ and by $|z_N - g(z')| < b$, such that a statement analogous to (5.5a, b) holds when $n(p) \cdot \nabla_x z_N < 0$.

(c) If the matrix B satisfies $B_{Nj} = -v_j(0)$ for $j = 1, \ldots, N$, then

$$z_N = -n(p) \cdot (x - p) \text{ for all } x \in \mathbb{R}^N, \quad (\partial_j g)(0) = 0 \text{ for } j = 1, \ldots, N-1, \tag{5.8}$$

so that the hyperplane $\{ z \mid z_N = 0 \}$ is tangent to $\partial\Omega$ at $x = p$ (at $z = 0$), as is shown in Figure 5.3.

[For a form of the implicit-function theorem that yields the C^{ℓ} property of g in (5.7), and not merely C^1 smoothness, see Dieudonné 1969, p.272, or Fleming 1965, p.117.] $\square$

Notation Until the contrary is stated, the unit vector k will be fixed; we adopt variables aligned with k (Definition 4.4). That is, we rotate the co-ordinate axes and use

$$\tilde{x}_1 := x \cdot k, \quad \tilde{x}'' := (\tilde{x}_2, \ldots, \tilde{x}_N),$$

$$\tilde{\Omega} := \{ \tilde{x} \in \mathbb{R}^N \mid x \in \Omega \}, \quad \tilde{Z}(\mu) := \{ \tilde{x} \in \tilde{\Omega} \mid \tilde{x}_1 > \mu \},$$

$$\widetilde{M} := \sup\{ \tilde{x}_1 \mid \tilde{x} \in \tilde{\Omega} \}$$

and so on. Again we *omit the tilde* of these aligned variables as long as there is no danger of confusion.

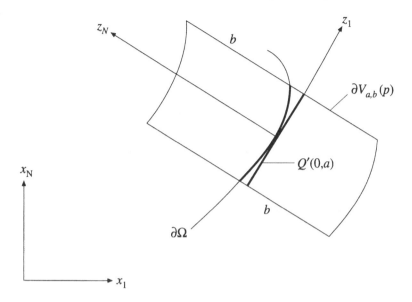

Fig. 5.3.

Lemma 5.6 *Let Ω be a bounded region with $\partial\Omega$ of class C^1. Define*

$$F := \left\{\, x \in \partial\Omega \mid x_1 = M \,\right\}.$$

Let $p \in F$ and let ξ_j be the local co-ordinates called z_j in Exercise 5.5, (c), so that the ξ_N-axis is along the inward normal at p, and $\partial\Omega$ is described near p by $\xi_N = g(\xi')$. Then

$$n(p) = e^1 := (1, 0, \ldots, 0) \quad and \quad g(\xi') \geq 0 \tag{5.9}$$

for all ξ' in the domain of g.

Proof Observe first that M is the supremum of the continuous function $x \mapsto x_1$ not merely over Ω, but also over the compact set $\overline{\Omega}$ (Exercise 1.4); this supremum is attained in $\overline{\Omega}$, but not in the open set Ω. Thus F is not empty and is closed in $\mathbb{R}^N$.

It is immediate that $n(p) = e^1$ implies $g(\xi') \geq 0$: if $n(p) = e^1$, then

$$x_1 - M = -\xi_N = -g(\xi') \quad \text{for } x \in \partial\Omega \text{ and } x \text{ near } p;$$

if $g(\xi') < 0$ for some ξ', then the definition of M is contradicted.

That $n(p) = e^1$ also seems self-evident: if $n(p) \neq e^1$ (Figure 5.4), then there are points x in Ω near p for which $x_1 > M$, so that the definition of

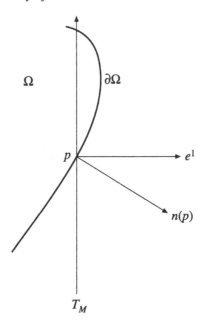

Fig. 5.4.

M is contradicted again. However, an analytical form of this argument is worthwhile.

Write the co-ordinate transformation as $\xi = C(x - p)$, so that C is an orthogonal $N \times N$ matrix. Then

$$n(p) = -\nabla_x \xi_N = -(C_{N1}, C_{N2}, \ldots, C_{NN}) \qquad (5.10)$$

and

$$x_1 - M = \sum_{i=1}^{N-1} C_{i1} \xi_i + C_{N1} g(\xi') \qquad \text{for } x \in \partial\Omega \quad \text{and} \quad x \text{ near } p. \quad (5.11)$$

Here $g(\xi') = o(|\xi'|)$ as $\xi' \to 0$, because

$$g(\xi') - g(0) = \int_0^1 \frac{dg(t\xi')}{dt}\, dt = \int_0^1 \xi' \cdot (\nabla g)(t\xi')\, dt$$

with $g(0) = 0$, $(\nabla g)(0) = 0$ by (5.8) and with ∇g continuous by (5.7).

Assume that $C_{m1} \neq 0$ for some $m \in \{1, \ldots, N-1\}$. Set $\xi' = (\ldots, 0, \xi_m, 0, \ldots)$ in (5.11), with $\xi_m > 0$ if $C_{m1} > 0$ or $\xi_m < 0$ if $C_{m1} < 0$. Since $g(\xi') = o(|\xi'|)$, we can choose $|\xi_m|$ to be so small that $x_1 - M > 0$. This is a contradiction, so that $C_{i1} = 0$ for $i = 1, \ldots, N-1$. The rows and columns of (C_{ij}) are unit vectors; therefore $C_{N1} = \pm 1$. If $C_{N1} = 1$, we have

$$x_1 - M = \sum_{i=1}^{N} C_{i1} \xi_i = \xi_N,$$

which is again a contradiction when $\xi' = 0$ and $\xi_N > 0$. Accordingly, $C_{N1} = -1$, and we see from (5.10), both sides of which are unit vectors, that $n(p) = e^1$. $\qquad\square$

Theorem 5.7 *Let Ω be a bounded region with $\partial\Omega$ of class C^1; we use variables aligned with a given unit vector k.*

There is a number $\delta > 0$ such that, if $\mu \in (M - \delta, M)$, then

(a) $Y(\mu) \subset \Omega$, which implies that Ω is admissible for the arbitrary direction k,

(b) $n_1(x) > 0$ for all $x \in \Gamma_\mu$.

Proof (i) Let F be as in Lemma 5.6. We shall use Exercise 5.5 again. The rhomboid neighbourhoods $U_{a,b}(p)$ of all points $p \in F$ form an open cover of the compact set F. [That is, each set $U_{a,b}(p)$ is open and F is a subset of the union of all of them.] We extract a finite subcover, say

$$\{ U_{a,b}(p^r) \mid r = 1, 2, \ldots, s \},$$

where $a = a(p^r)$ and $b = b(p^r)$. [When F is a finite set, this extraction is unnecessary.] Figure 5.5 shows one set of the finite subcover; the local co-ordinates ξ_j are as in Lemma 5.6. Now define

$$G := \bigcup_{r=1}^{s} U_{a,b}(p^r),$$

$$M - \varepsilon := \sup\{ x_1 \mid x \in \Omega \setminus G \} = \max\{ x_1 \mid x \in \overline{\Omega} \setminus G \},$$

$$\delta := \min\left\{ \varepsilon, \tfrac{1}{2}b(p^1), \ldots, \tfrac{1}{2}b(p^s) \right\}.$$

In describing $M - \varepsilon$, we have used Exercise 1.4 once more. Note that $\varepsilon > 0$, otherwise the definition of F would be contradicted, since G covers F.

(ii) Let $\mu \in (M - \delta, M)$. We wish to show that $Y(\mu) \subset \Omega$; equivalently, that

$$z \in \Omega \text{ and } z_1 > \mu \ \Rightarrow\ z^\mu \in \Omega.$$

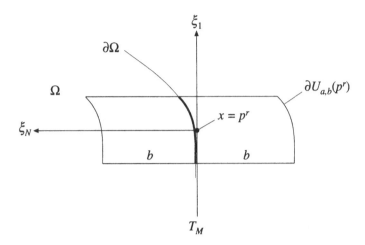

Fig. 5.5.

Now the conditions $z_1 > \mu$, $\mu > M - \delta$ and $\delta \leq \varepsilon$ imply that $z_1 > M - \varepsilon$, so that $z \notin \Omega \setminus G$. Therefore $z \in \Omega \cap G$, whence $z \in \Omega \cap U_{a,b}(p^r)$ for some $r \in \{1, \ldots, s\}$. Returning to the co-ordinate transformation $\xi = C(p)(x-p)$ used in Lemma 5.6, we define

$$\zeta := C(p^r)(z - p^r), \qquad \eta := C(p^r)(z^\mu - p^r). \tag{5.12}$$

We shall show that $g(\eta') < \eta_N < b$, where now $g = g(., p^r)$ and $b = b(p^r)$ for the particular r introduced before (5.12); then $z^\mu \in \Omega$.

By Lemma 5.6, $\xi_N = M - x_1$ when $p \in F$; therefore reflection in T_μ is reflection in $\left\{ \xi \mid \xi_N = M - \mu \right\}$. Accordingly,

$$\eta_N = 2(M - \mu) - \zeta_N, \tag{5.13}$$

in which $2(M - \mu) < 2\delta \leq b$ and $\zeta_N > g(\zeta') \geq 0$, so that $\eta_N < b$.

Next, by (5.13) and repeated use of the inequality $\zeta_N = M - z_1 < M - \mu$,

$$\eta_N > M - \mu > \zeta_N > g(\zeta') = g(\eta'),$$

since $\eta' = \zeta'$ by (5.12). The results $\eta_N < b$ and $\eta_N > g(\eta')$ now show (as we noted above) that $z^\mu \in \Omega$, and hence that $Y(\mu) \subset \Omega$.

(iii) It remains to consider $x \in \Gamma_\mu$ for $\mu \in (M - \delta, M)$ and to show that $n_1(x) > 0$. [Recall that $\Gamma_\mu := \partial Z(\mu) \setminus T_\mu$ and that the identity $Z(\mu) = \Omega \cap \left\{ x \mid x_1 > \mu \right\}$ implies that $\partial Z(\mu) \subset \partial \Omega \cup T_\mu$; therefore $\Gamma_\mu \subset \partial \Omega$ and here $x \in \partial \Omega$.] Since $\delta \leq \varepsilon$, we have $x_1 > \mu \geq M - \varepsilon$; then $x \notin \overline{\Omega} \setminus G$, by the definition of $M - \varepsilon$. Consequently $x \in \partial \Omega \cap G$, so that $x \in \partial\Omega \cap U_{a,b}(p^r)$ for some $r \in \{1,\ldots,s\}$. Equivalently, under the transformation $\xi = C(p^r)(x - p^r)$ we have $\xi \in V_{a,b}(p^r)$, $\xi_N = g(\xi')$ if we write $g = g(.,p^r)$, and

$$n_1(x) = -\nu_N(\xi') = \frac{1}{\left\{ (\partial_1 g)(\xi')^2 + \cdots + (\partial_{N-1} g)(\xi')^2 + 1 \right\}^{1/2}} > 0.$$

$\square$

The condition established in Theorem 5.7 that $n_1(x) > 0$ for all $x \in \Gamma_\mu$ is a useful property of the right-hand boundary of a cap. It prompts the following definition, which is phrased in terms of our original variables rather than in terms of those aligned with k.

Definition 5.8 Let Ω be a bounded region with $\partial\Omega$ of class C^1. For every unit vector $k \in \mathbb{R}^N$ we define

$$\lambda(k) := \inf\left\{ \sigma \mid \mu \in \big(\sigma, M(k)\big) \Rightarrow n(x) \cdot k > 0 \ \text{ for all } \ x \in \Gamma_\mu(k) \right\},$$

$$\beta(k) := \max\left\{ \alpha(k), \lambda(k) \right\},$$

and call $Z\big(\beta(k), k\big)$ the *beta cap for the direction* k. $\square$

Figure 5.6 illustrates this definition for $k = e^1$. The set in part (a) of the figure is

$$\Omega_1 := \left\{ x \in \mathbb{R}^2 \mid (x_1 - cx_1^2)^2 + x_2^2 < 1, \ x_1 < 1/2c \right\},$$
$$c = \text{ const.} \in (0, \tfrac{1}{4}), \quad (5.14)$$

and we define

$$\gamma := \frac{1}{4c}\left\{ 2 - \sqrt{1 - 4c} - \sqrt{1 + 4c} \right\}. \quad (5.15)$$

The reader should verify that, for the set Ω_1 and the direction e^1, the critical values of μ are $\alpha(e^1) = \gamma > 0$ and $\lambda(e^1) = 0$, so that $\beta(e^1) = \alpha(e^1)$.

The set in part (b) of Figure 5.6 is

$$\Omega_2 := \left\{ x \in \mathbb{R}^2 \mid (x_1 + cx_1^2)^2 + x_2^2 < 1, \ x_1 > -1/2c \right\},$$
$$c = \text{ const.} \in (0, \tfrac{1}{4}), \quad (5.16)$$

and now $\alpha(e^1) = 0$ and $\lambda(e^1) = 0$.

In part (c) of Figure 5.6 a small protuberance has been added to Ω_2 near the point $(M, 0)$, where $M := \sup\{ x_1 \mid x \in \Omega_2 \}$. This can be done

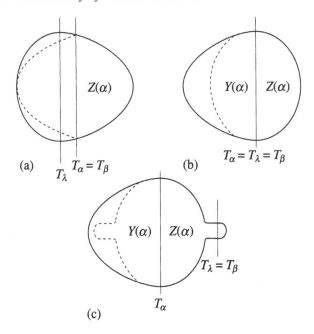

Fig. 5.6.

in such a way that $\alpha\left(e^1\right)$ is unchanged, $\lambda\left(e^1\right)$ is greatly increased, $\partial\Omega$ remains of class C^∞ and $\partial\Omega$ is unchanged for $x_1 \leq \frac{1}{2}M$.

Remark 5.9 *If Ω is a bounded region with $\partial\Omega$ of class C^1, then, for each unit vector k,*

(i) $n(x) \cdot k > 0$ *for all* $x \in \Gamma_{\lambda(k)}(k)$,

(ii) $n(x) \cdot k \geq 0$ *for all* $x \in \Gamma_{\alpha(k)}(k)$.

Proof We revert to variables aligned with k, so that $n(x) \cdot k$ and $\Gamma_{\lambda(k)}(k)$ become $n_1(x)$ and Γ_λ respectively.

(i) If $x \in \Gamma_\lambda$, then $x \in \partial\Omega$ and $x_1 > \lambda$. Hence $x \in \Gamma_\mu$ for some $\mu > \lambda$ [for example, $\mu = \frac{1}{2}(x_1 + \lambda)$]; the definition of λ now ensures that $n_1(x) > 0$.

(ii) Assume for contradiction that $p \in \Gamma_\alpha$ and $n_1(p) < 0$. By this last, $z := p + \varepsilon e^1 \in \Omega$ and $y := p - \varepsilon e^1 \notin \Omega$ for sufficiently small $\varepsilon > 0$. Choose $\mu = p_1$; then $\mu > \alpha$, $z \in Z(\mu)$ and $z^\mu = y \notin \Omega$. This contradicts the definition of α. □

Remark 5.9 implies that, when $\lambda(k) > \alpha(k)$, we have $n(x) \cdot k > 0$ for all $x \in \Gamma_{\beta(k)}(k)$ and $n(x) \cdot k \geq 0$ for all $x \in \Gamma_{\alpha(k)}(k)$. Although this distinction may seem slight, it makes certain proofs much easier for beta caps than for alpha caps.

5.3 Monotonicity in Ω

Our first monotonicity theorem, for sets Ω that may lack symmetry, is an extension of Theorem 3.6 that takes longer to state than to prove. The reader who finds the proof somewhat glib is invited (in Exercise 5.25) to spell out the corresponding extension of Theorem 3.3. This smaller task may illuminate the present proof.

Theorem 5.10 *Let* Ω *be a bounded region that is admissible for the direction* e^1 *(Definition 5.1); we abbreviate* $T_\mu \left(e^1 \right)$, $Z \left(\mu, e^1 \right)$, $\alpha \left(e^1 \right), \ldots$ *to* T_μ, $Z(\mu)$, $\alpha, \ldots$. *Assume that* $u : \overline{\Omega} \to \mathbb{R}$ *has the following properties.*
  (a) $u \in C(\overline{\Omega}) \cap C^1(\Omega)$, $u > 0$ *in* Ω, $u = 0$ *on* $\overline{\Gamma_\alpha}$.
  (b) *For all* $\varphi \in C_c^\infty(\Omega)$,

$$\int_\Omega \left\{ -\nabla\varphi \cdot \nabla u + \varphi f(u) \right\} = 0, \tag{5.17}$$

where f *has a decomposition* $f = f_1 + f_2$ *such that* $f_1 : [0,\infty) \to \mathbb{R}$ *is locally Lipschitz continuous, while* $f_2 : [0,\infty) \to \mathbb{R}$ *is non-decreasing and is identically 0 on* $[0,\kappa]$ *for some* $\kappa > 0$.
Then the conclusions of Theorem 3.6 hold in the alpha cap:

$$u(z) < u(z^\mu) \quad if \quad \mu \in (\alpha, M) \ and \ z \in Z(\mu), \tag{5.18}$$

$$\partial_1 u(x) < 0 \quad if \quad x_1 > \alpha \ and \ x \in \Omega. \tag{5.19}$$

Proof Scrutiny of the proof of Theorem 3.6 shows that there we used only conditions that have counterparts here: rather than the full Steiner symmetry of Ω relative to T_0, we used only inclusions implied here by Lemma 5.2; rather than the full boundary condition $u = 0$ on $\partial\Omega$, we used only that $u = 0$ on $\overline{\Gamma_0}$. (Of course, the full conditions were used for Corollary 3.9.) Therefore the proof of Theorem 3.6 extends to the present situation, provided that the earlier interval $(0, M)$ is replaced by the present (α, M). $\qquad\square$

Theorem 5.11 *Let* Ω *and* u *be as in Theorem 5.10. Suppose that, in addition,*

$$\left(\partial_1 u\right)(x_0) = 0 \quad at \ some \ point \ x_0 \in T_\alpha \cap \Omega. \tag{5.20}$$

Let $Z_0(\alpha)$ be the component of $Z(\alpha)$ that contains $x_0 + \varepsilon e^1$ for sufficiently small $\varepsilon > 0$, and let $Y_0(\alpha)$ be the reflection of $Z_0(\alpha)$ in T_α. Then

$$u(x^\alpha) \;=\; u(x) \qquad \text{for all } x \in \overline{Z_0(\alpha)}, \tag{5.21}$$

$$\overline{\Omega} \;=\; \overline{Y_0(\alpha) \cup Z_0(\alpha)}. \tag{5.22}$$

Proof (i) Once again let $w(.\,,\mu) := u - u_\mu$ on $\overline{Z(\mu)}$, this time for $\mu \in [\alpha, M)$. Fix $z \in Z(\alpha)$ and let $\mu \downarrow \alpha$. Since $u(z) < u(z^\mu)$ for $\mu \in (\alpha, z_1)$, by (5.18), and since $u(z^\mu)$ varies continuously for fixed z and varying μ, we have $w(z,\alpha) \le 0$. Lemma 3.8 now states that

$$\triangle w - Aw \ge 0 \qquad \text{in } Z(\alpha) \text{ in the generalized sense,} \tag{5.23}$$

where $w = w(z,\alpha)$ and the constant A is described in Lemma 3.8.

(ii) Suppose for contradiction that $w(z_0,\alpha) \ne 0$ at some point $z_0 \in Z_0(\alpha)$. Then $w(.\,,\alpha) < 0$ everywhere in $Z_0(\alpha)$ [because, if the supremum 0 of $w(.\,,\alpha)$ were attained in $Z_0(\alpha)$, then $w(.\,,\alpha)$ would be the zero function in $Z_0(\alpha)$, by (5.23) and the strong maximum principle, whereas our hypothesis is that $w(z_0,\alpha) \ne 0$].

Now consider the point x_0 in (5.20), observing that $w(x_0,\alpha) = 0$ because $x_0 \in T_\alpha$. We apply the boundary-point lemma for balls (Lemma 2.12) to the function $w(.\,,\alpha)$ and the ball $B := \mathscr{B}(x_0 + \rho e^1, \rho)$, with ρ so small that $B \subset Z_0(\alpha)$. It follows that $(\partial_1 w)(x_0,\alpha) < 0$, which contradicts (5.20) because $(\partial_1 w)(x_0,\alpha) = 2(\partial_1 u)(x_0)$. Accordingly, $w(z,\alpha) = 0$ first for all $z \in Z_0(\alpha)$, and then, by continuity, for all $z \in \overline{Z_0(\alpha)}$; thus we have proved (5.21).

(iii) It remains to prove (5.22); let $q \in \partial Y_0(\alpha) \setminus T_\alpha$. Then $q^\alpha \in \partial Z_0(\alpha) \setminus T_\alpha \subset \Gamma_\alpha$, so that $u(q^\alpha) = 0$ by hypothesis (a) of Theorem 5.10. It now follows from (5.21), with $x = q^\alpha$, that $u(q) = u(q^\alpha) = 0$. But then $q \in \partial\Omega$ [otherwise we would have $u(q) > 0$]; in other words,

$$\partial Y_0(\alpha) \setminus T_\alpha \subset \partial\Omega. \tag{5.24}$$

This implies, since Ω is connected, that $\overline{\Omega} = \overline{Y_0(\alpha) \cup Z_0(\alpha)}$. Some readers may regard this implication of (5.24) as self-evident; others may consider our claim to be too intuitive or too condensed. A detailed proof is offered in §5.4. □

We turn now to an analogue of Theorem 5.10 for the equation

$$\triangle u + b_1 \partial_1 u + f(u) = 0 \quad \text{in } \Omega, \tag{5.25}$$

in which $b_1 = $ const. ≥ 0. For equation (5.2) with $b \neq 0$, we have chosen variables aligned with $b/|b|$; in any case, the unit vector $k = e^1$. For the moment we suppose that f is locally Lipschitz continuous. The new feature is this: upon setting $w(.,\mu) := u - u_\mu$ on $\overline{Z(\mu)}$ for $\mu \in [\alpha, M)$, we obtain the equation

$$\triangle w - b_1 \partial_1 w + \gamma(z, \mu)w = -2b_1 (\partial_1 u) (z) \qquad \text{for } z \in Z(\mu), \qquad (5.26)$$

in place of (3.6); again $w = w(z, \mu)$ and γ is as in (3.5). When $b_1 > 0$, we can infer that $w(.,\mu)$ is a subsolution, relative to $\triangle - b_1 \partial_1 + \gamma$ and $Z(\mu)$, *only if we know that* $\partial_1 u \leq 0$ *in* $Z(\mu)$. It is sufficient in the first instance to know that $\partial_1 u \leq 0$ in some neighbourhood of Γ_α, but to prove even this is not easy if Γ_α is not smooth, or if Γ_α is smooth but contains points x at which $n_1(x) = 0$, or if $f(0) < 0$. Therefore we restrict attention to boundaries slightly smoother than those of class C^1, to beta caps and their right-hand boundaries Γ_β, and to functions f with $f(0) \geq 0$.

Theorem 5.12 *Let* Ω *be a bounded region with* $\partial\Omega$ *of class* C^1 *and with the interior-ball property (Definition 2.14) at every point of* Γ_β. *Here* Γ_β *is the right-hand boundary of the beta cap* $Z(\beta)$ *for the direction* e^1 *(Definitions 5.1 and 5.8) and* $T_\mu(e^1)$, $Z(\mu, e^1)$, $\beta(e^1)$, ... *are abbreviated to* T_μ, Z_μ, β, ... *Assume that* $u : \overline{\Omega} \to \mathbb{R}$ *has the following properties.*

(a) $u \in C^1(\overline{\Omega})$, $u > 0$ in Ω, $u = 0$ on $\overline{\Gamma_\beta}$.

(b) *For all* $\varphi \in C_c^\infty(\Omega)$,

$$\int_\Omega \{ -\nabla\varphi \cdot \nabla u + \varphi b_1 \partial_1 u + \varphi f(u) \} = 0, \qquad (5.27)$$

where $b_1 = $ const. ≥ 0, $f(0) \geq 0$ *and* f *is otherwise as in Theorems 3.6 and 5.10.*

Then the usual monotonicity conditions hold in the beta cap:

$$u(z) < u(z^\mu) \qquad if \qquad \mu \in (\beta, M) \quad and \quad z \in Z(\mu), \qquad (5.28)$$

$$\partial_1 u(x) < 0 \qquad if \qquad x_1 > \beta \quad and \quad x \in \Omega. \qquad (5.29)$$

Setting $w(.,\mu) := u - u_\mu$ on $\overline{Z(\mu)}$ for $\mu \in [\beta, M)$, we wish to prove that

$$z \in Z(\mu) \implies (\partial_1 u)(z) < 0 \text{ and } w(z, \mu) := u(z) - u(z^\mu) < 0 \qquad (**)$$

whenever $\mu \in (\beta, M)$. The proof of Theorem 3.6 involved a boundary strip $Z_h(\mu)$, a subset of $Z(\mu)$ having width h independent of μ, with the property that $u(x) < \kappa$ for $x \in Z_h(\mu)$. That h was independent of μ was a luxury rather than an essential ingredient of the proof. This time, we shall use a boundary strip $W_\eta(\mu)$, still within $Z(\mu)$, with the property that both $\partial_1 u(x) < 0$ and $u(x) < \kappa$ whenever $x \in W_\eta(\mu)$; as far as we know *a priori*, it might be that the width $\eta(\mu) \downarrow 0$ as $\mu \downarrow \beta$. This prospect may be uncomfortable, but we shall find that it is not catastrophic. In fact, Lemmas 5.13 and 5.14 will almost reduce the proof of Theorem 5.12 to that of Theorem 3.6.

Notation Once again

$$\gamma(.,\mu) := \begin{cases} \dfrac{f_1(u) - f_1(u_\mu)}{u - u_\mu} & \text{at points } z \text{ where } u(z) \neq u_\mu(z), \\ 0 & \text{at points } z \text{ where } u(z) = u_\mu(z). \end{cases} \quad (5.30)$$

Since u vanishes on $\overline{\Gamma}_\beta$ and is uniformly continuous on $\overline{\Omega}$,

$$\exists\, h > 0 \quad \text{such that} \quad \text{dist}\,(x, \overline{\Gamma}_\beta) < h \text{ and } x \in \overline{\Omega} \Rightarrow u(x) < \kappa. \quad (5.31)$$

We write

$$W_\eta(\mu) := Z(\mu) \cap \left\{ x \mid \text{dist}(x, \overline{\Gamma}_\mu) < \eta(\mu) \right\} \quad (5.32)$$

for the boundary strip, to be constructed in Lemma 5.13, with the additional property that $\partial_1 u < 0$ in $W_\eta(\mu)$.

Lemma 5.13 *For each* $\mu \in (\beta, M)$ *there is a number* $\eta(\mu) > 0$ *such that*

$$z \in W_\eta(\mu) \Rightarrow (\partial_1 u)(z) < 0 \quad \text{and} \quad u(z) < \kappa. \quad (5.33)$$

It follows that

$$\triangle w - b_1 \partial_1 w + \gamma(z, \mu)w \geq 0 \quad \text{in } W_\eta(\mu) \text{ in the generalized sense}, \quad (5.34)$$

where $w = w(z, \mu)$ *and* γ *is defined by* (5.30).

Proof (i) Let $\mu \in (\beta, M)$ and $p \in \overline{\Gamma}_\mu$. Then $p \in \Gamma_\beta$; since $\beta \geq \lambda$, Remark 5.9 shows that $n_1(p) > 0$. We shall use the boundary-point lemma to prove that $(\partial_1 u)(p) < 0$. By hypothesis there is a ball $B \subset \Omega$ such that $p \in \partial B$; then, for all $\varphi \in C_c^\infty(B)$ with $\varphi \geq 0$,

$$\int_B \{ -\nabla\varphi \cdot \nabla u + \varphi b_1 \partial_1 u \} = -\int_B \varphi \{ [f_1(u) - f_1(0)] + f_1(0) + f_2(u) \}$$

$$\leq \int_B \varphi \{ Au - f_1(0) - f_2(u) \},$$

where A is our usual Lipschitz constant for f_1 on the interval $[0, \sup_\Omega u]$, so that $A \geq 0$. Since $f_1(0) \geq 0$ and $f_2(u) \geq 0$ by hypothesis,

$$\int_B \{ -\nabla\varphi \cdot \nabla u + \varphi b_1 \partial_1 u - \varphi Au \} \leq 0,$$

so that u is a generalized supersolution relative to $\triangle + b_1\partial_1 - A$ and B. Moreover, $u(x) > u(p) = 0$ for all $x \in B$, and e^1 is an outward unit vector at p because $n_1(p) > 0$. Thus Lemma 2.12 implies that $(\partial_1 u)(p) < 0$.

(ii) Let $\mu \in [\alpha, M)$. Calculating as in the proof of Lemma 3.7, step (i), and abbreviating $Z(\mu)$ to Z, one finds that, for all $\varphi \in C_c^\infty(Z)$,

$$\int_Z \left\{ -\nabla\varphi \cdot \nabla w - \varphi b_1 \partial_1 w + \varphi\left[f_1(u) - f_1(u_\mu)\right] \right\} \, dz$$
$$= -\int_Z \varphi \left\{ 2b_1\partial_1 u + \left[f_2(u) - f_2(u_\mu)\right] \right\} \, dz, \qquad (5.35)$$

where $w = w(z, \mu)$, $\varphi = \varphi(z)$, $\partial_1 = \partial/\partial z_1$ and so on.

(iii) Let $\mu \in (\beta, M)$. We have shown in (i) that $\partial_1 u < 0$ on the compact set $\overline{\Gamma_\mu}$; by hypothesis, $\partial_1 u$ is (uniformly) continuous on $\overline{\Omega}$. Hence $\partial_1 u \leq -c(\mu)$ on $\overline{\Gamma_\mu}$ for some $c(\mu) > 0$, and then there is a number $\rho(\mu) > 0$ such that $(\partial_1 u)(z) < 0$ if $z \in \overline{Z(\mu)}$ and $\text{dist}(z, \overline{\Gamma_\mu}) < \rho(\mu)$. Choose $\eta(\mu)$ to be the smaller of $\rho(\mu)$ and the h in (5.31), observing that $\text{dist}(z, \overline{\Gamma_\beta}) \leq \text{dist}(z, \overline{\Gamma_\mu})$ because $\overline{\Gamma_\mu} \subset \overline{\Gamma_\beta}$. With this choice of $\eta(\mu)$, condition (5.33) holds.

Now restrict (5.35) to test functions $\varphi \in C_c^\infty(W_\eta(\mu))$ with $\varphi \geq 0$. On the right-hand side, $(\partial_1 u)(z) < 0$ and $f_2(u(z)) = 0$ for $z \in W_\eta(\mu)$, by (5.33), while $f_2(u_\mu(z)) \geq 0$ because $f_2 \geq 0$ on $[0, \infty)$. Thus the right-hand member of (5.35) is non-negative, which implies (5.34). $\qquad\square$

Lemma 5.14 *If* $\mu \in [\beta, M)$ *and if, for all* $z \in Z(\mu)$, *both* $(\partial_1 u)(z) \leq 0$ *and* $w(z, \mu) \leq 0$, *then*

$$\triangle w - b_1\partial_1 w - Aw \geq 0 \quad \text{in } Z(\mu) \text{ in the generalized sense;} \qquad (5.36)$$

here $w = w(z, \mu)$ *and* A *is as in Lemma 3.8.*

Proof If $\varphi \geq 0$ in (5.35), then $-\varphi b_1\partial_1 u \geq 0$ in $Z(\mu)$ by hypothesis. The rest follows from the proof of Lemma 3.8. $\qquad\square$

Proof of Theorem 5.12 (i) In order to prove that (∗∗) holds for all sufficiently small, positive $M - \mu$, we modify the proof of Theorem 3.6, step (i), as follows. First,

$$M - \eta\left(\frac{\beta + M}{2}\right) \leq \mu < M \Rightarrow \text{dist}(z, \overline{\Gamma_\mu}) < \eta\left(\frac{\beta + M}{2}\right) \text{ for all } z \in Z(\mu).$$

Second, we may and shall suppose that the width $\eta(\mu)$ of $W_\eta(\mu)$ does not decrease as μ increases [because, if $\mu < \nu$ and $z \in Z(\nu)$ and

$\text{dist}(z, \overline{\Gamma_\nu}) < \eta(\mu)$, then $\text{dist}(z, \overline{\Gamma_\mu}) < \eta(\mu)$ and so $(\partial_1 u)(z) < 0$ and $u(z) < \kappa]$. Consequently,

$$M - \eta \left(\frac{\beta + M}{2} \right) \leq \mu < M \quad \text{and} \quad \mu \geq \frac{\beta + M}{2}$$

$$\Longrightarrow \text{dist}(z, \overline{\Gamma_\mu}) < \eta \left(\frac{\beta + M}{2} \right) \leq \eta(\mu) \quad \text{for all } z \in Z(\mu),$$

so that $Z(\mu) = W_\eta(\mu)$ for such values of μ. Then Lemma 5.13 and the maximum principle for thin sets (Theorem 2.19, with δ as in the proof of Theorem 3.6) ensure that

$$M - \eta \left(\frac{\beta + M}{2} \right) \leq \mu < M \quad \text{and} \quad \mu \geq \frac{\beta + M}{2} \quad \text{and} \quad |Z(\mu)| < \delta$$

$$\Longrightarrow (\partial_1 u)(z) < 0 \quad \text{and} \quad w(z, \mu) \leq 0 \quad \text{for all } z \in Z(\mu).$$

That $(**)$ holds for these values of μ now follows from Lemma 5.14, from the strong maximum principle (Theorem 2.13) and from the existence of boundary points p, for each component of $Z(\mu)$, at which $w(p, \mu) < 0$.

(ii) Let (m, M) be the largest open interval of μ in which $(**)$ holds; assume (for contradiction) that $m > \beta$.

(a) First, $\partial_1 u < 0$ in $Z(m)$ because, if $z \in Z(m)$, then $z \in Z(\mu)$ for some $\mu > m$. Second, we obtain $w(z, m) \leq 0$ for $z \in Z(m)$ by fixing $z \in Z(m)$ and letting $\mu \downarrow m$. Then Lemma 5.14 and the strong maximum principle imply that

$$w(z, m) < 0 \qquad \text{for} \quad z \in Z(m). \tag{5.37}$$

The condition $\partial_1 u < 0$ now extends to $T_m \cap \Omega$ by application of the boundary-point lemma to $w(., m)$ and to balls in $Z(m)$ with boundary meeting T_m; therefore,

$$\partial_1 u < 0 \qquad \text{on} \quad \overline{Z(m)} \setminus \partial\Omega. \tag{5.38}$$

(b) Writing $w_\varepsilon := w(., m - \varepsilon)$ for $0 < \varepsilon \leq \varepsilon_0$, we shall obtain a contradiction by showing that $\partial_1 u < 0$ and $w_\varepsilon < 0$ in $Z(m - \varepsilon)$ if ε_0 is sufficiently small. Provided that $m - \varepsilon_0 > \beta$, we already have that $\partial_1 u < 0$ in $W_\eta(m - \varepsilon)$, by Lemma 5.13.

The only changes from the proof of Theorem 3.6 are that $\triangle w - b_1 \partial_1 w$ replaces $\triangle w$ in the two lemmas and that now we must so choose E_ε and F that

$$G_\varepsilon := Z(m - \varepsilon) \setminus (E_\varepsilon \cup F) \subset W_\eta(m - \varepsilon).$$

For this last, the main step is a suitable choice of the cross-section S of the cylindrical set E_ε. To make this choice, we refer to (3.29), replace the h there by $\eta\left(\frac{1}{2}(\beta + m)\right)$, and demand that the conditions on ε_0 include

$$m - \varepsilon_0 \geq \tfrac{1}{2}(\beta + m).$$

Then $\eta(\mu) \geq \eta\left(\frac{1}{2}(\beta + m)\right)$ for $m - \varepsilon_0 \leq \mu < M$ [by the second observation in (i)].

We already have $\partial_1 u < 0$ on F, by (5.38), and the earlier construction gives $\partial_1 u < 0$ on E_ε; that $w_\varepsilon < 0$ on $E_\varepsilon \cup F$ is argued as before. $\qquad\square$

Theorem 5.15 *Let Ω and u be as in Theorem 5.12. Suppose that, in addition,*

$$(\partial_1 u)(x_0) = 0 \qquad \text{at some point} \quad x_0 \in T_\beta \cap \Omega. \tag{5.39}$$

Let $Z_0(\beta)$ be the component of $Z(\beta)$ that contains $x_0 + \varepsilon e^1$ for sufficiently small $\varepsilon > 0$, and let $Y_0(\beta)$ be the reflection of $Z_0(\beta)$ in T_β. Then

$$u\left(x^\beta\right) = u(x) \quad \text{for all} \quad x \in \overline{Z_0(\beta)}, \tag{5.40}$$

$$\overline{\Omega} = \overline{Y_0(\beta)} \cup \overline{Z_0(\beta)}, \tag{5.41}$$

and the coefficient $b_1 = 0$.

Proof The proof of Theorem 5.11 remains valid here if we replace the interval (α, M) by (β, M), Theorem 5.10 by Theorem 5.12, and Lemma 3.8 by Lemma 5.14.

It remains to prove that $b_1 = 0$. Setting $\mu = \beta$ in (5.35), we observe from (5.40) and (5.41) that $u = u_\beta$ on $\overline{Z(\beta)}$ and hence $w(., \beta) = 0$ on $\overline{Z(\beta)}$; therefore (5.35) with $\mu = \beta$ reduces to

$$\int_{Z(\beta)} \varphi b_1 \partial_1 u = 0 \qquad \text{for all} \quad \varphi \in C_c^\infty\left(Z(\beta)\right).$$

Since $\partial_1 u < 0$ in $Z(\beta)$, by (5.29), it follows from a result of Exercise 1.16 that $b_1 = 0$. $\qquad\square$

5.4 A little topology

Remark 5.16 This section is concerned only with a full proof of (5.22) in Theorem 5.11. In addition to the basic fact that Ω is *open, bounded and connected*, we have the following data.

(A) $Z_0 := Z_0(\alpha)$ is a component of the alpha cap $Z := Z(\alpha)$.

(B) $Y_0 := Y_0(\alpha)$ is the reflection in T_α of Z_0 (so that $Y_0 \subset \Omega$ by Theorem 5.3).

(C) $\partial Y_0 \setminus T_\alpha \subset \partial \Omega$.

The question is whether these conditions imply that $\overline{\Omega} = \overline{Y_0 \cup Z_0}$.

We recall that, for sets A and B in a metric space,

$$\overline{A \cup B} = \overline{A} \cup \overline{B}, \qquad \overline{A \cap B} \subset \overline{A} \cap \overline{B},$$
$$\partial(A \cup B) \subset \partial A \cup \partial B, \quad \partial(A \cap B) \subset \partial A \cup \partial B.$$

$\square$

Lemma 5.17 *Let C be a connected, non-empty set in a metric space (M, d). If $A \subset C$ and $C \cap \partial A$ is empty, then either $A = \emptyset$ or $A = C$.*

Proof The pair $\left(C, d|_{C \times C}\right)$ is also a metric space, and

$$C = \mathrm{int}_C A \cup \partial_C A \cup \mathrm{ext}_C A,$$

where int_C, ∂_C and ext_C denote respectively the interior, boundary and exterior relative to C. Given the hint in Exercise 5.26, one checks without difficulty that $\partial_C A \subset C \cap \partial A$ (where ∂A is the boundary of A relative to M), so that $\partial_C A$ is empty. Since C is connected, it cannot be the union of two disjoint, non-empty sets that are open relative to C; one of $\mathrm{int}_C A$ and $\mathrm{ext}_C A$ must be empty.

If $\mathrm{int}_C A$ is empty, then

$$A \subset C = \mathrm{ext}_C A = \mathrm{int}_C (C \setminus A) \subset C \setminus A \;\Rightarrow\; A = \emptyset.$$

If $\mathrm{ext}_C A$ is empty, then

$$A \subset C = \mathrm{int}_C A \subset A \;\Rightarrow\; A = C.$$

$\square$

Lemma 5.18 *Let D and E be sets in a metric space, with D connected, non-empty and open. If D intersects E but not ∂E, then $D \subset E$.*

Proof Let $A := D \cap E$. Then $A \subset D$ and $D \cap \partial A$ is empty because

$$D \cap \partial A \subset D \cap (\partial D \cup \partial E) = D \cap \partial E = \emptyset.$$

Accordingly, we may apply Lemma 5.17 with $C = D$: either $D \cap E = \emptyset$ or $D \cap E = D$. The former is contrary to hypothesis; the latter implies that $D \subset E$. $\square$

Lemma 5.19 *If* $\{ G_t \mid t \in T \}$ *is a family of disjoint open sets in a metric space and* $G := \bigcup_{t \in T} G_t$, *then each* $\partial G_t \subset \partial G$.

Proof Let $x \in \partial G_s$ for some $s \in T$. Then $x \in \overline{G}$, $x \notin G_s$ and, since every open set containing x intersects G_s, we have $x \notin G_t$ if $t \neq s$. Thus $x \in \overline{G} \setminus G$. But $\overline{G} \setminus G = \partial G$ because G is open. $\qquad\square$

Proposition 5.20 *With* Ω, Z_0 *and* Y_0 *as in Remark 5.16, define*

$$A := Y_0 \cup Z_0 \cup (\Omega \cap \partial Z_0) \cup (\Omega \cap \partial Y_0). \tag{5.42}$$

Then $A = \Omega$, *which implies that* $\overline{\Omega} = \overline{A} = \overline{Y_0 \cup Z_0}$.

Proof (i) That $\overline{A} = \overline{Y_0 \cup Z_0}$ is immediate from the definition (5.42). We shall prove that $A = \Omega$ by means of Lemma 5.17 (with $C = \Omega$ and $M = \mathbb{R}^N$). First, $A \subset \Omega$ because $Y_0 \subset \Omega$ by (B) of Remark 5.16 and $Z_0 \subset \Omega$ by (A). Second, we shall prove in (ii) that A is open in $\mathbb{R}^N$; therefore,

$$
\begin{aligned}
\partial A = \overline{A} \setminus A &= \overline{Y_0 \cup Z_0} \setminus \{ Y_0 \cup Z_0 \cup (\Omega \cap \partial Z_0) \cup (\Omega \cap \partial Y_0) \} \\
&= \partial(Y_0 \cup Z_0) \setminus \{ (\Omega \cap \partial Z_0) \cup (\Omega \cap \partial Y_0) \} \\
&\subset (\partial Y_0 \cup \partial Z_0) \setminus \{ (\Omega \cap \partial Z_0) \cup (\Omega \cap \partial Y_0) \},
\end{aligned}
$$

which shows that $\Omega \cap \partial A$ is empty. Thus Lemma 5.17 applies; since A is not empty, we have $A = \Omega$.

(ii) To prove that A is open in $\mathbb{R}^N$, we consider separately three types of point in A.

(a) Let $x \in Y_0 \cup Z_0$. Then $x \in \text{int}\, A$ because $Y_0 \cup Z_0$ is open.

(b) Let $p \in \Omega \cap \partial Z_0$. Now $\partial Z_0 \subset \partial Z$ by Lemma 5.19, and $\partial Z \subset \partial \Omega \cup T_\alpha$ because $Z = \Omega \cap \{ x \mid x_1 > \alpha \}$; since Ω and $\partial \Omega$ are disjoint, $p \in \Omega \cap T_\alpha$. Hence there exists a radius $\rho = \rho(p) > 0$ such that $\mathscr{B}(p, \rho) \subset \Omega$. To show that $\mathscr{B}(p, \rho) \subset A$, we consider

$$
\begin{aligned}
B_+ &:= \mathscr{B}(p, \rho) \cap \{ x \mid x_1 > \alpha \}, \\
S &:= \mathscr{B}(p, \rho) \cap T_\alpha, \\
B_- &:= \mathscr{B}(p, \rho) \cap \{ x \mid x_1 < \alpha \}.
\end{aligned}
$$

Observe that $\mathscr{B}(p, \rho)$ intersects Z_0 because $p \in \partial Z_0$, and that S and B_- do not intersect Z_0; therefore B_+ intersects Z_0. Next, B_+ does not intersect ∂Z_0 because $\partial Z_0 \subset \partial \Omega \cup T_\alpha$. Therefore Lemma 5.18, with $D = B_+$ and $E = Z_0$, shows that $B_+ \subset Z_0$.

Then $B_- \subset Y_0$ (by reflection in T_α) and $S \subset \Omega \cap \partial Z_0$ (because $S \subset \Omega$ and points of S, being limit points of B_+, are also limit points of Z_0).

We have shown that B_+, B_- and S are each contained in a subset of A; consequently, $\mathscr{B}(p, \rho) \subset A$ and so $p \in \operatorname{int} A$.

(c) Let $q \in \Omega \cap \partial Y_0$. Then $q \in T_\alpha$, because $\partial Y_0 \setminus T_\alpha \subset \partial \Omega$ by (C) of Remark 5.16. In addition, there is a sequence (y_n) in Y_0 such that $y_n \to q$. But then $y_n^\alpha := (y_n)^\alpha \in Z_0$, and $y_n^\alpha \to q^\alpha = q$. Hence $q \in \Omega \cap \partial Z_0$, and $q \in \operatorname{int} A$ by step (b). $\qquad\square$

5.5 Exercises

Exercise 5.21 A boundary $\partial \Omega$ is *of class* $C^{0,1}$ iff, in each local representation $y_N = h(y', p)$ of $\partial \Omega$ [Chapter 0, (viii)], the function $h(., p)$ is uniformly Lipschitz continuous, so that $h(., p) \in C_b^{0,1}(\overline{G(p)})$ according to Definition A.10.

Show by means of an example that, for part (a) of Theorem 5.7 it is not enough to demand that Ω be a bounded region with $\partial \Omega$ of class $C^{0,1}$.

Exercise 5.22 The elliptic disk

$$E := \left\{ (x, y) \in \mathbb{R}^2 \mid Ax^2 - Bxy + Cy^2 < 1 \right\},$$

where

$$A := 1 + a \sin^2 \gamma, \quad B := a \sin 2\gamma, \quad C := 1 + a \cos^2 \gamma,$$

is characterized by the parameters $a > 0$ and $\gamma \in [0, \pi/2]$; the major axis has length 2 and slope $\tan \gamma$, while the minor axis has length $2(1+a)^{-1/2}$. Prove that, for the set E,

$$\alpha(e^1) = \lambda(e^1) = \tfrac{1}{2}(1+a)^{-1/2}(a \sin 2\gamma)(1 + a \sin^2 \gamma)^{-1/2},$$

according to Definitions 5.1 and 5.8.

Find the solution $u \in C(\overline{E}) \cap C^2(E)$ of

$$\triangle u + 1 = 0 \quad \text{in } E, \qquad u\big|_{\partial E} = 0.$$

For the direction e^1, how well does Theorem 5.10 describe the graph of u? Give separate answers for $\gamma = 0$, $0 < \gamma < \pi/2$ and $\gamma = \pi/2$.

[The function u can be found from Exercise 4.21.]

Exercise 5.23 Let Ω be the set called Ω_2 in (5.16); assume that a function u satisfies the hypotheses in Theorem 5.10, strengthened to $u = 0$ on $\partial \Omega$.

Show that $u(x_1, x_2) < u(-x_1, x_2)$ whenever $x_1 > 0$ and $x \in \Omega$, and that $u(x_1, x_2) = u(x_1, -x_2)$ for all $x \in \overline{\Omega}$. Establish the signs of $\partial_1 u$ and $\partial_2 u$ throughout Ω, except for the sign of $\partial_1 u$ in the open subset $\{ x \in \Omega \mid -\gamma < x_1 < 0 \}$, where γ is as in (5.15). Infer that the maximum of u can occur only in the line segment $(-\gamma, 0) \times \{0\}$.

Exercise 5.24 Let Ω and $\lambda = \lambda(e^1)$ be as in Definition 5.8. Prove that there exists a point $x_0 \in T_\lambda \cap \partial\Omega$ such that $n_1(x_0) \leq 0$; moreover, such that $n_1(x_0) = 0$ or $n_1(x_0) = -1$.

Exercise 5.25 State a result that extends Theorem 3.3 in the way that Theorem 5.10 extends Theorem 3.6. Prove your result by adjusting the notation in the proof of Theorem 3.3.

Exercise 5.26 Let $A \subset B \subset M$, where (M, d) is a metric space and B is not empty. Write ∂A and $\partial_B A$, respectively, for the boundary of A in M and the boundary of A relative to B. Prove that $\partial_B A \subset B \cap \partial A$, with equality if B is open in M. Give an example of inequality.

[A set $D \subset B$ is open relative to B if and only if $D = B \cap G$ for some G open in M. For a set S in a metric space (X, d), a point p belongs to ∂S if and only if every open set containing p intersects both S and $X \setminus S$.]

Exercise 5.27 Define an annulus or spherical shell by $A := \{ x \in \mathbb{R}^N \mid a < |x| < b \}$, $N \geq 2$, $0 < a < b$. Suppose that

$$u \in C(\overline{A}) \cap C^2(A), \quad u > 0 \text{ in } A \quad \text{and} \quad u = 0 \text{ on } \partial\mathscr{B}(0, b), \quad (5.43a)$$

$$\Delta u + f(u) = 0 \quad \text{in} \quad A, \quad (5.43b)$$

where $f : [0, \infty) \to \mathbb{R}$ is locally Lipschitz continuous. Let $\partial/\partial r := (x/|x|) \cdot \nabla$.

Prove that, if $x, y \in A$ with $|x| \in ((a+b)/2, b)$, $y/|y| - x/|x|$ and $|y| + |x| = a + b$, then

$$\frac{\partial u}{\partial r}(x) < 0 \quad \text{and} \quad u(x) < u(y); \quad (5.44a)$$

also that

$$\frac{\partial u}{\partial r}(x) < 0 \quad \text{if} \quad |x| = \frac{a+b}{2}. \quad (5.44b)$$

Exercise 5.28 Add to (5.43a, b) the condition

$$u = 0 \quad \text{on} \quad \partial\mathscr{B}(0, a). \quad (5.43c)$$

Display a spherically symmetric solution (depending only on $r := |x|$) of (5.43a, b, c) for the case in which

$$N = 3 \quad \text{and} \quad f(t) = \left(\frac{\pi}{b-a}\right)^2 t \quad \text{(for all } t \geq 0\text{)}.$$

Show that, for this case, (5.44a, b) are satisfied amply when a/b is small, but are satisfied only by a hair's breadth when a/b is very close to 1.

[A small part of Exercise 1.21 is relevant.]

Exercise 5.29 This exercise prepares for the description, in Exercise 5.30, of a solution of (5.43a, b, c) that is not spherically symmetric. It is to be hoped that the essence of the argument emerges, even though parts of it are beyond the range of this book.

With A as in Exercise 5.27, let H be the real Hilbert space formed by completion of the set $C_c^\infty(A)$ in the norm $\|\,.\,\|$ defined by

$$\|v\| := \sqrt{\langle v, v \rangle}, \quad \text{where} \quad \langle v, w \rangle := \int_A \nabla v \cdot \nabla w.$$

(i) Derive, or improve on, the following inequalities.

(a) If $v \in H$ and is spherically symmetric, say $r := |x|$ and $\tilde{v}(r) := v(x)$, then

$$\tilde{v}(r)^2 \leq \frac{\|v\|^2}{|\partial \mathscr{B}(0,a)|} q(r), \quad \text{where} \quad q(r) := \begin{cases} r - a & \text{if } a \leq r \leq \frac{1}{2}(a+b), \\ b - r & \text{if } \frac{1}{2}(a+b) \leq r \leq b. \end{cases}$$

(b) For all $v \in H$,

$$\int_A v^2 \leq \frac{1}{8}(b-a)^2 \left(\frac{a+b}{2a}\right)^{N-1} \|v\|^2.$$

(ii) Elements of H that are not spherically symmetric may have unbounded values. Verify that the following functions w satisfy $\|w\| < \infty$.

Given a point $p \in A$, choose δ so that $0 < \delta < \text{dist}(p, \partial A)$, and define

$$w(x) := h_N\left(\frac{|x-p|}{\delta}\right) \quad \text{for all} \quad x \in \overline{A} \setminus \{p\},$$

where, if $A \subset \mathbb{R}^2$, we use

$$h_2(t) := \begin{cases} \left(\log\frac{e}{t}\right)^\gamma - 1 & \text{if } 0 < t \leq 1, \quad \gamma = \text{const.} \in \left(0, \frac{1}{2}\right), \\ 0 & \text{if } t > 1, \end{cases}$$

or, if $A \subset \mathbb{R}^N$ with $N \geq 3$, we use

$$h_N(t) := \begin{cases} t^{-\gamma} - 1 & \text{if } 0 < t \leq 1, \quad \gamma = \text{const.} \in \left(0, \dfrac{N}{2} - 1\right), \\ 0 & \text{if } t > 1. \end{cases}$$

[The inequalities in (i) result from the identities

$$v(x_0) = \int_a^{r_0} \frac{\partial \tilde{v}}{\partial r}(r, \vartheta_0) \, dr = -\int_{r_0}^b \frac{\partial \tilde{v}}{\partial r}(r, \vartheta_0) \, dr,$$

in which $v \in C_c^\infty(A)$, $\vartheta := x/|x|$, $\tilde{v}(r, \vartheta) := v(x)$, and from the Buniakowsky–Schwarz inequality.

For each function w in (ii) one can show, by extensions of Exercises 1.23 and 1.25, that there is a sequence of functions $w_n \in C_c^\infty(A)$ such that $\|w - w_n\| \to 0$ as $n \to \infty$; therefore each w is indeed in H.]

Exercise 5.30 Here we shall describe a solution u of (5.43a, b, c) with $f : \mathbb{R} \to \mathbb{R}$ defined by

$$f(t) = \ell g(t) := \ell(t - c) f_H(t - c) \qquad \text{for all} \quad t \in \mathbb{R},$$

where f_H denotes the Heaviside function [Chapter 0, (v)], c is a given positive constant, and the positive constant ℓ will be calculated *a posteriori* because $\|u\|$ is prescribed (cf. Exercise 3.17). Let the Hilbert space H be as in Exercise 5.29 and let

$$G(t) := \tfrac{1}{2}(t - c)^2 f_H(t - c) \qquad \text{for all} \quad t \in \mathbb{R}.$$

For each $\rho > 0$, the variational problem of maximizing $\int_A G\big(v(x)\big) \, dx$ over the sphere $S_\rho := \{ v \in H \mid \|v\| = \rho \}$ has a solution $u \in S_\rho$ such that

$$\int_A G(u) > 0 \qquad \text{and} \qquad \langle u, \varphi \rangle = \ell \int_A g(u)\varphi \quad \text{for all } \varphi \in H,$$

where

$$\ell = \|u\|^2 \Big/ \int_A g(u)\, u \,.$$

Moreover, the equivalence class u (of functions equal almost everywhere in A) has a representative u_0 that satisfies (5.43a, b, c), for f and ℓ as above.

Given all this, show that the representative u_0 of u is not spherically symmetric if ρ is sufficiently small; in fact, if

$$0 < \left(\frac{b-a}{2|\partial \mathscr{B}(0,a)|} \right)^{1/2} \rho < c.$$

Appendix A. On the Newtonian Potential

A.1 Point sources in $\mathbb{R}^3$

(i) By the *potential of a unit source at the point* $c \in \mathbb{R}^3$ we mean the function Φ defined by

$$\Phi(x) := \frac{1}{4\pi} \frac{1}{|x - c|}, \qquad x \in \mathbb{R}^3 \setminus \{c\}. \tag{A.1}$$

The words 'potential' or 'potential function' imply that the vector field $-\nabla\Phi$ is the object of primary interest; the phrase 'unit source' or *source of unit strength* is intended to convey the result (A.3) below.

There is a good case for believing that God devised the function Φ on the first day of the Creation. Before considering a small part of the evidence for this belief, we note two properties of the function.

(a) Φ *satisfies the Laplace equation*:

$$\triangle\Phi := (\partial_1^2 + \partial_2^2 + \partial_3^2)\Phi = 0 \qquad \text{in } \mathbb{R}^3 \setminus \{c\}. \tag{A.2}$$

For, writing $R := |x - c| = \left\{ (x_1 - c_1)^2 + \cdots + (x_3 - c_3)^2 \right\}^{1/2}$, we have

$$\partial_i R = \frac{x_i - c_i}{R}, \qquad \partial_i \frac{1}{R} = -\frac{1}{R^2} \frac{x_i - c_i}{R},$$

$$\partial_i^2 \frac{1}{R} = \frac{3}{R^4} \frac{(x_i - c_i)^2}{R} - \frac{1}{R^3};$$

summing this last over i from 1 to 3, we obtain (A.2).

(b) *If Ω is a bounded open set that contains the point c, and $\partial\Omega$ is of class C^1, then*

$$-\int_{\partial\Omega} \frac{\partial\Phi}{\partial n} = 1. \tag{A.3}$$

Here n denotes the outward unit normal on $\partial\Omega$ (see Figure A.1), $\partial/\partial n := n \cdot \nabla$ and the element $\mathrm{d}S$ of surface area is implied.

167

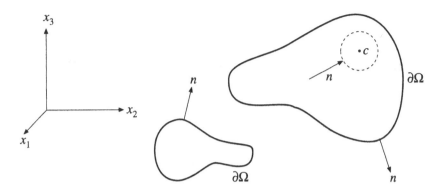

Fig. A.1.

To prove (A.3), we introduce a ball $\mathscr{B}(c,\rho)$ with ρ so small that $\overline{\mathscr{B}(c,\rho)} \subset \Omega$; on the sphere $\partial\mathscr{B}(c,\rho)$, the unit normal n outward from $\Omega' := \Omega \setminus \overline{\mathscr{B}(c,\rho)}$ points towards c. Hence (with $R = |x - c|$)

$$\left.\frac{\partial\Phi}{\partial n}\right|_{\partial\mathscr{B}(c,\rho)} = \left.-\frac{d}{dR}\frac{1}{4\pi R}\right|_{R=\rho} = \frac{1}{4\pi\rho^2},$$

so that

$$\int_{\partial\mathscr{B}(c,\rho)} \frac{\partial\Phi}{\partial n} = 1. \qquad (A.4)$$

Now apply the divergence theorem to the vector field $\nabla\Phi$ and the set Ω':

$$0 = \int_{\Omega'} \Delta\Phi = \int_{\partial\Omega} \frac{\partial\Phi}{\partial n} + \int_{\partial\mathscr{B}(c,\rho)} \frac{\partial\Phi}{\partial n}; \qquad (A.5)$$

then (A.3) follows from (A.4) and (A.5).

(ii) *Electrostatics* This may seem an absurd heading for the single rule that is stated here. But this rule is the basis of the whole subject, and leads to the electric fields of arbitrary distributions of electric charge. Consider a point charge Q (a particle having electric charge Q) at the point $c \in \mathbb{R}^3$. According to Coulomb's hypothesis of 1785, illustrated in Figure A.2, the force on a *test particle*, having unit electric charge and located at any point $x \neq c$, is

$$E(x) = \frac{Q}{4\pi\kappa}\frac{x-c}{|x-c|^3} = -\frac{Q}{\kappa}\nabla\Phi(x), \qquad (A.6)$$

where the positive constant κ is the dielectric constant of whatever material is assumed to occupy $\mathbb{R}^3$. The force on a particle at x having

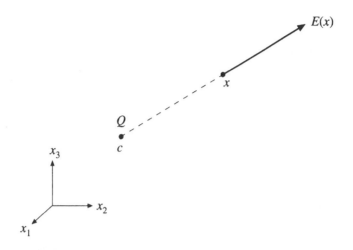

Fig. A.2.

charge q is $qE(x)$, and the vector field E is the *electric field* of the point charge Q. The formula (A.6) is an example of an *inverse-square law*, because $E(x)$ has direction along the line from c to x, and magnitude proportional to $|x - c|^{-2}$.

(iii) *Signs* No single convention, for the sign of potential functions and of source strength, is satisfactory for all applications. One reason is that positive point charges repel each other, whereas point masses attract each other. Nevertheless, we retain the definition and terminology in (i) for all three cases of our brief flirtation with physics.

(iv) *Gravitation* Here the basic hypothesis is traditionally attributed to Newton, who published it in his *Principia* of 1687 after perhaps twenty years' contemplation. However, it may be that Hooke has not had his share of the credit (Arnold 1990, Fauvel *et al.* 1988). Consider a point mass M (a particle having mass M) at the point $c \in \mathbb{R}^3$. According to the Newtonian hypothesis, the force on a test particle, having unit mass and located at any point $x \neq c$, is

$$F(x) = -\frac{M}{4\pi k_g} \frac{x - c}{|x - c|^3} = \frac{M}{k_g} \nabla \Phi(x), \qquad (A.7)$$

where k_g is a positive gravitational constant. The force on a particle at x having mass m is $mF(x)$. (Our signs are unfortunate here in that the *potential energy* of this particle is $-Mm\Phi(x)/k_g$ apart from an additive constant.) Again we have an inverse-square law.

(v) *Point sources in hydrodynamics* In this subject it is the Laplace equation that is basic, and point sources can be used to construct explicit solutions. For the flow of a highly idealized fluid past a compact set (or body) E in $\mathbb{R}^3$, one seeks a *velocity potential* φ such that $-\nabla\varphi$ represents the fluid velocity and such that

$$\left. \begin{array}{l} \triangle\varphi = 0 \ \text{in} \ \mathbb{R}^3 \setminus E, \qquad \dfrac{\partial\varphi}{\partial n} = 0 \ \text{on} \ \partial E, \\[2mm] -\nabla\varphi(x) \to (U,0,0) \ \text{as} \ |x| \to \infty. \end{array} \right\} \tag{A.8a}$$

Here U is a given positive constant, so that the velocity 'at infinity' is prescribed. Let $\Omega := \mathbb{R}^3 \setminus E$. By a *solution of the problem* (A.8) we mean a function φ that satisfies not only (A.8a) but also the conditions

$$\left. \begin{array}{l} \varphi \in C^1(\overline{\Omega}) \cap C^2(\Omega), \\[2mm] \text{as} \ r := |x| \to \infty, \quad \nabla\varphi(x) + (U,0,0) = O(r^{-2}) \ \text{and} \ \varphi(x) + Ux_1 \to 0. \end{array} \right\}$$
$$\tag{A.8b}$$

Theorem A.1 *If $\partial\Omega$ is of class C^1 and Ω is connected, then the problem (A.8) has at most one solution.*

Proof Let $w := \varphi_1 - \varphi_2$ be the difference of two solutions, so that $w \in C^1(\overline{\Omega}) \cap C^2(\Omega)$ and

$$\triangle w = 0 \ \text{in} \ \Omega, \qquad \frac{\partial w}{\partial n} = 0 \ \text{on} \ \partial\Omega,$$

$$\text{as} \ r \to \infty, \qquad \nabla w(x) = O(r^{-2}) \ \text{and} \ w(x) \to 0.$$

Then $w(x) = O(r^{-1})$ as $r \to \infty$, by Remark 4 before Theorem A.2 below. Substituting $V = w\nabla w$ into the divergence theorem

$$\int_{\Omega'} \nabla \cdot V = \int_{\partial\Omega'} n \cdot V,$$

where V is to be in $C^1(\overline{\Omega'}, \mathbb{R}^3)$ and a suitable approximation Ω' to Ω will be chosen presently, we obtain the *energy identity*

$$\int_{\Omega'} |\nabla w|^2 = \int_{\partial\Omega'} w\frac{\partial w}{\partial n} - \int_{\Omega'} w\triangle w,$$

in which the left-hand member is twice a kinetic energy, for a fluid of density 1. We cannot begin with the energy identity for the set Ω, for two reasons: Ω is unbounded and w need not be in $C^2(\overline{\Omega})$. (It would be artificial to demand in (A.8) that $\varphi \in C^2(\overline{\Omega})$.)

To define Ω', we first set $\Omega_A := \Omega \cap \mathcal{B}(0,A)$ with A so large that $\partial\Omega = \partial E \subset \mathcal{B}(0,A)$, and then choose $\Omega' := \Omega_{A,m}$, the mth approximation to Ω_A in the sense of Theorem D.9 (Appendix D). Since $\overline{\Omega}_{A,m} \subset \Omega_A \subset \Omega$, we have $w \in C^2(\overline{\Omega}_{A,m})$ and $\triangle w = 0$ in $\Omega_{A,m}$, so that

$$\int_{\Omega_{A,m}} |\nabla w|^2 = \int_{\partial\Omega_{A,m}} w \frac{\partial w}{\partial n}.$$

Let $m \to \infty$; by Theorem D.9 and because $\partial w/\partial n = 0$ on $\partial\Omega$,

$$\int_{\Omega_A} |\nabla w|^2 = \int_{\partial\mathcal{B}(0,A)} w \frac{\partial w}{\partial n}.$$

Now let $A \to \infty$; we have $w\partial w/\partial n = O(A^{-3})$ for $r = A$, while the surface area $|\partial\mathcal{B}(0,A)| = 4\pi A^2$, so that finally $\int_\Omega |\nabla w|^2 = 0$. This implies that $|\nabla w| = 0$ (the zero function) because $|\nabla w|$ is continuous; then $w(x)$ is constant in Ω because Ω is connected, and, in fact, $w = 0$ in Ω because $w(x) \to 0$ as $r \to \infty$. $\square$

If Ω is not connected, we still obtain $|\nabla w| = 0$ in the foregoing proof, so that the problem (A.8) admits at most one velocity field $-\nabla\varphi$.

For the construction of a solution of (A.8) by an elementary method, it is usual and profitable to cheat by writing

$$\left.\begin{array}{c} \varphi(x) = -Ux_1 + \dfrac{q}{4\pi}\left\{\dfrac{1}{|x+c|} - \dfrac{1}{|x-c|}\right\}, \\[2mm] q = \text{const.} > 0, \qquad c = (\gamma,0,0) \text{ with } \gamma > 0, \end{array}\right\} \qquad \text{(A.9)}$$

and by then showing that this φ is the velocity potential of flow past a certain body E called a *Rankine solid* (Figure A.3). Note that the three terms in (A.9) are the potentials, respectively, of a uniform stream with velocity $(U,0,0)$, of a point source at $-c$ of strength q and of a point source at c of strength $-q$. Thus φ is defined on $\mathbb{R}^3 \setminus \{-c,c\}$, indeed, $\varphi \in C^\infty(\mathbb{R}^3 \setminus \{-c,c\})$, but we are interested mainly in the velocity $-\nabla\varphi$ on and outside ∂E. It is far from obvious that the formula (A.9) implies the picture in Figure A.3; before taking up this question in (vi), we consider a limiting case.

Set $q = k/2\gamma$, and let $\gamma \to 0$ with k fixed and positive, and with $x \neq 0$. Then

$$\frac{q}{4\pi}\left\{\frac{1}{|x+c|} - \frac{1}{|x-c|}\right\} = \frac{1}{4\pi}\frac{k}{2\gamma}\left\{\left(\partial_1 \frac{1}{|x|}\right)2\gamma + O(\gamma^2)\right\}$$

$$\to \frac{k}{4\pi}\partial_1\frac{1}{|x|} = -\frac{k}{4\pi}\frac{x_1}{|x|^3}. \qquad \text{(A.10)}$$

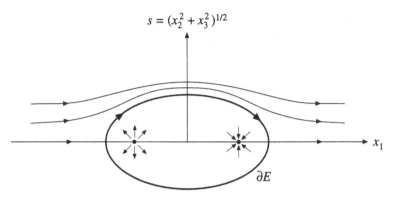

$$s = (x_2^2 + x_3^2)^{1/2}$$

Fig. A.3.

This is the potential of a *dipole* (or doublet) at the origin, of *strength k and type* $(-1,0,0)$, or of strength $-k$ and type $(1,0,0)$; the type states the direction from negative source to positive source before the limit is taken. This function satisfies the Laplace equation in $\mathbb{R}^3 \setminus \{0\}$, because Δ and ∂_1 commute when acting on C^3 functions.

Now define spherical co-ordinates r, θ, λ by

$$\left. \begin{array}{l} x = (x_1, x_2, x_3) = (r \cos \theta, \ r \sin \theta \cos \lambda, \ r \sin \theta \sin \lambda), \\[4pt] r \geq 0, \ \ 0 \leq \theta \leq \pi, \ \ -\pi < \lambda \leq \pi. \end{array} \right\} \qquad \text{(A.11)}$$

Let $k = 2\pi U a^3$; then the limiting form of (A.9) is

$$\varphi_0(x) = -U \cos \theta \left(r + \frac{1}{2} \frac{a^3}{r^2} \right) \qquad (r > 0), \qquad \text{(A.12)}$$

and the reason for our choice of k is that

$$\frac{\partial \varphi_0}{\partial r}(x) = -U \cos \theta \left(1 - \frac{a^3}{r^3} \right),$$

which vanishes if $r = a$. Thus φ_0 is the *unique solution of the problem* (A.8) *when the body is a closed ball*: $E = \overline{\mathscr{B}(0, a)}$.

(vi) *The Stokes stream function* In addition to the spherical co-ordinates r, θ, λ in (A.11), we introduce cylindrical co-ordinates x_1, s, λ (**Figure A.4**):

$$\left. \begin{array}{l} x = (x_1, \ s \cos \lambda, \ s \sin \lambda), \\[4pt] s = \left(x_2^2 + x_3^2 \right)^{1/2}, \ \ -\pi < \lambda \leq \pi. \end{array} \right\} \qquad \text{(A.13)}$$

Suppose that a potential function φ satisfies *the Laplace equation*

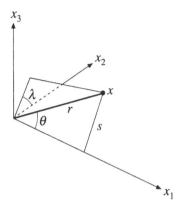

Fig. A.4.

$\triangle\varphi = 0$ in Ω and is *cylindrically symmetric* about the x_1-axis. (This last means that Ω is a figure of revolution about the x_1-axis and that

$$\frac{\partial\varphi}{\partial\lambda} := x_2\frac{\partial\varphi}{\partial x_3} - x_3\frac{\partial\varphi}{\partial x_2} = 0 \quad\text{in } \Omega.)$$

If both these conditions hold, we write $\varphi(x) = \tilde{\varphi}(x_1, s)$ and, apart from an additive constant, define the *Stokes stream function* $\psi(x) = \tilde{\psi}(x_1, s)$ corresponding to the potential φ by

$$\left(\frac{\partial\tilde{\psi}}{\partial x_1}, \frac{\partial\tilde{\psi}}{\partial s}\right) = s\left(-\frac{\partial\tilde{\varphi}}{\partial s}, \frac{\partial\tilde{\varphi}}{\partial x_1}\right). \tag{A.14}$$

A better definition will be given in Exercise A.24, but the present one is legitimate because (by a formula for $\triangle$ in Exercise 1.18)

$$\frac{\partial}{\partial x_1}\frac{\partial\tilde{\psi}}{\partial s} - \frac{\partial}{\partial s}\frac{\partial\tilde{\psi}}{\partial x_1} = s\left(\frac{\partial^2\tilde{\varphi}}{\partial x_1^2} + \frac{\partial^2\tilde{\varphi}}{\partial s^2} + \frac{1}{s}\frac{\partial\tilde{\varphi}}{\partial s}\right) = s\triangle\varphi = 0.$$

We now show that *the vector field* $\nabla\varphi$ *is tangential to the level sets* $\{x \mid \psi(x) = \text{const.}\}$ *of* ψ; it is this fact that makes ψ a splendid companion to φ. Here we adopt the convention that the zero vector is both perpendicular and parallel to every vector.

First, the vector fields $\nabla\varphi$ and $\nabla\psi$ are orthogonal (are perpendicular to each other):

$$\nabla\varphi \cdot \nabla\psi = \frac{\partial\tilde{\varphi}}{\partial x_1}\frac{\partial\tilde{\psi}}{\partial x_1} + \frac{\partial\tilde{\varphi}}{\partial s}\frac{\partial\tilde{\psi}}{\partial s} = s\left(-\frac{\partial\tilde{\varphi}}{\partial x_1}\frac{\partial\tilde{\varphi}}{\partial s} + \frac{\partial\tilde{\varphi}}{\partial s}\frac{\partial\tilde{\varphi}}{\partial x_1}\right) = 0.$$

Second, at any point $x_0 \in \Omega$, the vector $(\nabla\psi)(x_0)$ is normal to the level set $\{ x \in \Omega \mid \psi(x) = \text{const.} = \psi(x_0) \}$. Together, these two facts show that $(\nabla\varphi)(x_0)$ is tangential at x_0 to the level set of ψ containing x_0.

Accordingly, a picture of the level sets of ψ (which are surfaces of revolution in $\mathbb{R}^3$ wherever $|\nabla\psi| \neq 0$) or of the level sets of $\tilde{\psi}$ (which are curves in the half-plane $\mathbb{R} \times [0, \infty)$ wherever $|\nabla\tilde{\psi}| \neq 0$) indicates the directions of the vector field $\nabla\varphi$. Such a picture also indicates, very approximately, the magnitudes $|\nabla\varphi| = |\nabla\psi|/s$. For, let $\delta(x)$ be the distance from a point x on a level set to an adjacent level set; if $\delta(x)$ is small, then

$$|\nabla\psi(x)| \approx |\psi_2 - \psi_1|/\delta(x),$$

where ψ_1 and ψ_2 are the constant values of ψ on the level sets in question.

For the velocity potentials in (A.9) and (A.12), the definition (A.14) yields

$$\psi(x) = -\frac{1}{2}Us^2 + \frac{q}{4\pi}\left[\frac{x_1 + \gamma}{\{(x_1 + \gamma)^2 + s^2\}^{1/2}} - \frac{x_1 - \gamma}{\{(x_1 - \gamma)^2 + s^2\}^{1/2}}\right], \quad \text{(A.15)}$$

$$\psi_0(x) = -\frac{1}{2}U\sin^2\theta\left(r^2 - \frac{a^3}{r}\right). \quad \text{(A.16)}$$

To calculate ψ_0 one first writes (A.14) in terms of spherical co-ordinates (Exercise A.18). The boundary ∂E of the Rankine solid is a part of the level set $\{x \mid \psi(x) = 0\}$. Detailed description of ∂E still requires work (Exercise A.26), but the Stokes stream function enables us to establish the main properties of this boundary before resorting to numerical calculation.

A.2 The Newtonian potential: first steps

The Newtonian kernel This is another name for the potential of a unit source at the origin, but now the setting is $\mathbb{R}^N$ with $N \in \{1, 2, 3, \ldots\}$. [In (A.21) below, K is said to be the *kernel* of the integral operator acting on f.]

It is a practical outlook, not a hollow desire for generality, that prompts our interest in the Laplace operator $\partial_1^2 + \cdots + \partial_N^2$ for all N. The case $N = 1$ has the virtue that easy arguments (as in Remark 1.5 and Exercise 1.17) may point the way to results that are true in higher dimensions. In $\mathbb{R}^2$, there is the golden result that the real and imaginary parts of a holomorphic or analytic function $u+iv$, of the complex variable $z = x_1 + ix_2$, satisfy the Laplace equation; this leads to many explicit solutions, some of which are even relevant to physical situations. (Such situations involve long cylindrical bodies with axis normal to the plane being considered, and of arbitrary cross-section.) The prevalence of the Laplace operator in $\mathbb{R}^3$ has been illustrated, in a small way, in §A.1. The story does not end there; wholly concrete problems, originally set in $\mathbb{R}^3$, can lead to the Laplace equation in $\mathbb{R}^5$ (Exercise A.25).

Let $\mathscr{B}_N(c, \rho)$ denote, for the moment, the ball in $\mathbb{R}^N$ with centre c and radius ρ, and let $|\partial \mathscr{B}_N(c, \rho)|$ denote the surface area of its spherical boundary. [Note that $\mathscr{B}_1(c, \rho)$ is the interval $(c - \rho, c + \rho)$ of length or 'volume' 2ρ; hence the 'surface area' $|\partial \mathscr{B}_1(c, \rho)|$ of the corresponding sphere is $\mathrm{d}(2\rho)/\mathrm{d}\rho = 2$.] Our extension $K(x - c; N)$ of the $\Phi(x)$ in (A.1) is the result of three conditions: $K(x; N)$ is to depend only on $|x|$ and N; we demand that

$$-\frac{\mathrm{d}K(x; N)}{\mathrm{d}r} = \frac{1}{|\partial \mathscr{B}_N(0, r)|} \qquad \text{whenever} \ \ r := |x| > 0; \qquad \text{(A.17)}$$

and $K(x; N)$ is to vanish for $r = 0$ if $N = 1$, for $r = 1$ if $N = 2$, and as $r \to \infty$ if $N \geq 3$. Accordingly,

$$K(x; N) := \begin{cases} -\frac{1}{2}|x| & \text{in } \mathbb{R}, \\[2mm] \frac{1}{2\pi} \log \frac{1}{|x|} & \text{in } \mathbb{R}^2 \setminus \{0\}, \\[2mm] \kappa_N \frac{1}{|x|^{N-2}} & \text{in } \mathbb{R}^N \setminus \{0\}, \quad N \geq 3, \end{cases} \qquad \text{(A.18a)}$$

where

$$\kappa_N = \frac{1}{(N-2)|\partial \mathscr{B}_N(0, 1)|} \ \ (N \neq 2), \qquad |\partial \mathscr{B}_N(0, 1)| = \frac{N\pi^{N/2}}{(N/2)!}. \ \ \text{(A.18b)}$$

The formula for the surface area of the unit sphere is a result of Exercise 1.14. Note that the formula stated for $N \geq 3$, in (A.18a), is also valid for $N = 1$. The basic properties (A.2) and (A.3) of Φ are unchanged, and are proved exactly as before.

$$\Delta K(x - c; N) = 0 \qquad \text{for } x \in \mathbb{R}^N \setminus \{c\}. \qquad \text{(A.19)}$$

If Ω is a bounded open set in $\mathbb{R}^N$ that contains the point c, and $\partial \Omega$ is of class C^1, then

$$-\int_{\partial \Omega} \frac{\partial K(x - c; N)}{\partial n} = 1; \qquad \text{(A.20)}$$

here n denotes the outward unit normal on $\partial \Omega$.

The Newtonian potential Let G be a bounded open set in $\mathbb{R}^N$ and let

$f : G \to \mathbb{R}$ be a given function. Provided that the integral exists for all $x \in \mathbb{R}^N$, we call

$$u(x) := \int_G K(x - \xi; N) f(\xi) \, d\xi \qquad (x \in \mathbb{R}^N, \quad d\xi = d\xi_1 \dots d\xi_N) \quad (A.21)$$

the *Newtonian potential of the density function* f. In classical physics the following hypothesis is made. The rule that

$$\frac{1}{\kappa} \sum_{j=1}^m K(x - \xi^j; 3) \, Q_j, \qquad x \in \mathbb{R}^3 \setminus \{\xi^1, \dots, \xi^m\},$$

is the electric potential at x of charges Q_j at the points ξ^j, extends to the rule that $(1/\kappa)u(x)$ is the potential, *at all* $x \in \mathbb{R}^3$, of charge density f. Here f is measured in units of charge per volume. Essentially the same hypothesis is made for the gravitational potential of a density f (now measured in units of mass per volume). One often calls ξ^j or ξ a *source point*, and x a *field point*.

The new feature, that the potential is defined at points where there is charge or mass, prompts the question: is there a simple relationship between u and f at points of G? A short answer is that the *Poisson equation* $-\Delta u = f$ holds in G, provided that f is sufficiently smooth; this will be proved in §A.5. In this section we shall give an answer that is less direct but is both more general and easier to prove: if f has a certain integrability property, then a weak form of the Poisson equation holds in G.

Notation Henceforth we omit the label N when its absence causes no confusion. For some purposes the notation in

$$u(x_0) = \int_G K(x_0 - x) f(x) \, dx, \qquad x_0 \in \mathbb{R}^N, \qquad (A.22a)$$

is more convenient than that in (A.21); accompanying symbols are

$$r := |x|, \quad r_0 := |x_0|, \quad R := |x_0 - x|, \qquad (A.22b)$$

so that, for example, $K(x - x_0) = \kappa_N R^{-N+2}$ if $N \neq 2$.

There are a few cases for which the integral defining the Newtonian potential can be evaluated explicitly (for which the potential can be found 'in closed form'). Of these, the case of constant density in a ball is the simplest and perhaps the most valuable; we state the result and then explain its derivation in a sequence of remarks.

The potential of unit density in a ball Let $B := \mathscr{B}(0, a)$; then

$$u_B(x) := \int_B K(x - \xi)\, d\xi = \begin{cases} -\dfrac{r^2}{2N} + c_N(a), & r \le a, \\ |B| K(x), & r \ge a, \end{cases} \tag{A.23a}$$

where $|B| = \pi^{N/2} a^N / (N/2)!$ denotes the volume of B, and

$$c_N(a) = \begin{cases} -\frac{1}{2}a^2 & \text{if } N = 1, \\ \frac{1}{2}a^2 \left(\log \dfrac{1}{a} + \frac{1}{2} \right) & \text{if } N = 2, \\ \frac{1}{2}a^2 \dfrac{1}{N - 2} & \text{if } N \ge 3. \end{cases} \tag{A.23b}$$

Remarks 1. For $N = 1$ the result (A.23) follows from any easy exercise in integration. The same is true for $N = 2$ if we use the notation in (A.22) and the series

$$\log \frac{1}{|z - z_0|} = \operatorname{Re} \log \frac{1}{z - z_0} = \log \frac{1}{r_0} + \sum_{n=1}^{\infty} \frac{1}{n} \left(\frac{r}{r_0} \right)^n \cos n(\theta - \theta_0), \left.\begin{array}{c} \\ \\ \\ \end{array}\right\} \\ r < r_0, \tag{A.24}$$

where $z = x_1 + ix_2 = re^{i\theta}$. [If $r > r_0$, we interchange z and z_0.] For $N = 3$ we choose, for source points x, spherical co-ordinates r, θ, λ such that the axis $\{\theta = 0\}$ passes through the field point x_0 (Figure A.5). The Fubini theorem and a theorem about co-ordinate transformations (Apostol 1974, p.421; Weir 1973, p.158) allow us to write

$$u_B(x_0) = \frac{1}{4\pi} \int_0^a r^2\, dr \int_0^\pi \frac{\sin \theta\, d\theta}{(r^2 + r_0^2 - 2rr_0 \cos \theta)^{1/2}} \int_{-\pi}^\pi d\lambda\,,$$

and this is an elementary integral.

These integrations yield (A.23) for $N = 1, 2, 3$ (the properly sceptical reader is invited to verify this). The function u_B for $N = 3$ is shown in Figure A.6. One easily checks that, for all $N \le 3$, both u_B and du_B / dr are continuous at $r = a$; hence $u_B \in C^1(\mathbb{R}^N)$ for $N \le 3$.

2. Observe that for $r \ge a$ the potential u_B is merely that of a point source at the origin of strength $|B|$, and that

$$-\Delta u_B = \begin{cases} (\partial_1^2 + \cdots + \partial_N^2) \dfrac{x_1^2 + \cdots + x_N^2}{2N} = 1 & \text{if } r < a, \\ 0 & \text{if } r > a, \end{cases} \tag{A.25}$$

at least for $N \le 3$.

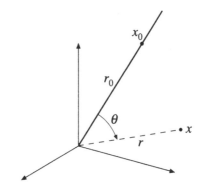

Fig. A.5.

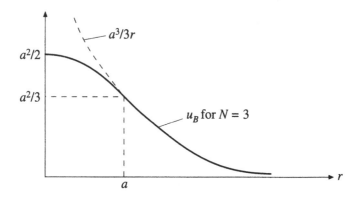

Fig. A.6.

We now proceed backwards. The result (A.25) can be regarded as an ordinary differential equation in terms of r, and one can use this equation to construct, with relatively little labour, a tentative formula for all values of N. However, we shall exclude the case $N = 1$, because a third condition 'at infinity' would be required in what follows.

Using a formula in Exercise 1.18 for $\triangle$ acting on functions of r alone, we seek $V(r)$ such that

$$\triangle V = \frac{1}{r^{N-1}} \frac{\mathrm{d}}{\mathrm{d}r} \left(r^{N-1} \frac{\mathrm{d}}{\mathrm{d}r} \right) V = \begin{cases} -1, & 0 < r < a, \\ 0, & r > a, \end{cases}$$

$$V(r) - \frac{\pi a^2}{2\pi} \log \frac{1}{r} \to 0 \text{ as } r \to \infty \text{ if } N = 2,$$

$$V(r) \to 0 \text{ as } r \to \infty \text{ if } N \geq 3,$$

and such that $V \in C^1[0, \infty) \cap C^2([0, \infty) \setminus \{a\})$. This last condition fixes certain constants that are unknown initially and rules out singular behaviour at the origin. An elementary calculation now yields

$$
\text{for } N = 2: \quad V(r) = \begin{cases} -\dfrac{r^2}{4} + \dfrac{1}{2}a^2 \left(\log \dfrac{1}{a} + \dfrac{1}{2}\right), & r \le a, \\[3mm] \dfrac{1}{2}a^2 \log \dfrac{1}{r}, & r \ge a; \end{cases}
$$

$$
\text{for } N \ge 3: \quad V(r) = \begin{cases} -\dfrac{r^2}{2N} + \dfrac{1}{2}a^2 \dfrac{1}{N-2}, & r \le a, \\[3mm] \dfrac{a^N}{N(N-2)} \dfrac{1}{r^{N-2}}, & r \ge a. \end{cases}
$$

These are the functions on the right-hand side of (A.23a).

3. It remains to prove that $u_B(x) = V(r)$ for $N \ge 4$, where $u_B(x)$ denotes the integral in (A.23a). This follows from Theorem A.2 below, which gives a second characterization of the Newtonian potential when f and ∂G are exceptionally smooth. One easily verifies that, with $G = B$ and $v(x) = V(r)$, the hypotheses of Theorem A.2 are satisfied. The theorem has other applications, because one may well ask: when is a given function v the Newtonian potential of $-\Delta v|_G$?

4. We note the following in connection with condition (c) of Theorem A.2. If, as $r := |x| \to \infty$, $|\nabla w(x)| = O(r^{-1-m})$ and $w(x) \to 0$, where the constant $m > 0$, then $w(x) = O(r^{-m})$ as $r \to \infty$.

The reason is this: there are positive constants C_1 and C_2 such that, for $r \ge C_1$,

$$
|w(x)| = \left| \int_1^\infty \frac{dw(tx)}{dt} \, dt \right| = \left| \int_1^\infty x \cdot (\nabla w)(tx) \, dt \right|
$$
$$
\le \int_1^\infty |x| C_2 t^{-1-m} |x|^{1-m} \, dt = \frac{C_2 r^{-m}}{m}.
$$

5. In Theorem A.2 the condition $\Delta v \in L_p(G)$, $p > N/2$, is needed because condition (a) does not restrict sufficiently the behaviour of $(\Delta v)(x)$ as x approaches ∂G.

Theorem A.2 *Let G be a bounded open subset of $\mathbb{R}^N$ with $N \ge 2$; suppose that either ∂G is of class C^1 or G is listed in Remark D.4. Consider a function $v : \mathbb{R}^N \to \mathbb{R}$ with the following properties.*

(a) $v \in C^1(\mathbb{R}^N) \cap C^2(\mathbb{R}^N \setminus \partial G)$.

(b) $\triangle v \in L_p(G)$ *for some* $p > N/2$, *and* $\triangle v = 0$ *in* $\mathbb{R}^N \setminus \overline{G}$.

(c) *As* $r := |x| \to \infty$,

$$\nabla v(x) = O(r^{-N+1}) \quad and \quad v(x) \to 0 \quad if \ N \geq 3,$$

$$\nabla\{v(x) - cK(x)\} = O(r^{-2}) \quad and \quad v(x) - cK(x) \to 0 \quad if \ N = 2,$$

for some constant c.

Then v *is the Newtonian potential of* $-\triangle v|_G$.

Proof (i) In this first step, v need not be the function specified above. Let Ω be a bounded open set with $\partial\Omega$ of class C^1, or else a set listed in Remark D.4, and let v and w be in $C^2(\overline{\Omega})$. Then the divergence theorem for the vector field $v\nabla w$ states that

$$\int_\Omega \{\nabla v \cdot \nabla w + v\triangle w\} = \int_{\partial\Omega} v\frac{\partial w}{\partial n}.$$

Interchanging v and w, then subtracting the result, we obtain the *Green identity*

$$\int_\Omega \{v\triangle w - w\triangle v\} = \int_{\partial\Omega} \left\{ v\frac{\partial w}{\partial n} - w\frac{\partial v}{\partial n} \right\}. \qquad (A.26)$$

(ii) Let x_0 be a given field point in $\mathbb{R}^N \setminus \partial G$, arbitrary apart from $x_0 \notin \partial G$, but fixed henceforth. If $x_0 \in G$, we define (Figure A.7)

$$G_\varepsilon := G \setminus \overline{\mathscr{B}(x_0, \varepsilon)}, \quad H_A := \mathscr{B}(0, A) \setminus \overline{G},$$

where ε is so small that $\overline{\mathscr{B}(x_0, \varepsilon)} \subset G$ and A is so large that $\overline{G} \subset \mathscr{B}(0, A)$. If $x_0 \in \mathbb{R}^N \setminus \overline{G}$, we define H_A as before, but with A now so large that $\overline{G} \cup \{x_0\} \subset \mathscr{B}(0, A)$, and use the sets G and

$$H_{A,\varepsilon} := H_A \setminus \overline{\mathscr{B}(x_0, \varepsilon)},$$

where ε is so small that $\overline{\mathscr{B}(x_0, \varepsilon)} \subset H_A$. From now until step (v), we suppose that $x_0 \in G$; for the other case, the argument is changed only slightly.

(iii) Let $G_{\varepsilon,m}$ and $H_{A,m}$ be mth approximations to G_ε and H_A, respectively, in the sense of Theorem D.9. We apply the Green identity (A.26) to $G_{\varepsilon,m}$ and $H_{A,m}$, choosing v to be as in the statement of the present theorem, and choosing $w(x) = K(x - x_0) =: K_0(x)$, say. [We cannot apply the Green identity immediately to G_ε and H_A, because v need not

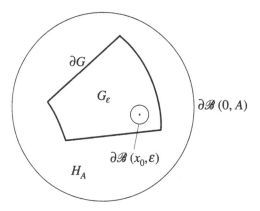

Fig. A.7.

be in $C^2(\overline{G}_\varepsilon)$ and in $C^2(\overline{H}_A)$.] Since $\triangle K_0 = 0$ in $\mathbb{R}^N \setminus \{x_0\}$ and $\triangle v = 0$ in $\mathbb{R}^N \setminus \overline{G}$, there results

$$-\int_{G_{\varepsilon,m}} K_0 \triangle v = \int_{\partial G_{\varepsilon,m}} \left\{ v\frac{\partial K_0}{\partial n} - K_0 \frac{\partial v}{\partial n} \right\}, \qquad (A.27a)$$

$$0 = \int_{\partial H_{A,m}} \left\{ v\frac{\partial K_0}{\partial n} - K_0 \frac{\partial v}{\partial n} \right\}. \qquad (A.27b)$$

In order to apply Theorem D.9, we must verify that

(α) $v\partial_j K_0 - K_0 \partial_j v \in C(\overline{G}_\varepsilon) \cap C(\overline{H}_A)$ for each $j \in \{1,\ldots,N\}$,

(β) $K_0 \triangle v \in L_1(G_\varepsilon)$.

Now, K_0 is infinitely smooth outside $\mathscr{B}(x_0, \varepsilon)$, so that the hypothesis $v \in C^1(\mathbb{R}^N)$ implies (α), and hypothesis (b) implies more than (β) [namely, that $K_0 \triangle v$ is in $L_p(G_\varepsilon)$ with $p > N/2$, which is better than membership of $L_1(G_\varepsilon)$ because G_ε is bounded; see Chapter 0, (xiv)].

Letting $m \to \infty$, using Theorem D.9, and adding the two identities which result from (A.27a,b), we find that the two contributions of ∂G cancel each other, because v and K_0 are in $C^1(\mathbb{R}^N \setminus \{x_0\})$ and because the outward unit normals n to ∂G_ε and ∂H_A are in opposite directions at each point of ∂G where they exist. Accordingly,

$$-\int_{G_\varepsilon} K_0 \triangle v = \int_{\partial B_\varepsilon} + \int_{S_A} \left\{ v\frac{\partial K_0}{\partial n} - K_0 \frac{\partial v}{\partial n} \right\}, \qquad (A.28)$$

where $B_\varepsilon := \mathscr{B}(x_0, \varepsilon)$, $S_A := \partial\mathscr{B}(0, A)$ and n points towards x_0 on ∂B_ε.

(iv) Now let $\varepsilon \to 0$. Define χ_ε to be the characteristic function of G_ε

[equal to 1 on G_ε, and to 0 on $\mathbb{R}^N \setminus G_\varepsilon$]; then

$$\int_{G_\varepsilon} K_0 \Delta v = \int_G \chi_\varepsilon K_0 \Delta v \to \int_G K_0 \Delta v$$

by the Lebesgue dominated convergence theorem, since, with $\delta := \frac{1}{2} \operatorname{dist}(x_0, \partial G)$, we have K_0 integrable and Δv bounded on B_δ, while K_0 is bounded and Δv is integrable on $G \setminus B_\delta$. Next,

$$\left\{ \int_{\partial B_\varepsilon} v \frac{\partial K_0}{\partial n} \right\} - v(x_0) = \int_{\partial B_\varepsilon} \{v(x) - v(x_0)\} \frac{\partial K_0(x)}{\partial n} \, dS(x) \to 0,$$

because $\partial K_0/\partial n = 1/|\partial B_\varepsilon|$ on ∂B_ε by the definitive property (A.17) of the Newtonian kernel and because $v(x) \to v(x_0)$ when $x \in \partial B_\varepsilon$. Finally,

$$\int_{\partial B_\varepsilon} K_0 \frac{\partial v}{\partial n} \to 0,$$

because $K_0 \partial v/\partial n = o(\varepsilon^{-N+1})$ uniformly on ∂B_ε, while $|\partial B_\varepsilon| = \text{const}.\varepsilon^{N-1}$. Thus (A.28) yields

$$-\int_G K_0 \Delta v = v(x_0) + \int_{S_A} \left\{ v \frac{\partial K_0}{\partial n} - K_0 \frac{\partial v}{\partial n} \right\}. \tag{A.29}$$

(v) We now show that the last integral in (A.29) tends to zero as $A \to \infty$, for each fixed $x_0 \in \mathbb{R}^N \setminus \partial G$. First, $K_0(x)$ is close to $K(x)$ when r is sufficiently large; in fact, for $r \geq 2r_0$ and for each j we have

$$\left| (\partial_j K)(x - x_0) - (\partial_j K)(x) \right| \leq |x_0| \max_{0 \leq t \leq 1} \left| (\nabla \partial_j K)(x - tx_0) \right| = O(r^{-N}),$$

because $|x - tx_0| \geq \frac{1}{2}r$. Thus $\partial_j K_0(x)$ is $O(r^{-N+1})$ [the same order as $\partial_j K(x)$] and, for $N \geq 3$, $K_0(x)$ is $O(r^{-N+2})$ by Remark 4. We proceed to apply hypothesis (c), Remark 4 and the observation just made about K_0.

For $N \geq 3$ and $r = A \to \infty$, both terms $v \partial K_0/\partial r$ and $K_0 \partial v/\partial r$ are $O(A^{-2N+3})$, while the surface area $|S_A|$ is $O(A^{N-1})$; then the integral over S_A is $O(A^{-N+2})$ and tends to zero. For $N = 2$ and $r = A \to \infty$,

$$v \frac{\partial K_0}{\partial r} - K_0 \frac{\partial v}{\partial r} = \left\{ cK(x) + O(r^{-1}) \right\} \left\{ \frac{dK(x)}{dr} + O(r^{-2}) \right\}$$
$$- \left\{ K(x) + O(r^{-1}) \right\} \left\{ c \frac{dK(x)}{dr} + O(r^{-2}) \right\}$$
$$= O(A^{-2} \log A),$$

because the dominant terms cancel. Since $|S_A| = 2\pi A$, the integral over S_A is $O(A^{-1} \log A)$ and tends to zero. Therefore (A.29) reduces to

$$v(x_0) = -\int_G K_0 \Delta v \qquad (x_0 \in \mathbb{R}^N \setminus \partial G). \tag{A.30}$$

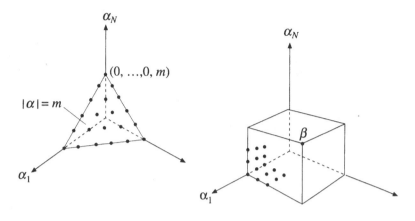

Fig. A.8.

(vi) Finally, we use continuity to extend the result (A.30) to points $x_0 \in \partial G$. The function v is more than continuous on $\mathbb{R}^N$, by hypothesis. Theorem A.6 will show (independently of the present theorem) that the condition $\triangle v \in L_p(G)$, $p > N/2$, makes the integral in (A.30) a continuous function of x_0 on $\mathbb{R}^N$. Therefore the function $v + \int_G K_0 \triangle v$, being zero outside ∂G and continuous on $\mathbb{R}^N$, must vanish also on ∂G.

□

To make further progress, we need (a) a notation that enables us to write partial derivatives of arbitrary order without undue labour, (b) bounds for all derivatives of the Newtonian kernel K.

Definition A.3 We write

$$\partial^\alpha := \partial_1^{\alpha_1} \partial_2^{\alpha_2} \dots \partial_N^{\alpha_N}, \qquad x^\alpha := x_1^{\alpha_1} x_2^{\alpha_2} \dots x_N^{\alpha_N};$$

the N-tuple $\alpha := (\alpha_1, \alpha_2, \dots, \alpha_N)$ of integers $\alpha_j \geq 0$ is called a *multi-index* of length N; its *order* is $|\alpha| := \alpha_1 + \alpha_2 + \dots + \alpha_N$. (Figure A.8 shows a set of multi-indices of fixed order, and also illustrates the partial ordering $\alpha \leq \beta$ introduced in Exercise A.23.) It is to be understood that $x_j^0 = 1$ even when $x_j = 0$. The label $|\alpha| = m$, when appended to $\sum$, means summation over all multi-indices of order m, the length N being implied by the context. For example, if $N = 2$,

$$\sum_{|\alpha|=3} c_\alpha x^\alpha = c_{(3,0)} x_1^3 + c_{(2,1)} x_1^2 x_2 + c_{(1,2)} x_1 x_2^2 + c_{(0,3)} x_2^3.$$

□

Lemma A.4 *For all multi-indices β if $N \neq 2$, or for $|\beta| \geq 1$ if $N = 2$,*

$$\left| \partial^\beta K(x; N) \right| \leq C_\beta |x|^{-N+2-|\beta|}, \qquad x \neq 0, \qquad (A.31)$$

where the constant C_β depends only on β (the length of which specifies N).

Proof We shall prove by induction that (for β as above and $r := |x|$)

$$\partial^\beta K(x; N) = r^{-N+2-2|\beta|} P_\beta(x), \qquad x \neq 0, \qquad (A.32a)$$

where P_β is a homogenous polynomial in $x_1, \ldots, x_N$ of degree $|\beta|$ (possibly the zero polynomial):

$$P_\beta(x) = \sum_{|\alpha| = |\beta|} c_{\beta, \alpha} \, x^\alpha. \qquad (A.32b)$$

[Many of the coefficients $c_{\beta, \alpha}$ are zero; for example,

$$P_{(1)}(x) = -\tfrac{1}{2}x_1 = -\tfrac{1}{2}x, \qquad P_{(2)}(x) = 0,$$
$$P_{(1,0)}(x) = -x_1/2\pi, \qquad P_{(1,1)}(x) = x_1 x_2/\pi,$$
$$P_{(0,1,0)}(x) = -x_2/4\pi, \qquad P_{(0,0,2)}(x) = (-x_1^2 - x_2^2 + 2x_3^2)/4\pi.]$$

Assume that (A.32) is true for $|\beta| = m$ (it is true for $|\beta| = 0$ if $N \neq 2$, or for $|\beta| = 1$ if $N = 2$). Then

$$\partial_j \partial^\beta K(x; N)$$
$$= (-N + 2 - 2m) r^{-N+1-2m} (x_j/r) P_\beta(x) + r^{-N+2-2m} \partial_j P_\beta(x)$$
$$= r^{-N-2m} \left\{ (-N + 2 - 2m) x_j P_\beta(x) + (x_1^2 + \cdots + x_N^2) \partial_j P_\beta(x) \right\},$$

where $\partial_j P_\beta$ is a homogenous polynomial in $x_1, \ldots, x_N$ of degree $m - 1$. Hence the factor of r^{-N-2m} is a homogeneous polynomial in $x_1, \ldots, x_N$ of degree $m + 1$. This proves (A.32) for $|\beta| = m + 1$ and hence for all β (except $\beta = 0$ if $N = 2$). Finally, (A.32) implies (A.31) because $|P_\beta(x)| \leq \text{const.} |x|^{|\beta|}$. $\qquad \square$

We now extend the density function f in (A.21) to have domain $\mathbb{R}^N$ by setting

$$f(x) := 0 \quad \text{if} \quad x \in \mathbb{R}^N \setminus G.$$

The *support* of f, written $\operatorname{supp} f$ and defined fully in Chapter 0, (iv), is the smallest closed set outside which f equals zero.

The next theorem concerns the set $\mathbb{R}^N \setminus \operatorname{supp} f$, which always contains $\mathbb{R}^N \setminus \overline{G}$; it also contains any open subset of G on which f happens to be zero. For the moment, we relax the condition that the potential be defined at all points of $\mathbb{R}^N$.

Theorem A.5 *If* $f \in L_1(G)$, *then the Newtonian potential* u *of* f *has the properties*

$$u \in C^\infty(\mathbb{R}^N \setminus \operatorname{supp} f) \quad and \quad \triangle u = 0 \quad in \quad \mathbb{R}^N \setminus \operatorname{supp} f.$$

Proof (i) The essence of the matter is the formula

$$\left.\begin{array}{c} \left(\partial^\alpha u\right)(x_0) = \int_G \left(\partial^\alpha K\right)(x_0 - x) f(x) \, \mathrm{d}x \\[4pt] \text{for all} \quad x_0 \in \mathbb{R}^N \setminus \operatorname{supp} f \quad \text{and all} \quad \alpha. \end{array}\right\} \tag{A.33}$$

Since $\mathbb{R}^N \setminus \operatorname{supp} f$ is open, we have (for $x_0 \in \mathbb{R}^N \setminus \operatorname{supp} f$)

$$\delta_0 := \operatorname{dist}(x_0, \operatorname{supp} f) > 0,$$

$$\delta_0 \le |x_0 - x| \le \delta_0 + \operatorname{diam} G \qquad \text{for all} \quad x \in \operatorname{supp} f,$$

so that nothing can go wrong in merely formal, repeated differentiation. However, as preparation for more dangerous cases, we shall justify the formula (A.33). The symbol $\left(\partial^\alpha K\right)(x_0 -.)$ will denote the function with values $\left(\partial^\alpha K\right)(x_0 - x)$; the point x_0 is a parameter, and is outside $\operatorname{supp} f$ throughout the proof.

(ii) First, does the integral in (A.33) always exist? Since $f \in L_1(G)$, it is sufficient that $\left(\partial^\alpha K\right)(x_0 -.)$ be bounded and measurable on $\operatorname{supp} f$, and this is the case because (A.32) shows [when we replace the x there by $x_0 - x$] that $\left(\partial^\alpha K\right)(x_0 -.)$ is continuous on the compact set $\operatorname{supp} f$.

(iii) Next, is the integral in (A.33) the result of operating on u with ∂^α? Assume that (A.33) is true for $|\alpha| = m$ (it is true for $|\alpha| = 0$); we prove it for $|\alpha| = m + 1$ by calculating $\left(\partial_j \partial^\alpha u\right)(x_0)$. Let

$$h := (h_j \delta_{ji})_{i=1}^N = (0, \ldots, 0, h_j, 0, \ldots, 0) \qquad \text{with} \quad h_j \neq 0,$$

and define the difference kernel

$$k(x_0, x, h) \quad := \quad \frac{1}{h_j} \big\{ \left(\partial^\alpha K\right)(x_0 + h - x) - \left(\partial^\alpha K\right)(x_0 - x)$$

$$- h_j \left(\partial_j \partial^\alpha K\right)(x_0 - x) \big\}, \tag{A.34}$$

because then, if $|h| \leq \frac{1}{2}\delta_0$,

$$\left| \frac{(\partial^\alpha u)(x_0 + h) - (\partial^\alpha u)(x_0)}{h_j} - \int_G (\partial_j \partial^\alpha K)(x_0 - x) f(x) \, dx \right|$$

$$= \left| \int_G k(x_0, x, h) f(x) \, dx \right|$$

$$\leq \|k(x_0, ., h) \mid L_\infty\| \int_G |f|, \tag{A.35}$$

where, in the present case,

$$\|k(x_0, ., h) \mid L_\infty\| = \max_{x \in \mathrm{supp} f} |k(x_0, x, h)|.$$

If this norm of k tends to zero as $h_j \to 0$, then the inequality (A.35) justifies the formula (A.33) for $|\alpha| = m + 1$. Now define

$$g(t) := (\partial^\alpha K)(x_0 + th - x), \qquad 0 \leq t \leq 1,$$

because then

$$k(x_0, x, h) = \frac{1}{h_j} \{ g(1) - g(0) - g'(0) \} = \frac{1}{h_j} \int_0^1 (1 - t) \, g''(t) \, dt.$$

Referring to (A.31), we let $\Gamma_n := \max_{|\beta|=n} C_\beta$, and bound k as follows.

$$|g''(t)| = \left| h_j^2 (\partial_j^2 \partial^\alpha K)(x_0 + th - x) \right| \leq h_j^2 \Gamma_{|\alpha|+2} \left(\tfrac{1}{2}\delta_0 \right)^{-N-|\alpha|}$$

because $|x_0 + th - x| \geq \frac{1}{2}\delta_0$; also $|\alpha| = m$. Accordingly,

$$\|k(x_0, ., h) \mid L_\infty\| \leq \tfrac{1}{2} |h_j| \, \Gamma_{m+2} \left(\tfrac{1}{2}\delta_0 \right)^{-N-m} \to 0 \quad \text{as} \quad h_j \to 0. \tag{A.36}$$

This completes the inductive proof of (A.33).

(iv) The continuity of $\partial^\alpha u$ in $\mathbb{R}^N \setminus \mathrm{supp} f$ follows from the continuity of $\partial^\alpha K$ away from its singular point. Again let $h \in \mathbb{R}^N$, with magnitude $|h| \leq \frac{1}{2}\delta_0$; now let h have arbitrary direction. By application of (A.31) to $\nabla \partial^\alpha K$, or else by (A.32),

$$\left| (\partial^\alpha K)(x_0 + h - x) - (\partial^\alpha K)(x_0 - x) \right| \leq A_\alpha \delta_0^{-N+1-|\alpha|} |h|,$$

where the constant A_α depends only on α. Consequently

$$\left| (\partial^\alpha u)(x_0 + h) - (\partial^\alpha u)(x_0) \right| \leq M(x_0, \alpha, f) \, |h|,$$

where

$$M(x_0, \alpha, f) := A_\alpha \delta_0^{-N+1-|\alpha|} \int_G |f|.$$

(v) That $\triangle u = 0$ in $\mathbb{R}^N \setminus \mathrm{supp} f$ follows from (A.33) and the fact that $(\triangle K)(x_0 - x) = 0$ whenever $x_0 \neq x$. $\qquad \square$

The small-ball technique This is a reliable, although slightly laborious, method that we use repeatedly henceforth. Consider the function in (A.34) with $|\alpha| = 0$ and $h := (h_1, 0, \ldots, 0)$, $h_1 \neq 0$, so that

$$k(x_0, x, h) := \frac{1}{h_1} \left\{ K(x_0 + h - x) - K(x_0 - x) - h_1(\partial_1 K)(x_0 - x) \right\},$$

$$(A.37)$$

where now x_0 may be in the support of f. We bound some specified norm of k (or of a function like k) as follows.

(a) For $x \in \mathscr{B}(x_0, 2|h|)$, the three terms of k are treated separately; each makes a small contribution.

(b) For $x \notin \mathscr{B}(x_0, 2|h|)$, the function k as a whole is bounded, as it was in the proof of Theorem A.5, by means of a formula for the remainder of a short Taylor series.

In this context we write

$$R := |x - x_0|, \qquad R_h := |x - x_0 - h| \qquad (A.38)$$

(Figure A.9). Also, Γ will be used as a symbol for any positive constant independent of x_0, x, h, f and G (but possibly depending on N and, in Theorem A.6, the exponent p). Thus Γ may have different values at successive appearances within a single line.

The estimates of the method are simple applications of the following inequality, usually with $c = x_0$ or $c = x_0 + h$. Suppose that $\Omega \subset \mathscr{B}(c, M)$ in $\mathbb{R}^N$, that $\varphi : \Omega \to \mathbb{R}$ is measurable and that

$$|\varphi(x)| \leq \psi(|x - c|) \quad \text{almost everywhere in } \Omega,$$

where ψ is defined and non-negative on $(0, M)$, and the function with values $R^{N-1}\psi(R)$ belongs to $L_1(0, M)$. Then

$$\int_\Omega |\varphi(x)| \, \mathrm{d}x \leq \int_{\mathscr{B}(c, M)} \psi(|x - c|) \, \mathrm{d}x = |\partial\mathscr{B}(0, 1)| \int_0^M \psi(R) R^{N-1} \, \mathrm{d}R.$$

$$(A.39)$$

Equality of the last two integrals follows, as it did in Remark 1 after (A.23), from Fubini's theorem and a theorem about co-ordinate transformations (Apostol 1974, p.421; Weir 1973, p.158).

Theorem A.6 *Let G be a bounded open subset of $\mathbb{R}^N$ with $N \geq 2$. If $f \in L_p(G)$ for some $p > N/2$, then the Newtonian potential u of f is uniformly continuous on $\mathbb{R}^N$.*

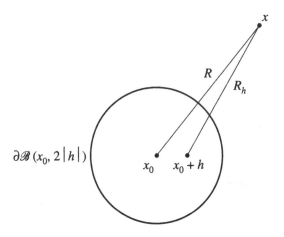

Fig. A.9.

Proof (i) Let q be the Hölder conjugate of p [that is, $1/p + 1/q = 1$]; since $p > N/2$, we have $-Nq + 2q + N > 0$.

The integral in the definition

$$u(x_0) := \int_G K(x_0 - x) f(x) \, \mathrm{d}x, \qquad x_0 \in \mathbb{R}^N,$$

exists because the integrand is measurable and the integral can be bounded by means of the Hölder inequality; the calculations in step (ii) imply that $K(x_0-.) \in L_q(G)$.

(ii) To prove the continuity of u, we introduce the difference kernel

$$a(x_0, x, h) := K(x_0 + h - x) - K(x_0 - x),$$

so that

$$
\begin{aligned}
\left| u(x_0 + h) - u(x_0) \right| &= \left| \int_G a(x_0, x, h) f(x) \, \mathrm{d}x \right| \\
&\le \left\| a(x_0, ., h) \mid L_q(G) \right\| \, \left\| f \mid L_p(G) \right\|.
\end{aligned}
\tag{A.40}
$$

If, as $|h| \to 0$, this norm of a tends to zero independently of $x_0 \in \mathbb{R}^N$, then the uniform continuity of u follows. To this end we apply the small-ball technique; in addition to the symbols R, R_h and Γ [defined by (A.38) and the remark following it], we use a notation that is abbreviated but self-explanatory.

(a) $R < 2|h|$, $N \ge 3$. Note that $R < 2|h|$ implies $R_h < 3|h|$. Since

$K(x_0 + h - x) = \Gamma R_h^{-N+2}$, the definition of a and the triangle inequality for the norm of L_q give

$$\left\{ \int_{R<2|h|} |a|^q \, dx \right\}^{1/q} \leq \Gamma \left\{ \int_0^{3|h|} R_h^{-Nq+2q} R_h^{N-1} \, dR_h \right\}^{1/q}$$

$$+ \Gamma \left\{ \int_0^{2|h|} R^{-Nq+2q} R^{N-1} \, dR \right\}^{1/q}$$

$$= \Gamma \left\{ |h|^{-Nq+2q+N} \right\}^{1/q} \qquad (N \geq 3); \qquad (A.41)$$

the integrals converge at $R_h = 0$ and $R = 0$ because $-Nq + 2q + N > 0$, which we noted in (i). A similar integration, but over the ball

$$\left\{ x \mid R < \operatorname{diam} G + \operatorname{dist}(x_0, G) \right\},$$

shows that $K(x_0 - .) \in L_q(G)$ if $N \geq 3$.

(a') $R < 2|h|$, $N = 2$. Since now $K(x_0 + h - x) = \Gamma \log(1/R_h)$,

$$\left\{ \int_{R<2|h|} |a|^q \, dx \right\}^{1/q} \leq \Gamma \left\{ \int_0^{3|h|} |\log R_h|^q R_h \, dR_h \right\}^{1/q} + \quad \text{similar}$$

$$\leq \Gamma \left\{ |h|^2 \mid \log |h| \mid^q + |h|^2 \right\}^{1/q} \qquad (N = 2). \qquad (A.42)$$

(b) $R \geq 2|h|$, $N \geq 2$. Define $b(t) := K(x_0 + th - x)$, $0 \leq t \leq 1$, because then

$$a(x_0, x, h) = b(1) - b(0) = \int_0^1 b'(t) \, dt.$$

Now,

$$|h| \leq \tfrac{1}{2} R \Rightarrow |x_0 + th - x| \geq |x_0 - x| - |h| \geq \tfrac{1}{2} R,$$

whence, by (A.31),

$$|b'(t)| = \left| \sum_{j=1}^N h_j (\partial_j K)(x_0 + th - x) \right| \leq \Gamma |h| R^{-N+1},$$

so that

$$|a(x_0, x, h)| \leq \Gamma |h| R^{-N+1}.$$

We may suppose that $N/2 < p < N$, because the volume $|G|$ is finite, whence a function in $L_p(G)$ is also in $L_r(G)$ for $1 \leq r \leq p$ [as explained in Chapter 0, (xiv)]. The following constants Γ may then depend on $|G|$. With $p < N$, we have $-Nq + q + N < 0$, so that the next integral will converge at infinity, thereby simplifying the estimate because we need not mention diam G and dist(x_0, G). In fact,

$$\left\{ \int_{R \geq 2|h|} |a|^q \, dx \right\}^{1/q} \leq \Gamma \left\{ |h|^q \int_{2|h|}^{\infty} R^{-Nq+q} \, R^{N-1} \, dR \right\}^{1/q}$$

$$= \Gamma \left\{ |h|^{-Nq+2q+N} \right\}^{1/q}, \qquad (A.43)$$

and the exponent of $|h|$ is the same as the final one in (A.41) and (A.42).

It is legitimate to add the inequality (A.43) to (A.41) or to (A.42) [because $(\alpha + \beta)^{1/p} \leq \alpha^{1/p} + \beta^{1/p}$ for real numbers $\alpha \geq 0$, $\beta \geq 0$ and $p \geq 1$]. The uniform continuity of u follows from (A.40), either (A.41) or (A.42), and (A.43). $\qquad \square$

Theorem A.6 may seem unexciting, but the result is a good one in the following sense. If the condition $p > N/2$ is widened slightly to $p \geq N/2$, then not only may u fail to be continuous, but it may fail to exist on a set that is dense in $\overline{G}$ [that has closure equal to $\overline{G}$]; Exercise A.27 demonstrates this.

We turn now to the question of how the Newtonian potential is related to the Poisson equation $-\Delta u = f$ in G, even when the density function f is so bad that the potential has no derivative. One method is not to explore in detail the behaviour of u and f at each point of G, but to ask how u and f combine with arbitrary *test functions* $\varphi \in C_c^{\infty}(G)$. These are differentiable any number of times and have compact support within G; the set $C_c^{\infty}(\Omega)$ is defined fully in Chapter 0, (vii). In the following definition, Ω need not be bounded and v need not be a Newtonian potential.

Definition A.7 As always, let Ω be an open non-empty subset of $\mathbb{R}^N$. We shall say

(a) that v is a *C^2-solution* of $-\Delta v = g$ in Ω iff $v \in C^2(\Omega)$ and the equation holds pointwise:

$$-\Delta v(x) = g(x) \qquad \text{at each } x \in \Omega; \qquad (A.44)$$

(b) that v is a *generalized solution* of $-\Delta v = g$ in Ω iff $v \in C^1(\Omega)$ and

$$\int_{\Omega} \nabla \varphi \cdot \nabla v = \int_{\Omega} \varphi g \qquad \text{whenever } \varphi \in C_c^{\infty}(\Omega); \qquad (A.45)$$

(c) that v is a *distributional solution* of $-\Delta v = g$ in Ω iff v is locally integrable in Ω (integrable on each compact subset of Ω) and

$$-\int_\Omega (\triangle\varphi)v = \int_\Omega \varphi g \qquad \text{whenever } \varphi \in C_c^\infty(\Omega). \qquad (A.46)$$

Regarding the condition $\varphi \geq 0$, which appears in some statements resembling (A.45) and (A.46), see Remark 8 below. ☐

Remarks 6. An explanation of Definition A.7 seems desirable. Note first that, when $V \in C^2(\Omega)$ and $g \in C(\Omega)$, equation (A.44) is equivalent to the statement

$$\int_\Omega \varphi\{\triangle v + g\} = 0 \qquad \text{whenever } \varphi \in C_c^\infty(\Omega). \qquad (A.44')$$

For, to pass from (A.44) to (A.44'), we multiply (A.44) by φ and integrate over Ω; to pass from (A.44') to (A.44), we use a result in Exercise 1.16.

Second, (A.45), (A.46) and (A.44') are related by integration by parts:

$$\int_\Omega \nabla\varphi \cdot \nabla v = -\int_\Omega \varphi \triangle v \quad \text{if } \varphi \in C_c^\infty(\Omega) \text{ and } v \in C^2(\Omega), \quad (A.47)$$

$$-\int_\Omega (\triangle\varphi)v = \int_\Omega \nabla\varphi \cdot \nabla v \quad \text{if } \varphi \in C_c^\infty(\Omega) \text{ and } v \in C^1(\Omega); \quad (A.48)$$

these formulae are derived from first principles in Remark 7. If v is a C^2-solution of $-\triangle v = g$ in Ω [which implies that $g \in C(\Omega)$], then v is also a generalized solution, by (A.44') and (A.47). Similarly, a generalized solution is also a distributional solution.

Conversely, if a generalized solution should happen to be in $C^2(\Omega)$, then it is also a C^2-solution when $g \in C(\Omega)$; if a distributional solution should happen to be in $C^1(\Omega)$, then it is also a generalized solution.

To sum up, we may say that (A.45) and (A.46) are as close to (A.44) as is possible for functions in the classes that house generalized and distributional solutions.

7. To derive (A.47) and (A.48) we do not need a divergence theorem, because there is no boundary integral in these identities. Define φ to be zero not merely in $\Omega \setminus \operatorname{supp}\varphi$ but in $\mathbb{R}^N \setminus \operatorname{supp}\varphi$; then $\varphi \in C_c^\infty(\mathbb{R}^N)$ and it is sufficient to prove that, for each $j \in \{1,\ldots,N\}$,

$$\int_{\mathbb{R}^N} \partial_j(\varphi\partial_j v) = 0 \quad \text{if } \varphi \in C_c^\infty(\mathbb{R}^N) \text{ and } v \in C^2(\operatorname{supp}\varphi), \quad (A.49)$$

$$\int_{\mathbb{R}^N} \partial_j\{(\partial_j\varphi)v\} = 0 \quad \text{if } \varphi \in C_c^\infty(\mathbb{R}^N) \text{ and } v \in C^1(\operatorname{supp}\varphi); \quad (A.50)$$

summation over j yields (A.47) and (A.48). Fubini's theorem allows us to replace each volume integral in (A.49) and (A.50) by a repeated integral and to integrate first with respect to x_j. Then, if b is so large that the open cube $(-b, b)^N$ contains supp φ,

$$\int_{-\infty}^{\infty} \partial_j(\varphi \partial_j v) \, dx_j = \int_{-b}^{b} \partial_j(\varphi \partial_j v) \, dx_j = \left[\varphi \partial_j v\right]_{x_j=-b}^{b} = 0,$$

and similarly for (A.50).

8. Definition A.7 resembles a part of Definition 2.10, and in Remark 6 we noted the relevance of Exercise 1.16. But there seems to be an inconsistency: in Exercise 1.16 and Definition 2.10 only non-negative test functions are used, $\varphi \in C_c^{\infty}(\Omega)$ and $\varphi \geq 0$.

The answer is that the condition $\varphi \geq 0$ is needed for the inequalities in Exercise 1.16 and for the subsolutions in Definition 2.10, but not for the definition of generalized and distributional *solutions*. In fact, let $(A.46)_+$ denote the statement which results from adding the condition $\varphi \geq 0$ to (A.46); *if v and g are locally integrable in Ω, then (A.46) and $(A.46)_+$ are equivalent statements*; we sketch the proof.

It is obvious that (A.46) implies $(A.46)_+$; to prove that $(A.46)_+$ implies (A.46), one uses the decomposition $\varphi = \varphi^+ + \varphi^-$ of a given function $\varphi \in C_c^{\infty}(\Omega)$ and applies the smoothing operation in Exercise 1.23 to form approximations α_n to φ^+, and $-\beta_n$ to φ^-, that allow use of $(A.46)_+$ for α_n, v, g and for β_n, v, g. Then a limiting procedure yields (A.46): if α_n and β_n are formed with the same smoothing kernel, say a kernel s_n with smoothing radius 2^{-n}, we have

$$
\begin{aligned}
\triangle(\alpha_n - \beta_n)(x) &= \int_{\Omega} (\triangle s_n)(x - y)\left\{\varphi^+(y) + \varphi^-(y)\right\} dy \\
&= \int_{\Omega} s_n(x - y)(\triangle \varphi)(y) \, dy \\
&\rightarrow \triangle \varphi(x) \quad \text{as} \quad n \rightarrow \infty,
\end{aligned}
$$

uniformly over $x \in \Omega$. Thus $\triangle(\alpha_n - \beta_n) \rightarrow \triangle \varphi$ in $L_\infty(\Omega)$ and, similarly, $\alpha_n - \beta_n \rightarrow \varphi$ in $L_\infty(\Omega)$; consequently, $(A.46)_+$ for α_n, v, g and for β_n, v, g implies (A.46) for φ, v, g.

Similar remarks apply to (A.45).

9. The proof of the next theorem involves the identity

$$\varphi(x_0) = -\int_{\mathbb{R}^N} K(x_0 - x) \triangle \varphi(x) \, dx \quad \text{if} \quad \varphi \in C_c^{\infty}(\mathbb{R}^N) \quad \text{and} \quad x_0 \in \mathbb{R}^N.$$

$$(A.51)$$

This is implied by Theorem A.2 if the set G in that theorem is chosen to be a ball (or a cube) containing the support of φ; then the hypotheses of the theorem are satisfied with much room to spare. Indeed, we need only steps (i) and (iv) of the proof of Theorem A.2, and it may be worthwhile to demonstrate this.

Let $B_\varepsilon := \mathscr{B}(x_0, \varepsilon)$, let n be the unit normal on ∂B_ε outward from $\mathbb{R}^N \setminus B_\varepsilon$, hence pointing towards x_0, and let $K_0(x) := K(x - x_0)$. Since φ is infinitely smooth everywhere, and identically zero outside the compact set supp φ,

$$
\int_{\mathbb{R}^N} K(x_0 - x) \triangle \varphi(x) \, dx
$$

$$
= \lim_{\varepsilon \to 0} \int_{\mathbb{R}^N \setminus B_\varepsilon} K_0 \triangle \varphi
$$

$$
= \lim_{\varepsilon \to 0} \left\{ \int_{\partial B_\varepsilon} \left(K_0 \frac{\partial \varphi}{\partial n} - \frac{\partial K_0}{\partial n} \varphi \right) + \int_{\mathbb{R}^N \setminus B_\varepsilon} (\triangle K_0) \varphi \right\}
$$

$$
= -\varphi(x_0)
$$

by step (iv) of the earlier proof and because $\triangle K_0 = 0$ in $\mathbb{R}^N \setminus B_\varepsilon$.

Theorem A.8 *Let G be a bounded open subset of $\mathbb{R}^N$ with $N \geq 2$. If $f \in L_p(G)$ for some $p > N/2$, and $f := 0$ on $\mathbb{R}^N \setminus G$, then the Newtonian potential u of f is a distributional solution of $-\triangle u = f$ in $\mathbb{R}^N$.*

Proof In view of Definition A.7, we must prove that u is locally integrable in $\mathbb{R}^N$, which is certainly true because $u \in C(\mathbb{R}^N)$ by Theorem A.6, and that

$$
-\int_{\mathbb{R}^N} (\triangle \varphi) u = \int_{\mathbb{R}^N} \varphi f \qquad \text{whenever} \quad \varphi \in C_c^\infty(\mathbb{R}^N).
$$

Now, by Fubini's theorem and (A.51),

$$
-\int_{\mathbb{R}^N} (\triangle \varphi) u \quad - \quad -\int_{\mathbb{R}^N} \triangle \varphi(x) \left\{ \int_G K(x - \xi) f(\xi) \, d\xi \right\} dx
$$

$$
= -\int_G f(\xi) \left\{ \int_{\mathbb{R}^N} K(\xi - x) \triangle \varphi(x) \, dx \right\} d\xi
$$

$$
= \int_G f(\xi) \, \varphi(\xi) \, d\xi,
$$

and this last equals $\int_{\mathbb{R}^N} \varphi f$ because f vanishes outside G. $\qquad \square$

Of course, u is also a distributional solution of $-\triangle u = f$ in G [because local integrability in $\mathbb{R}^N$ implies local integrability in G, and because an identity that holds for all test functions $\varphi \in C_c^\infty(\mathbb{R}^N)$ certainly holds for all $\varphi \in C_c^\infty(G)$].

A.3 Continuity of the force field ∇u

In the remainder of this appendix it is to be understood, unless the contrary is stated, that *G is bounded and open in* $\mathbb{R}^N$, that *u is the Newtonian potential of* $f : G \to \mathbb{R}$, and that *formulae like* (A.52) *below hold for all j in* $\{1, \dots, N\}$.

The pointwise form of the Poisson equation $-\triangle u = f$ in G, which will be established in §A.5, is conceptually important and often useful. But in many contexts, such as that of §4.1, we need consider only first derivatives of the potential. It is remarkable that this can be done without reference to the boundary ∂G; Theorem A.11 below remains true when ∂G is pathological. Note also that our daily lives, as particles on the boundary of the earth, would be very different if the gravitational force field were discontinuous at ∂G.

Lemma A.9 *If* $f \in L_p(G)$ *for some* $p > N$, *then*

$$(\partial_j u)(x_0) = \int_G (\partial_j K)(x_0 - x) f(x) \, dx \qquad \text{for all} \quad x_0 \in \mathbb{R}^N. \qquad (A.52)$$

Proof (i) We proceed very much as in the proof of Theorem A.6. Since now $p > N$, the Hölder conjugate q of p [such that $1/p + 1/q = 1$] now satisfies $-Nq + q + N > 0$.

The integral in (A.52) exists because the integrand is measurable and because the Hölder inequality provides a bound; this bound can be inferred from the calculations in step (ii).

(ii) To justify the formula (A.52), we may set $j = 1$ without any real loss of generality, as the proof of Theorem A.5 shows. As in (A.37), let $h := (h_1, 0, \dots, 0)$, $h_1 \neq 0$, and define k by

$$k(x_0, x, h) := \frac{1}{h_1} \left\{ K(x_0 + h - x) - K(x_0 - x) - h_1 (\partial_1 K)(x_0 - x) \right\}.$$

Then

$$\left| \frac{u(x_0 + h) - u(x_0)}{h_1} - \int_G (\partial_1 K)(x_0 - x) f(x) \, dx \right| = \left| \int_G k(x_0, x, h) f(x) \, dx \right|$$

$$\leq \left\| k(x_0, ., h) \mid L_q(G) \right\| \left\| f \mid L_p(G) \right\|,$$

and the lemma will follow if this norm of k tends to zero as $h_1 \to 0$. In fact, we shall prove more: that the norm of $k(x_0, ., h)$ in $L_q(\mathbb{R}^N)$ tends to zero, uniformly over $x_0 \in \mathbb{R}^N$. We use the small-ball technique.

(a) $R < 2|h|$, $N \neq 2$. Since $K(x_0 + h - x) = \mp \Gamma R_h^{-N+2}$ (minus only for $N = 1$), the definition of k and the triangle inequality for the norm of L_q yield

$$
\left\{ \int_{R<2|h|} |k|^q \, dx \right\}^{1/q} \leq \Gamma \left\{ |h|^{-q} \int_0^{3|h|} R_h^{-Nq+2q} R_h^{N-1} \, dR_h \right\}^{1/q}
$$

$$
+ \Gamma \left\{ |h|^{-q} \int_0^{2|h|} R^{-Nq+2q} R^{N-1} \, dR \right\}^{1/q}
$$

$$
+ \Gamma \left\{ \int_0^{2|h|} R^{-Nq+q} R^{N-1} \, dR \right\}^{1/q}
$$

$$
\leq \Gamma \left\{ |h|^{-Nq+q+N} \right\}^{1/q},
$$

where $-Nq + q + N > 0$.

(a') $R < 2|h|$, $N = 2$. Since $K(x_0 + h - x) = \Gamma \log(1/R_h)$, the first integrand becomes $\left| \log R_h \right|^q R_h$, the second becomes $|\log R|^q R$, the third is unchanged. Consequently,

$$
\left\{ \int_{R<2|h|} |k|^q \, dx \right\}^{1/q} \leq \Gamma \left\{ |h|^{2-q} \left| \log |h| \right|^q + |h|^{2-q} \right\}^{1/q},
$$

where $q < 2$.

(b) $R \geq 2|h|$, $N \geq 1$. Let $g(t) := K(x_0 + th - x)$, $0 \leq t \leq 1$; as before,

$$
|h| \leq \tfrac{1}{2} R \Rightarrow |x_0 + th - x| \geq |x_0 - x| - |h| \geq \tfrac{1}{2} R,
$$

so that, by (A.31),

$$
|g''(t)| = \left| h_1^2 (\partial_1^2 K)(x_0 + th - x) \right| \leq h_1^2 \left\{ \max_{|\beta|=2} C_\beta \right\} \left(\tfrac{1}{2} R \right)^{-N},
$$

and

$$
|k(x_0, x, h)| = \left| \frac{1}{h_1} \{ g(1) - g(0) - g'(0) \} \right|
$$

$$
= \left| \frac{1}{h_1} \int_0^1 (1-t) g''(t) \, dt \right| \leq \Gamma |h| R^{-N}.
$$

Just as we assumed in the proof of Theorem A.6 that $N/2 < p < N$, so we may suppose here that $N < p < \infty$; the following constants Γ may then depend on $|G|$. With $p < \infty$, hence $q > 1$, the next integral will converge at infinity and yield a simple estimate. In fact,

$$\left\{ \int_{R \geq 2|h|} |k|^q \, dx \right\}^{1/q} \leq \Gamma \left\{ |h|^q \int_{2|h|}^{\infty} R^{-Nq} R^{N-1} \, dR \right\}^{1/q} \quad (-Nq + N < 0)$$

$$= \Gamma \left\{ |h|^{q-Nq+N} \right\}^{1/q},$$

and the exponent of $|h|$ is the same as the final one in (a) and (a').

Thus $\|k(x_0,.,h) \mid L_q(\mathbb{R}^N)\| \to 0$ independently of x_0, as $h_1 \to 0$. $\qquad \square$

It is now appropriate to distinguish certain continuous functions on $\mathbb{R}^N$ that are decidedly better than arbitrary members of $C(\mathbb{R}^N)$. First, observing that functions in the set $C(\overline{\Omega})$ can be unbounded when Ω is unbounded, we shall define the normed linear space $C_b(\overline{\Omega})$ of *bounded* continuous functions on $\overline{\Omega}$. Second, we introduce Hölder continuity. A function $v : \mathbb{R}^N \to \mathbb{R}$ is *Hölder continuous at the point* c if and only if there are constants $\lambda \in (0,1]$, A and $\delta > 0$ such that

$$|v(x) - v(c)| \leq A|x - c|^{\lambda} \qquad \text{whenever} \ \ x \in \mathscr{B}(c, \delta).$$

The number λ is a *Hölder exponent* and A is a *Hölder constant*. When $\lambda = 1$, the name *Lipschitz* often replaces the name Hölder. Here is an example. If, for some $\alpha \in \mathbb{R}$,

$$w(x) := \begin{cases} \dfrac{1 + \alpha r}{1 + |\log r|}, & r := |x| > 0, \\ 0, & r = 0, \end{cases}$$

then $w \in C(\mathbb{R}^N)$. But $w \in C_b(\mathbb{R}^N)$ only if $\alpha = 0$, and w is not Hölder continuous at the origin, whatever α may be. In the following definition, we demand *uniform* Hölder continuity on $\overline{\Omega}$. [We take suprema over Ω, rather than $\overline{\Omega}$, because this is sufficient in the definition and is traditional.]

Definition A.10 (a) The normed linear space $C_b(\overline{\Omega})$ consists of the norm defined by

$$\|v \mid C_b(\overline{\Omega})\| := \sup_{x \in \Omega} |v(x)|$$

and of the set

$$\left\{ v : \overline{\Omega} \to \mathbb{R} \mid v \in C(\overline{\Omega}), \ \ \|v \mid C_b(\overline{\Omega})\| < \infty \right\}.$$

(b) For given $\lambda \in (0,1]$, *the Hölder constant* of $w : \Omega \to \mathbb{R}$ is

$$[w]_\lambda := \sup \left\{ \frac{|w(x) - w(y)|}{|x - y|^\lambda} \,\middle|\, x, y \in \Omega, \ 0 < |x - y| \le 1 \right\},$$

in which the final constant 1 may be replaced by any positive constant characteristic of Ω or of the problem in hand.

(c) The normed linear space $C_b^{0,\lambda}(\overline{\Omega})$ consists of the norm defined by

$$\left\| v \,\middle|\, C_b^{0,\lambda}(\overline{\Omega}) \right\| := \sup_{x \in \Omega} |v(x)| + [v]_\lambda$$

and of the set

$$\left\{ v \in C_b(\overline{\Omega}) \,\middle|\, [v]_\lambda < \infty \right\}.$$

$\square$

It is seldom necessary to know *the* Hölder constant $[w]_\lambda$ of w; usually one is content with *a* Hölder constant A such that $|w(x) - w(y)| \le A|x - y|^\lambda$ for all x and y as above. The spaces $C_b(\overline{\Omega})$ and $C_b^{0,\lambda}(\overline{\Omega})$ are Banach spaces [are complete]. We shall use neither this fact nor the considerable machinery to which the definition leads. But the definition gives us a convenient way of stating the next result and those in §A.5.

Theorem A.11 *If $f \in L_p(G)$ for some $p \in (N, \infty)$ and $f := 0$ on $\mathbb{R}^N \setminus G$, then the first derivatives of u are bounded and uniformly Hölder continuous:*

$$\partial_j u \in C_b^{0,\lambda}(\mathbb{R}^N), \qquad \text{with } \lambda = 1 - N/p,$$

and u is a generalized solution of $-\triangle u = f$ in $\mathbb{R}^N$.

Proof As in the proof of Lemma A.9, the Hölder conjugate q of p satisfies

$$0 < -Nq + q + N < 1, \tag{A.53}$$

with equality on the right only if $N = 1$. Again we may take $j = 1$.

(i) First we bound $\partial_1 u$. By Lemma A.9,

$$\begin{aligned} |(\partial_1 u)(x_0)| &= \left| \int_G (\partial_1 K)(x_0 - x) f(x) \, dx \right| \\ &\le \left\| (\partial_1 K)(x_0 - \cdot) \,\middle|\, L_q(G) \right\| \left\| f \,\middle|\, L_p(G) \right\| ; \end{aligned}$$

to bound this norm of $\partial_1 K$, we consider two cases. Let $d_0 := \text{dist}(x_0, G)$ and $D := \text{diam}\, G$.

(a) $d_0 \leq 1$. Then $|x - x_0| \leq D + 1$ when $x \in G$; by the left-hand inequality in (A.53),

$$\int_G |\partial_1 K|^q \leq \Gamma \int_0^{D+1} R^{-Nq+q} R^{N-1} \, dR = \Gamma (D+1)^{-Nq+q+N}.$$

(b) $d_0 \geq 1$. Since $d_0 \leq |x - x_0| \leq D + d_0$ when $x \in G$, we have $R \geq 1$ and use the right-hand inequality in (A.53):

$$\int_G |\partial_1 K|^q \leq \Gamma \int_{d_0}^{d_0+D} R^{-Nq+q} R^{N-1} \, dR \qquad (-Nq + q + N - 1 \leq 0)$$

$$\leq \Gamma \int_{d_0}^{d_0+D} dR = \Gamma D.$$

(ii) Now we establish the uniform Hölder continuity of $\partial_1 u$. Let $h \in \mathbb{R}^N$ have magnitude $|h| \leq 1$ and arbitrary direction. Define

$$a(x_0, x, h) := (\partial_1 K)(x_0 + h - x) - (\partial_1 K)(x_0 - x);$$

then

$$\left| (\partial_1 u)(x_0 + h) - (\partial_1 u)(x_0) \right| = \left| \int_G a(x_0, x, h) f(x) \, dx \right|$$

$$\leq \left\| a(x_0, ., h) \mid L_q(G) \right\| \left\| f \mid L_p(G) \right\|. \quad \text{(A.54)}$$

We apply the small-ball technique.

(a) $R < 2|h|$. Since $\left| (\partial_1 K)(x_0 + h - x) \right| \leq \Gamma R_h^{-N+1}$ by (A.31), we have

$$\left\{ \int_{R < 2|h|} |a|^q \, dx \right\}^{1/q} \leq \Gamma \left\{ \int_0^{3|h|} R_h^{-Nq+q} R_h^{N-1} \, dR_h \right\}^{1/q} + \quad \text{similar}$$

$$= \Gamma \left\{ |h|^{-Nq+q+N} \right\}^{1/q}.$$

(b) $R \geq 2|h|$. Let $b(t) := (\partial_1 K)(x_0 + th - x)$, $0 \leq t \leq 1$. As before, we have $|x_0 + th - x| \geq \frac{1}{2}R$, and, by (A.31),

$$|b'(t)| = \left| \sum_{i=1}^{N} h_i (\partial_i \partial_1 K)(x_0 + th - x) \right| \leq \Gamma |h| R^{-N},$$

$$|a(x_0, x, h)| = |b(1) - b(0)| = \left| \int_0^1 b'(t) \, dt \right| \leq \Gamma |h| R^{-N}.$$

Then, exactly as in the proof of Lemma A.9, step (b),

$$\left\{\int_{R\geq 2|h|} |a|^q \, dx\right\}^{1/q} \leq \Gamma\left\{|h|^q \int_{2|h|}^{\infty} R^{-Nq} \, R^{N-1} \, dR\right\}^{1/q} \quad (-Nq+N<0)$$

$$= \Gamma\left\{|h|^{q-Nq+N}\right\}^{1/q},$$

and

$$\frac{q-Nq+N}{q} = 1-N+N\left(1-\frac{1}{p}\right) = \lambda.$$

Accordingly, (A.54) implies that, whenever $x_0 \in \mathbb{R}^N$ and $|h| \leq 1$,

$$\left|(\partial_1 u)(x_0+h) - (\partial_1 u)(x_0)\right| \leq \Gamma \left\|f \mid L_p(G)\right\| \, |h|^{\lambda}.$$

(iii) The potential u is certainly in $C^1(\mathbb{R}^N)$, by what we have just proved. Integration by parts, as in (A.48) or (A.50), gives

$$\int_{\mathbb{R}^N} \nabla\varphi \cdot \nabla u = -\int_{\mathbb{R}^N} (\Delta\varphi)u \qquad \text{for all } \varphi \in C_c^{\infty}(\mathbb{R}^N),$$

and the proof of Theorem A.8 shows that

$$-\int_{\mathbb{R}^N} (\Delta\varphi)u = \int_{\mathbb{R}^N} \varphi f.$$

Therefore the potential is a generalized solution of $-\Delta u = f$ in $\mathbb{R}^N$. ☐

A.4 Multipoles and the far field

In (A.10) we encountered the potential of a dipole in $\mathbb{R}^3$; inevitably, this is one of a hierarchy of point singularities. Let α be a multi-index of length N (Definition A.3); then

$$\varphi_\alpha(x) := \sigma(-1)^{|\alpha|} \partial^\alpha K(x;N), \qquad x \in \mathbb{R}^N \setminus \{0\}, \qquad (A.55)$$

is the potential of a *multipole* at the origin, *of strength σ and type α*. As before, $\Delta\varphi_\alpha = 0$ in $\mathbb{R}^N \setminus \{0\}$, because Δ and ∂^α commute when acting on functions with continuous derivatives of order $2+|\alpha|$. For $N=1$, multipoles are not exciting: $\varphi_{(1)}(x)$ is constant in $(-\infty,0)$ and in $(0,\infty)$; $\varphi_{(m)} = 0$ if $m \geq 2$.

A multipole is called a *dipole* if $|\alpha| = 1$, or a *quadrupole* if $|\alpha| = 2$. Figure A.10 shows the field lines of three such potentials; these are curves to which $-\nabla\varphi_\alpha$ is tangential. In the case of $\mathbb{R}^2$, the field lines are drawn with the help of a stream function (Exercises A.19 and A.21); in the case of $\mathbb{R}^3$ and cylindrical symmetry, with the help of a Stokes stream function (§A.1, item (vi)).

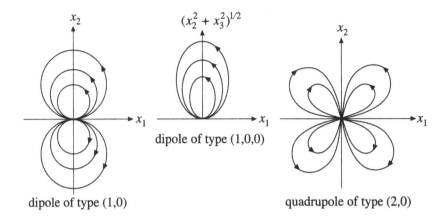

dipole of type (1,0)   quadrupole of type (2,0)

Fig. A.10.

Equation (A.23a) showed that, for constant density in a ball B, the potential at points outside B is that of a point source at the centre of B. One cannot expect so simple a result in general, but Theorem A.12 provides a corresponding result that is not complicated for an observer sufficiently far from G. This will be discussed after the proof, but it should be noted now that the terms of the series in Theorem A.12 decrease in order of magnitude for large $|x|$ as $|\alpha|$ increases.

The theorem is stated only for $N \geq 2$ because its analogue for one dimension is immediate: if $N = 1$ and $(a, b) \supset G$, then the definition of u implies that

$$u(x) = \begin{cases} -\dfrac{1}{2}x \displaystyle\int_G f(\xi)\,d\xi + \dfrac{1}{2}\displaystyle\int_G \xi f(\xi)\,d\xi, & x \geq b, \\[4mm] \dfrac{1}{2}x \displaystyle\int_G f(\xi)\,d\xi - \dfrac{1}{2}\displaystyle\int_G \xi f(\xi)\,d\xi, & x \leq a. \end{cases} \qquad \text{(A.56)}$$

Once again we relax the condition that the potential be defined at all points of $\mathbb{R}^N$.

Theorem A.12 *Let* $a := \sup\{\, |x| \mid x \in G \,\}$. *If* $N \geq 2$, $f \in L_1(G)$ *and* $|x| \geq 2a$, *then*

$$u(x) = \sum_{0 \leq |\alpha| \leq k-1} \sigma_\alpha (-1)^{|\alpha|} \partial^\alpha K(x) + E_k(x), \qquad \text{(A.57a)}$$

where

$$\sigma_\alpha := \int_G \frac{\xi^\alpha}{\alpha!} f(\xi) \, d\xi, \qquad \alpha! := \alpha_1! \, \alpha_2! \ldots \alpha_N!, \tag{A.57b}$$

and

$$\left| \partial^\beta E_k(x) \right| \le \text{const.} \, |x|^{-N+2-k-|\beta|} \tag{A.57c}$$

for all multi-indices β. The constant depends on k, β, f and G, but not on x.

Proof We have

$$u(x) = \int_G K(x - \xi) f(\xi) \, d\xi, \qquad |\xi| < a, \quad |x| \ge 2a,$$

and set $g(t) := K(x - t\xi)$, $0 \le t \le 1$. Then

$$g^{(m)}(t) := \left(\frac{d}{dt} \right)^m g(t) = \left(-\sum_{j=1}^N \xi_j \partial_j \right)^m K(x - t\xi),$$

where x and ξ are fixed, and ∂_j means, as always, differentiation with respect to the jth argument of the operand. By the multinomial theorem (Exercise A.22),

$$\left(\sum_{j=1}^N \xi_j \partial_j \right)^m = \sum_{|\alpha|=m} \frac{m!}{\alpha!} \xi^\alpha \partial^\alpha,$$

whence

$$g^{(m)}(t) = (-1)^m \sum_{|\alpha|=m} \frac{m!}{\alpha!} \xi^\alpha \left(\partial^\alpha K \right)(x - t\xi).$$

Accordingly [Taylor's theorem or integration by parts],

$$
\begin{aligned}
K(x - \xi) = g(1) &= \sum_{m=0}^{k-1} \frac{g^{(m)}(0)}{m!} + \int_0^1 \frac{(1-t)^{k-1}}{(k-1)!} g^{(k)}(t) \, dt \\
&= \sum_{0 \le |\alpha| \le k-1} (-1)^{|\alpha|} \frac{\xi^\alpha}{\alpha!} \partial^\alpha K(x) + A_k(x, \xi),
\end{aligned}
$$

where

$$A_k(x, \xi) := (-1)^k \int_0^1 \frac{(1-t)^{k-1}}{(k-1)!} \left\{ \sum_{|\alpha|=k} \frac{k!}{\alpha!} \xi^\alpha \left(\partial^\alpha K \right)(x - t\xi) \right\} dt.$$

Substituting for $K(x - \xi)$ into the integral defining u, we obtain (A.57a,b) with

$$E_k(x) = \int_G A_k(x, \xi) f(\xi) \, d\xi.$$

It remains to bound $\partial^\beta E_k$. The result (A.33) allows us to differentiate $K(x - \xi)$ with respect to x under the integral sign $\int_G$, and it is certainly legitimate to differentiate the part $A_k(x, \xi)$ of $K(x - \xi)$ under the integral sign $\int_0^1$, because that integrand is C^∞ by the condition $|x - t\xi| \geq a$. Now (A.31) and the inequality $|x - t\xi| \geq \frac{1}{2}|x|$ imply that

$$\left| (\partial^{\alpha + \beta} K)(x - t\xi) \right| \leq C_{k,\beta} \left| \tfrac{1}{2} x \right|^{-N + 2 - k - |\beta|}, \quad \text{where} \quad C_{k,\beta} := \max_{|\alpha| = k} C_{\alpha + \beta};$$

then

$$\left| \left(\frac{\partial}{\partial x} \right)^\beta A_k(x, \xi) \right| \leq \text{const.} \, a^k |x|^{-N + 2 - k - |\beta|},$$

and

$$\left| \partial^\beta E_k(x) \right| \leq \text{const.} \, a^k \, \big\| f \mid L_1(G) \big\| \, |x|^{-N + 2 - k - |\beta|},$$

where the constant depends only on k and β.   □

Remarks 1. The theorem states that, as $|x| \to \infty$, the dominant part of the potential, and of all its derivatives, is due to a source ($|\alpha| = 0$) at the origin, of strength equal to the total charge or mass $\int f$. (We assume for purposes of description that, at each order $|\alpha| \leq k - 1$, at least one coefficient $\sigma_\alpha \neq 0$.) Next in order of magnitude come the dipole terms ($|\alpha| = 1$); their strengths are moments $\int \xi_j f$ of the density function. Then, still smaller, come the quadrupole terms ($|\alpha| = 2$); their strengths are quadratic moments $\int \xi_i \xi_j f(\xi) \, d\xi$, apart from a numerical factor. The hierarchy of multipole terms, becoming smaller (for large $|x|$) as $|\alpha|$ increases, continues in this way.

2. *Provided that $\int_G f \neq 0$, we can so choose co-ordinates that the dipole terms vanish.* (This suggests the first step in the proof of Theorem 4.2.) Define the centre c of charge or mass by

$$c_j := \frac{\int_G \xi_j f(\xi) \, d\xi}{\int_G f} \quad (j = 1, \ldots, N),$$

and translate the origin to c, in effect, by setting $v(x) := u(x + c)$. Multi-indices are not always convenient; we now write

$$Q := \sigma_{(0, \ldots, 0)} = \int_G f,$$

$$a_1 := \sigma_{(1,0,\ldots,0)} = \int_G \xi_1 f(\xi)\, d\xi, \quad \ldots, \quad a_N := \sigma_{(0,\ldots,0,1)} = \int_G \xi_N f(\xi)\, d\xi,$$

so that $c_j = a_j/Q$. Then for $|x| \geq 2(a + |c|)$, say,

$$
\begin{aligned}
v(x) &= \sum_{0 \leq |\alpha| \leq 1} \sigma_\alpha (-1)^{|\alpha|} (\partial^\alpha K)(x + c) + O\left(|x|^{-N}\right) \\
&= Q\left\{ K(x) + \sum_{j=1}^N c_j \partial_j K(x) + O\left(|x|^{-N}\right) \right\} \\
&\quad - \left\{ \sum_{j=1}^N a_j \partial_j K(x) + O\left(|x|^{-N}\right) \right\} + O\left(|x|^{-N}\right) \\
&= Q K(x) + O\left(|x|^{-N}\right)
\end{aligned}
\tag{A.58}
$$

because $Q c_j = a_j$.

3. Let $N = 2$ and fix x with $|x| > a$. Then one can use (A.24) to prove that the series in the theorem, and the series resulting from it by repeated differentiation, all converge as $k \to \infty$. For $N \geq 3$, a corresponding result is true (but more difficult to prove). However, convergence of these series is not the object of our interest here; rather, we have sought an ordered picture of the field at large distances from G.

A.5 Second derivatives of u at points in G

For $N \geq 2$, the condition $f \in C(\overline{G})$ fails to imply that $u \in C^2(G)$; this is shown by Exercise A.29. The theme of this section is that *Hölder continuity* of f on $\overline{G}$ ensures not merely that $u \in C^2(G)$, but that second derivatives of u are Hölder continuous in G with the same Hölder exponent as f. In symbols,

$$
\left.\begin{aligned}
f \in C_b^{0,\mu}(\overline{G}) &\Rightarrow \partial_i \partial_j u \in C_b^{0,\mu}(\overline{\Omega}) \\
\text{whenever} \quad \mu \in (0,1) \quad &\text{and} \quad \Omega \subset G.
\end{aligned}\right\}
$$

The reason for the condition $\mu < 1$ will be mentioned after the condition has been used. [Of course, if $f \in C_b^{0,1}(\overline{G})$, then $f \in C_b^{0,\mu}(\overline{G})$ for each $\mu \in (0,1)$.] Except when G is a ball, we shall not pursue the more difficult question of *regularity up to the boundary*: if $f \in C_b^{0,\mu}(\overline{G})$, does $\partial_i \partial_j u\big|_G$ have an extension to $\overline{G}$ that is in $C(\overline{G})$ or perhaps in $C_b^{0,\mu}(\overline{G})$? Exercise A.30 shows that the answer can be No when f is constant on $\overline{G}$ and ∂G has a corner; a positive answer requires a condition on the smoothness of ∂G and more analysis than is appropriate here.

For $N = 1$, the analogue of Theorems A.15 and A.16 is again very simple; the definition of u implies that

$$-u''(x) = f(x) \quad \text{whenever} \ x \in G \ \text{ and } \ f \in C(\overline{G}). \tag{A.59}$$

Notation Following convention, we often *omit the subscript b from* $C_b(\Omega)$ and $C_b^{0,\lambda}(\overline{\Omega})$ when Ω is *bounded*, because then $C(\overline{\Omega})$ contains only bounded functions. The ambiguity, that $C(\overline{\Omega})$ may or may not be not be normed, causes little difficulty.

Lemma A.13 *Assume that* $N \geq 2$. *Let* $B := \mathscr{B}(0, a)$, *let* $f \in C^{0,\mu}(\overline{B})$ *for some* $\mu \in (0, 1)$, *and let*

$$v(x_0) := \int_B K(x_0 - x) f(x) \, dx, \qquad x_0 \in \mathbb{R}^N.$$

Extend f *to be zero outside* $\overline{B}$. *Then* [*for all i and j in* $\{1, \ldots, N\}$]

$$(\partial_i \partial_j v)(x_0) = \int_B (\partial_i \partial_j K)(x_0 - x) \{ f(x) - f(x_0) \} \, dx - \delta_{ij} \frac{f(x_0)}{N} \tag{A.60}$$

whenever $x_0 \in \mathbb{R}^N \setminus \partial B$.

Proof If $x_0 \in \mathbb{R}^N \setminus \overline{B}$, then $f(x_0) = 0$ and (A.60) is a particular case of (A.33). Therefore we suppose in the remainder of this proof that $x_0 \in B$. The existence of the integral in (A.60) will be implied once again by the calculations that follow. It is enough to prove the result for $i = 1$; let

$$h = (h_1, 0, \ldots, 0), \qquad 0 < |h| \leq 1, \qquad |h| < \text{dist}(x_0, \partial B).$$

Note first that, by Lemma A.9 and (A.23a),

$$\int_B \left\{ (\partial_j K)(x_0 + h - x) - (\partial_j K)(x_0 - x) \right\} \, dx$$

$$= (\partial_j u_B)(x_0 + h) - (\partial_j u_B)(x_0) = -\frac{1}{N} \left\{ (x_0 + h)_j - x_{0j} \right\}$$

$$= -\delta_{1j} \frac{h_1}{N}.$$

Hence, by a second application of Lemma A.9,

$$(\partial_j v)(x_0 + h) - (\partial_j v)(x_0) = \int_B \left\{ (\partial_j K)(x_0 + h - x) - (\partial_j K)(x_0 - x) \right\}$$

$$\times \{ f(x) - f(x_0) \} \, dx - \delta_{1j} \frac{h_1}{N} f(x_0).$$

We are now on a familiar path. Fix j and define

$$k(x_0, x, h) := \frac{1}{h_1} \left\{ (\partial_j K)(x_0 + h - x) - (\partial_j K)(x_0 - x) \right.$$
$$\left. - h_1 (\partial_1 \partial_j K)(x_0 - x) \right\};$$

then

$$\frac{(\partial_j v)(x_0 + h) - (\partial_j v)(x_0)}{h_1}$$
$$- \int_B (\partial_1 \partial_j K)(x_0 - x) \left\{ f(x) - f(x_0) \right\} \, dx + \delta_{1j} \frac{f(x_0)}{N}$$
$$= \int_B k(x_0, x, h) \left\{ f(x) - f(x_0) \right\} \, dx.$$

If this last integral tends to zero as $h_1 \to 0$, then the lemma will be proved. Let $\|f\| := \left\| f \mid C^{0,\mu}(\overline{B}) \right\|$ and $R = |x - x_0|$; the following coarse bound will be useful.

$$|f(x) - f(x_0)| \leq \left\{ \begin{array}{ll} [f]_\mu R^\mu & \text{if } R \leq 1, \\ 2\|f \mid C(\overline{B})\| R^\mu & \text{if } R \geq 1 \end{array} \right\} \leq 2\|f\| R^\mu. \quad (A.61)$$

In the inevitable application of the small-ball technique, constants Γ now depend on N and μ. We recall the notation $R_h = |x - x_0 - h|$.

(a) $R < 2|h|$. By the definition of k, by (A.31) and by (A.61),

$$\left| \int_{R < 2|h|} k \{ f(x) - f(x_0) \} \, dx \right|$$
$$\leq \Gamma |h|^{-1} \int_0^{3|h|} R_h^{-N+1} \|f\| \, |h|^\mu \, R_h^{N-1} \, dR_h + \quad \text{similar}$$
$$+ \Gamma \int_0^{2|h|} R^{-N} \|f\| R^\mu \, R^{N-1} \, dR$$
$$= \Gamma \|f\| \, |h|^\mu.$$

(b) $R \geq 2|h|$. Let $g(t) := (\partial_j K)(x_0 + th - x)$, $0 \leq t \leq 1$. Since $|x_0 + th - x| \geq \frac{1}{2} R$, we have from (A.31)

$$|g''(t)| = \left| h_1^2 (\partial_1^2 \partial_j K)(x_0 + th - x) \right| \leq \Gamma |h|^2 R^{-N-1},$$

$$|k(x_0, x, h)| = \left| \frac{1}{h_1} \int_0^1 (1 - t) g''(t) \, dt \right| \leq \Gamma |h| R^{-N-1};$$

accordingly,

$$\left| \int_{R \geq 2|h|} k\{ f(x) - f(x_0) \} \, dx \right| \leq \Gamma |h| \int_{2|h|}^{\infty} R^{-N-1} \|f\| R^{\mu} R^{N-1} \, dR$$

$$= \Gamma \|f\| \, |h|^{\mu}.$$

$\square$

Remarks 1. At the last step of the foregoing proof, the integrand was $\|f\| R^{-2+\mu}$. If $\mu = 1$ had been allowed, then for this case we would have been driven to some estimate like

$$\left| \int_{R \geq 2|h|} k\{ f(x) - f(x_0) \} \, dx \right| \leq \Gamma |h| \int_{2|h|}^{2a} \|f\| R^{-1} \, dR = \Gamma \|f\| \, |h| \log \frac{a}{|h|}.$$
(A.62)

At this stage, the logarithm would do no harm, but in Theorems A.15 and A.16 the effect of $\mu = 1$ and of a similar inequality would be that the Hölder exponent 1 is not conserved.

2. Again let $B := \mathscr{B}(0, a)$ in $\mathbb{R}^N$, $N \geq 2$; denote the usual small ball by $H := \mathscr{B}(x_0, 2|h|)$ with $0 < |h| \leq a/4$, say. It will be desirable to bound the integral

$$I_{ij}(x_0, h) := \int_{B \setminus H} (\partial_i \partial_j K)(x_0 - x) \, dx, \quad \text{where} \quad x_0 \in B,$$
(A.63)

independently of x_0 and h. Our standard method gives

$$|I_{ij}(x_0, h)| \leq \Gamma \int_{2|h|}^{2a} R^{-N} R^{N-1} \, dR = \Gamma \log \frac{a}{|h|},$$

which is not good enough for our purpose. The logarithm can be avoided by a more careful calculation, which unfortunately runs to some length; this is the subject of Lemma A.14.

3. If Lemma A.14 and Theorem A.15 are thought to be of no interest, then one can follow a shorter path to Theorem A.16. This requires a bound for I_{ij} only when the balls B and H are concentric. Now, with the notation $B_0 := \mathscr{B}(x_0, \rho)$, $\rho > 2|h|$, we have

$$\left| \int_{B_0 \setminus H} (\partial_i \partial_j K)(x_0 - x) \, dx \right| = \left| \int_{\partial(B_0 \setminus H)} n_i(x)(\partial_j K)(x - x_0) \, dS(x) \right|$$

$$\leq \int_{\partial B_0 \cup \partial H} \left| (\nabla K)(x - x_0) \right| \, dS(x)$$

$$= 2$$
(A.64)

in view of the definitive property (A.17) of the Newtonian kernel.

Lemma A.14 *Let I_{ij} be as in (A.63). Then*

$$\sum_{i,j=1}^{N} I_{ij}(x_0, h)^2 \le \text{const.,}$$

where the constant is independent of x_0 and h; indeed, it depends only on N.

Proof (i) It will be convenient to rotate the co-ordinate frame: let $x = T(x_0)\tilde{x}$ and $x_0 = T(x_0)\tilde{x}_0$, where $T(x_0)$ is an orthogonal $N \times N$ matrix depending on x_0. Now,

$$K_{ij}(x_0 - x) := (\partial_i \partial_j K)(x_0 - x) = \frac{N(x_{0i} - x_i)(x_{0j} - x_j) - \delta_{ij}|x_0 - x|^2}{\sigma_N |x_0 - x|^{N+2}},$$
(A.65)

where $\sigma_N := |\partial \mathscr{B}_N(0, 1)|$. Since the transformation of co-ordinates depends on x_0, we use the explicit form of K_{ij}, rather than the rule for transforming partial derivatives, to infer that

$$K_{ij}(x_0 - x) = \sum_{p,q=1}^{N} T_{ip} T_{jq} K_{pq}(\tilde{x}_0 - \tilde{x}), \quad \text{where } T = T(x_0). \quad (A.66)$$

Let $\tilde{H} := \mathscr{B}(\tilde{x}_0, 2|h|)$ and let $\tilde{I}_{pq}(\tilde{x}_0, h)$ denote the integral with respect to $\tilde{x}$ of $K_{pq}(\tilde{x}_0 - \tilde{x})$ over $B \setminus \tilde{H}$; then

$$I_{ij}(x_0, h) = \sum_{p,q=1}^{N} T_{ip} T_{jq} \tilde{I}_{pq}(\tilde{x}_0, h),$$

and, by the orthogonality of T,

$$\sum_{i,j=1}^{N} I_{ij}(x_0, h)^2 = \sum_{p,q=1}^{N} \tilde{I}_{pq}(\tilde{x}_0, h)^2. \quad (A.67)$$

(ii) Given $x_0 \in B$, we so choose T that $\tilde{x}_0 = (|x_0|, 0, \dots, 0)$; we shall *use these special co-ordinates $\tilde{x}_j$ and (A.67) to prove the Lemma*, but henceforth we *omit the tilde* for ease of writing. Two configurations are possible (Figure A.11): case I if $\overline{H} \subset B$, or case II if $\overline{H}$ intersects ∂B; we shall write formulae and inequalities that are valid for both. It is immaterial whether the integration is over $B \setminus H$ or over $B \setminus \overline{H}$; in the rest of this proof we shall use $B \setminus \overline{H}$ because it is open. The divergence theorem is valid for both configurations because $\partial(B \setminus \overline{H})$ is of class C^∞ in case I, while the set $B \setminus \overline{H}$ of case II is listed in Remark D.4.

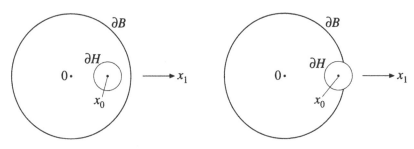

Fig. A.11.

(iii) If $i \neq j$, then $I_{ij} = 0$. For, at least one of i and j must differ from 1; suppose that $j \neq 1$. Then $x_{0j} = 0$ and $\delta_{ij} = 0$, so that $K_{ij}(x_0 - x)$ is an odd function of x_j, while $B \setminus \overline{H}$ is symmetrical about the hyperplane $\{ x \mid x_j = 0 \}$. Also,

$$\sum_{j=1}^{N} K_{jj}(x_0 - x) = 0 \quad \text{for } x \neq x_0 \Rightarrow I_{11}(x_0, h) = -\sum_{j=2}^{N} I_{jj}(x_0, h);$$

therefore it is sufficient to bound $|I_{22}|, \ldots, |I_{NN}|$, and these are all equal, by cylindrical symmetry about the x_1-axis.

(iv) Let $n = n(x)$ be the unit normal on $\partial(B \setminus \overline{H})$ outward from $B \setminus \overline{H}$, and let $dS = dS(x)$. Then

$$|I_{jj}(x_0, h)| = \left| \int_{B \setminus \overline{H}} (\partial_j^2 K)(x - x_0) \, dx \right| = \left| \int_{\partial(B \setminus \overline{H})} n_j (\partial_j K)(x - x_0) \, dS \right|$$

$$\leq \left| \int_{\partial B \setminus H} n_j (\partial_j K)(x - x_0) \, dS \right| + 1 \qquad (A.68)$$

because $|(\nabla K)(x - x_0)| = 1/|\partial H|$ when $x \in \partial H$. Now let $j \neq 1$. Then $n_j(x) = x_j/a$ on ∂B and $x_j - x_{0j} = x_j$, so that

$$\left| \int_{\partial B \setminus H} n_j (\partial_j K)(x - x_0) \, dS \right| = \int_{\partial B \setminus H} \frac{x_j^2}{a \sigma_N |x - x_0|^N} \, dS \qquad (j \neq 1)$$

$$= \frac{1}{(N-1) a \sigma_N} \int_{\partial B \setminus H} \frac{x_2^2 + \cdots + x_N^2}{|x - x_0|^N} \, dS$$

by cylindrical symmetry about the x_1-axis. Transforming to polar coordinates, we set

$$x_1 = a \cos \theta,$$
$$(x_2^2 + \cdots + x_N^2)^{1/2} = a \sin \theta \qquad (x \in \partial B, \ 0 \leq \theta \leq \pi);$$

we also write $x_{01} = |x_0| =: a(1 - \delta)$, so that $0 < \delta \leq 1$. Then

$$
\begin{aligned}
|x - x_0|^2 &= a^2 \left\{ 1 - 2(1 - \delta)\cos\theta + (1 - \delta)^2 \right\} \\
&= a^2 \left\{ \delta^2 + 4(1 - \delta)\sin^2\frac{\theta}{2} \right\},
\end{aligned}
$$

and, for some angle β in $[0, \pi/4)$ that depends on $|x_0|$ and $|h|$,

$$
\left| \int_{\partial B \backslash H} n_j(\partial_j K)(x - x_0)\, dS \right|
$$

$$
= \frac{1}{(N-1)a\sigma_N} \int_\beta^\pi \frac{a^2 \sin^2\theta}{a^N \left\{ \delta^2 + 4(1-\delta)\sin^2(\theta/2) \right\}^{N/2}} \sigma_{N-1}(a\sin\theta)^{N-2}\, a\, d\theta
$$

$$
= \frac{\sigma_{N-1}}{(N-1)\sigma_N} \int_\beta^\pi \left[\frac{4\cos^2(\theta/2)\sin^2(\theta/2)}{\delta^2 + 4(1-\delta)\sin^2(\theta/2)} \right]^{N/2} d\theta. \tag{A.69a}
$$

We replace β by 0 in this integral, discard the δ^2 if $\delta \leq \frac{1}{2}$, and discard the $4(1 - \delta)\sin^2(\theta/2)$ if $\delta > \frac{1}{2}$; accordingly,

$$
\int_\beta^\pi \left[\frac{4\cos^2(\theta/2)\sin^2(\theta/2)}{\delta^2 + 4(1-\delta)\sin^2(\theta/2)} \right]^{N/2} d\theta
$$

$$
\leq \begin{cases} 2^{N/2} \int_0^\pi (\cos(\theta/2))^N\, d\theta & \text{if } 0 < \delta \leq \frac{1}{2}, \\ 2^N \int_0^\pi (\sin\theta)^N\, d\theta & \text{if } \frac{1}{2} < \delta \leq 1, \end{cases}
$$

$$
\leq 2^{N+1} \int_0^{\pi/2} (\sin\theta)^N\, d\theta \quad \text{for } 0 < \delta \leq 1, \tag{A.69b}
$$

because

$$
\int_0^\pi \left(\cos\frac{\theta}{2} \right)^N d\theta = 2 \int_0^{\pi/2} (\sin\varphi)^N\, d\varphi \quad \left[\frac{\theta}{2} = \frac{\pi}{2} - \varphi \right].
$$

The results (A.68) and (A.69) yield a bound for $|I_{jj}(x_0, h)|$, $j \neq 1$, that depends only on N; in view of (A.67) and remarks in (iii), this bound proves the lemma. ☐

Remark 4. When proving uniform Hölder continuity, with the supremum of the function already bounded, one need consider only points x and y satisfying $0 < |x - y| < \delta$ for some $\delta > 0$, because

$$
|x - y| \geq \delta \implies \frac{|w(x) - w(y)|}{|x - y|^\lambda} \leq \frac{2\sup_z |w(z)|}{\delta^\lambda}.
$$

Theorem A.15 *As in Lemma* A.13, *let* $B := \mathscr{B}(0,a)$ *in* $\mathbb{R}^N$, $N \geq 2$, *let* $f \in C^{0,\mu}(\overline{B})$ *for some* $\mu \in (0,1)$ *and let* v *be the Newtonian potential of* $f|_B$. *Then each second derivative* $\partial_i \partial_j v|_B$ *has an extension to* $\overline{B}$ *that is in* $C^{0,\mu}(\overline{B})$, *and* v *is a* C^2*-solution of* $-\Delta v = f$ *in* B.

Proof In this proof $(.)_{ij} := \partial_i \partial_j (.)$ for functions (but not for the Kronecker delta), and $\|f\| := \|f \mid C^{0,\mu}(\overline{B})\|$.

(i) To bound $v_{ij}(x_0)$ independently of $x_0 \in B$, we use the formula (A.60) for $v_{ij}(x_0)$ and the bound (A.61) for $|f(x) - f(x_0)|$; accordingly,

$$
\begin{aligned}
|v_{ij}(x_0)| &= \left| \int_B K_{ij}(x_0 - x)\{f(x) - f(x_0)\}dx - \delta_{ij}\frac{f(x_0)}{N} \right| \\
&\leq \Gamma \int_0^{2a} R^{-N} \|f\| R^\mu R^{N-1} \, dR + \frac{\|f\|}{N} \\
&\leq \Gamma \|f\| (a^\mu + 1).
\end{aligned}
\tag{A.70}
$$

(ii) To prove uniform Hölder continuity of v_{ij}, we consider field points x_0 and $x_0 + h$ in B, with $0 < |h| \leq a/4$ [recall Remark 4]. The corresponding small ball is $H := \mathscr{B}(x_0, 2|h|)$. This time we do not separate the two steps of the small-ball technique, but proceed from the decomposition

$$
\begin{aligned}
v_{ij}&(x_0 + h) - v_{ij}(x_0) \\
&= \int_{H \cap B} K_{ij}(x_0 + h - x)\{f(x) - f(x_0 + h)\} \, dx \\
&\quad - \int_{H \cap B} K_{ij}(x_0 - x)\{f(x) - f(x_0)\} \, dx \\
&\quad + \int_{B \setminus H} \{K_{ij}(x_0 + h - x) - K_{ij}(x_0 - x)\}\{f(x) - f(x_0 + h)\} \, dx \\
&\quad + \{f(x_0) - f(x_0 + h)\} \int_{B \setminus H} K_{ij}(x_0 - x) \, dx \\
&\quad - \frac{1}{N}\delta_{ij}\{f(x_0 + h) - f(x_0)\}.
\end{aligned}
\tag{A.71}
$$

The inequality (A.61), with x_0 replaced by $x_0 + h$ [hence with R replaced by $R_h := |x - x_0 - h|$], yields

$$
|f(x) - f(x_0 + h)| \leq 2\|f\| R_h^\mu
\tag{A.72}
$$

for all x and $x_0 + h$ in B. Then the modulus of the first integral in (A.71) is bounded by

$$
\Gamma \int_0^{3|h|} R_h^{-N} \|f\| R_h^\mu R_h^{N-1} \, dR_h = \Gamma \|f\| \, |h|^\mu.
$$

The second integral is similar. The first factor of the third integrand is estimated by our usual procedure for $R \geq 2|h|$; the second factor by means of (A.72), in which $R_h \leq (3/2)R$ for $R \geq 2|h|$. Then the modulus of the third integral is bounded by

$$\Gamma \int_{2|h|}^{\infty} R^{-N-1} |h| \, \|f\| \, R^{\mu} R^{N-1} \, \mathrm{d}R = \Gamma \|f\| \, |h|^{\mu}.$$

The fourth integral has been bounded in Lemma A.14. Hence the fourth and fifth terms on the right of (A.71) also contribute $\Gamma \|f\| \, |h|^{\mu}$, so that

$$\left| v_{ij}(x_0 + h) - v_{ij}(x_0) \right| \leq \Gamma \|f\| \, |h|^{\mu} \tag{A.73}$$

for all x_0 and $x_0 + h$ in B with $0 < |h| \leq a/4$. Therefore v_{ij} has a unique continuous extension to $\overline{B}$ (Dieudonné 1969, pp. 56–57) and this extension belongs to $C^{0,\mu}(\overline{B})$.

(iii) That $v \in C^2(B)$ follows from what has just been proved; that $-(\Delta v)(x_0) = f(x_0)$ at each $x_0 \in B$ is implied by (A.60), since $(\Delta K)(x_0 - x) = 0$ for $x \neq x_0$. Therefore v is a C^2-solution of the Poisson equation in B. □

Theorem A.16 *Let G be bounded and open in $\mathbb{R}^N$, $N \geq 2$, and let $f \in C^{0,\mu}(\overline{G})$ for some $\mu \in (0,1)$. Then each second derivative of the Newtonian potential u of f has the property*

$$\partial_i \partial_j u \in C^{0,\mu}(\overline{\Omega}) \quad \text{whenever} \quad \overline{\Omega} \subset G, \tag{A.74}$$

and u is a C^2-solution of $-\Delta u = f$ in G.

Proof In this proof $(.)_{ij} := \partial_i \partial_j(.)$ for functions (but not for the Kronecker delta), and $\|f\| := \left\| f \mid C^{0,\mu}(\overline{G}) \right\|$.

(i) Let an (open) set Ω, such that $\overline{\Omega} \subset G$, be given. Then $\mathrm{dist}(\overline{\Omega}, \partial G) > 0$ because $\overline{\Omega}$ and ∂G are compact and disjoint; let $\rho := \frac{1}{2}\,\mathrm{dist}(\overline{\Omega}, \partial G)$. For each $c \in \overline{\Omega}$ and all $x_0 \in \mathbb{R}^N$, define $B := \mathcal{B}(c, \rho)$ and

$$v(x_0) \quad := \quad \int_B K(x_0 - x) f(x) \, \mathrm{d}x, \tag{A.75a}$$

$$w(x_0) \quad := \quad \int_{G \setminus \overline{B}} K(x_0 - x) f(x) \, \mathrm{d}x. \tag{A.75b}$$

[The symbols $v(x_0, c)$ and $w(x_0, c)$ would be more accurate but more clumsy.] Then $\overline{B} \subset G$ and $u = v + w$. Application of (A.60) to v and of

(A.33) to w shows that, whenever $c \in \overline{\Omega}$ and $x_0 \in B$,

$$v_{ij}(x_0) = \int_B K_{ij}(x_0 - x)\{ f(x) - f(x_0) \}\, dx - \delta_{ij}\frac{f(x_0)}{N}, \quad (A.76a)$$

$$w_{ij}(x_0) = \int_{G\setminus\overline{B}} K_{ij}(x_0 - x) f(x)\, dx. \quad (A.76b)$$

In particular, when $x_0 \in \overline{\Omega}$ is given, we may choose $c = x_0$, so that the field point is at the centre of B. [This is legitimate because (A.76) holds for every $c \in \overline{\Omega}$ and every $x_0 \in \mathcal{B}(c,\rho)$, but it must be remembered that the differentiation with respect to x_{0j} and x_{0i} was for fixed c and before the choice $c = x_0$ was made.]

(ii) To bound $u_{ij}(x_0)$ independently of $x_0 \in \overline{\Omega}$, we choose $c = x_0$, adapt (A.70) to the present situation, obtaining

$$\left| v_{ij}(x_0) \right| \le \Gamma \|f\| \left(\rho^\mu + 1 \right),$$

and infer from (A.76b) that, with $D := \operatorname{diam} G$,

$$\left| w_{ij}(x_0) \right| \le \Gamma \int_\rho^D R^{-N} \|f\| R^{N-1}\, dR = \Gamma \|f\| \log \frac{D}{\rho}.$$

(iii) To prove uniform Hölder continuity of u_{ij} on $\overline{\Omega}$, we consider field points x_0 and $x_0 + h$ in $\overline{\Omega}$, with $0 < |h| \le \rho/4$ [recall Remark 4]; again we choose $c = x_0$. The identity (A.71) and estimates like those leading from (A.71) to (A.73), again with $H := \mathcal{B}(x_0, 2|h|)$ but now with $B = \mathcal{B}(x_0, \rho)$, with $0 < |h| \le \rho/4$ and with (A.64) replacing Lemma A.14, yield

$$\left| v_{ij}(x_0 + h) - v_{ij}(x_0) \right| \le \Gamma \|f\|\, |h|^\mu$$

whenever $x_0, x_0 + h \in \overline{\Omega}$ and $0 < |h| \le \rho/4$.

Next, our standard method of estimating for $R \ge 2|h|$ gives

$$\left| w_{ij}(x_0 + h) - w_{ij}(x_0) \right| = \left| \int_{G\setminus\overline{B}} \{ K_{ij}(x_0 + h - x) - K_{ij}(x_0 - x) \} f(x)\, dx \right|$$

$$\le \Gamma \int_\rho^\infty R^{-N-1} |h| \|f\| R^{N-1}\, dR$$

$$= \Gamma \|f\| |h|/\rho$$

whenever $x_0, x_0 + h \in \overline{\Omega}$ and $0 < |h| \le \rho/4$.

Thus $u_{ij} \in C^{0,\mu}(\overline{\Omega})$.

(iv) The result (A.74) implies that $u \in C^2(G)$ [because for each point $x_0 \in G$ there is an open set $\Omega = \Omega(x_0)$ such that $x_0 \in \Omega$ and $\overline{\Omega} \subset G$]. To prove that $-(\triangle u)(x_0) = f(x_0)$ at a given point $x_0 \in G$, choose any ball

B such that $x_0 \in B$ and $\overline{B} \subset G$; in (A.75) and (A.76), let B denote this new ball. It follows that $u = v + w$, that $-(\Delta v)(x_0) = f(x_0)$ and that $(\Delta w)(x_0) = 0$. Thus u is a C^2-solution of the Poisson equation in G. $\qquad \square$

A.6 Exercises

Exercise A.17 Considering the Newtonian potential in one dimension $(N = 1)$,

(i) derive the formula (A.23) for the potential of unit density in $(-a, a)$;

(ii) prove that, if $f \in C(\overline{G})$, then $-u'' = f$ in G.

Exercise A.18 Show that in terms of spherical co-ordinates r, θ, λ the definition (A.14) of the Stokes stream function becomes

$$\left(\psi_r, \frac{1}{r} \psi_\theta \right) = r \sin \theta \left(-\frac{1}{r} \varphi_\theta, \; \varphi_r \right).$$

Derive (modulo an additive constant) the Stokes stream function ψ_0 in (A.16) from the potential φ_0 in (A.12).

Exercise A.19 Let $z = x + iy = re^{i\theta}$ denote an arbitrary point of the complex plane $\mathbb{C}$. [The point $(x, y) \in \mathbb{R}^2$ and the point $x + iy \in \mathbb{C}$ can be considered one and the same for many purposes, such as the definition of balls, of open sets and of continuity.] If χ is holomorphic (or analytic) in an open set $\Omega \subset \mathbb{C}$, if $\chi = \varphi + i\psi$, and if φ is regarded as a potential function, then χ is called a *complex potential,* ψ a *stream function,* and $- \, d\chi(z)/\,dz$ a *complex velocity* (or *complex electric field* or ...).

(i) How is $d\chi(z)/\,dz$ related to $\nabla \varphi(x, y)$ and to $\nabla \psi(x, y)$? Show that the vector field $\nabla \varphi$ is tangential to the level sets of ψ. [Recall that the zero vector is both normal and parallel to every vector.]

(ii) The complex potential of a unit source at the origin is $-(\log z)/2\pi$, after some suitable restriction of θ. Use this to find the analogues in the plane ($\mathbb{R}^2$ or $\mathbb{C}$) of the φ_0 and ψ_0 in §A.1 for 'flow past a closed ball'.

Exercise A.20 In the analogue for $\mathbb{R}^2$ of the problem (A.8), one allows any bad behaviour of φ that is not passed on to $\nabla \varphi$.

(i) Referring to Exercise A.19, (ii), exhibit infinitely many solutions for flow past a closed ball in $\mathbb{R}^2$ under the condition $\nabla \varphi(x, y) + (U, 0) = O(r^{-1})$ as $r \to \infty$. [Contemplate the complex potential $i \log z$, $z \neq 0$.]

(ii) State and prove an analogue, for flow in $\mathbb{R}^2$, of the uniqueness result in Theorem A.1.

Exercise A.21 (i) Show that in the plane a multipole at the origin, of unit strength and of type α with $|\alpha| \geq 1$, has complex potential

$$\chi_\alpha(z) = \frac{i^{\alpha_2}}{2\pi} \left(-\frac{d}{dz} \right)^{|\alpha|} \log \frac{1}{z} = \frac{i^{\alpha_2}}{2\pi} (|\alpha| - 1)!\, z^{-|\alpha|} \quad (z \neq 0).$$

(ii) For $\alpha = (1, 0)$, find equations describing the field lines; sketch these level sets of $\psi_{(1,0)}$.

(iii) Explain, and demonstrate by an example, how your picture in (ii) can be combined with a picture of the conformal map $\zeta = z^n$, where $n \in \{2, 3, 4, \ldots\}$ and $0 \leq \arg z < 2\pi/n$, to yield a sketch of any multipole in the plane with $|\alpha| \geq 2$. [To draw a picture of the map $\zeta = z^n$, write $z = re^{i\theta}$ and draw corresponding level sets of r and of θ in the z-plane and in the ζ-plane.]

Exercise A.22 Derive the multinomial theorem: for any positive integer m,

$$(x_1 + \cdots + x_N)^m = \sum_{|\alpha|=m} \frac{m!}{\alpha!} x^\alpha, \quad \text{where } \alpha! := \alpha_1! \ldots \alpha_N! \,.$$

[One method is induction on N; a second is a combinatorial argument; a third proceeds from

$$(1 - x_1 - \cdots - x_N) \sum_{m=0}^{\infty} (x_1 + \cdots + x_N)^m = 1 \quad (|x_1| + \cdots + |x_N| < 1).]$$

Exercise A.23 For multi-indices of length N, the partial ordering $\alpha \leq \beta$ means that $\alpha_j \leq \beta_j$ for $j = 1, \ldots, N$. As before, $\alpha! := \alpha_1! \ldots \alpha_N!$. Derive the Leibniz rule for differentiating the product of sufficiently smooth functions f and g, namely,

$$\partial^\beta(fg) = \sum_{\alpha \leq \beta} \frac{\beta!}{(\beta - \alpha)!\,\alpha!} (\partial^{\beta-\alpha} f)(\partial^\alpha g).$$

Exercise A.24 The Stokes stream function ψ can exist (and be useful) without an accompanying potential function. Write $\mathbf{x} = (x_1, x_2, x_3) = (x, s\cos\lambda, s\sin\lambda)$ for points of $\mathbb{R}^3$, with $s \geq 0$ and $-\pi < \lambda \leq \pi$. Suppose that we are given a cylindrically symmetric vector field V of the form

$$V(\mathbf{x}) = \big(V_1(\mathbf{x}), V_2(\mathbf{x}), V_3(\mathbf{x})\big) = \big(u(x, s), v(x, s)\cos\lambda, v(x, s)\sin\lambda\big),$$

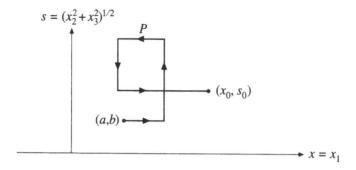

Fig. A.12.

and that $V \in C^1(\mathbb{R}^3, \mathbb{R}^3)$. Suppose also that V is *solenoidal* (or divergence-free):

$$\nabla \cdot V = \frac{\partial u}{\partial x} + \frac{1}{s} \frac{\partial}{\partial s}(sv) = 0 \quad \text{in } \mathbb{R}^3,$$

limiting values being taken for $s = 0$. Define, for all $(x_0, s_0) \in \mathbb{R} \times [0, \infty)$,

$$\psi(x_0, s_0; P) := \int_{(a,b)}^{(x_0, s_0)} \{ sv \ dx - su \ ds \},$$

where (a, b) is a fixed 'reference point'; the path P of integration is in the half-plane $\mathbb{R} \times [0, \infty)$, is connected, and consists of finitely many straight-line segments, each parallel to the x-axis or to the s-axis (Figure A.12).

(i) Prove that for fixed (x_0, s_0) the value $\psi(x_0, s_0; P)$ is independent of the choice of P, so that $\psi(x_0, s_0; P) = \psi(x_0, s_0)$. [The interior of the set bounded by two different paths is, or can be made, a finite union of open rectangles; the divergence theorem applies.]

(ii) Show that

$$\left(\frac{\partial \psi}{\partial x}, \frac{\partial \psi}{\partial s} \right) = s(v, -u),$$

which is a generalization of (A.14). Deduce that the vector field (u, v) is tangential to the level sets of ψ.

Exercise A.25 The equation

$$\left(\frac{\partial^2}{\partial x^2} + \frac{\partial^2}{\partial s^2} - \frac{1}{s} \frac{\partial}{\partial s} \right) \psi(x, s) = s^2 f(x, s) \tag{A.77}$$

can be used to find a vector field V as in Exercise A.24 for which curl V is prescribed by

$$\frac{\partial v}{\partial x} - \frac{\partial u}{\partial s} = sf(x, s).$$

(In hydrodynamics, the equations of motion allow f, but not sf, to be piecewise constant.)

Show that, if we set $\psi(x, s) =: s^2 \chi(x, s)$ and interpret x as $x = y_1$ and s as $s = (y_2^2 + \cdots + y_5^2)^{1/2}$, then (A.77) becomes

$$(\Delta \chi)(y) := \left\{ \left(\frac{\partial}{\partial y_1} \right)^2 + \cdots + \left(\frac{\partial}{\partial y_5} \right)^2 \right\} \chi(y_1, s) = f(y_1, s).$$

Exercise A.26 This question concerns the Stokes stream function in (A.15) for flow past Rankine's solid E. Write x for the previous x_1 and consider

$$\begin{aligned} F(x, s) \quad &:= \quad -\frac{4\pi}{q} \tilde{\psi}(x, s) \quad = \quad As^2 - \frac{x+\gamma}{R_1} + \frac{x-\gamma}{R_2} \\ &= \quad As^2 - \cos\theta_1 + \cos\theta_2, \quad \text{say,} \end{aligned}$$

where $R_1, R_2 := \{ (x \pm \gamma)^2 + s^2 \}^{1/2}$ respectively, $R_1 > 0$, $R_2 > 0$, $s \geq 0$, and A, γ are given positive constants. In order to describe ∂E, prove the following.

(a) If $x > 0$ and $s > 0$, then $F_x(x, s) > 0$.

(b) If $-\gamma \leq x \leq \gamma$ and $s > 0$, then $F_s(x, s) > 0$.

(c) For fixed $x \in [-\gamma, \gamma]$, there is exactly one number $g(x)$ such that $g(x) > 0$ and $F(x, g(x)) = 0$; for fixed $s \in (0, g(0)]$, there is exactly one number $h(s)$ such that $h(s) \geq 0$ and $F(h(s), s) = 0$.

(d) If $s > 0$ and $F(x, s) = 0$, then $F_s(x, s) \neq 0$. [Assume the contrary and observe that $3\cos\theta - \cos^3\theta$ decreases on $(0, \pi)$.]

(e) There is exactly one number a such that $a > \gamma$ and $F(a, s)/s^2 \to 0$ as $s \to 0$.

Deduce that ∂E has a representation $s = g(x), -a \leq x \leq a$. What can be said about the function g?

Exercise A.27 In this exercise we relax the condition that the Newtonian potential is to exist at all points of $\mathbb{R}^N$. Let $B := \mathscr{B}(0, 1)$ in $\mathbb{R}^3$; Theorem A.6 states that, if $f \in L_p(B)$ with $p > 3/2$, then the Newtonian potential of f is uniformly continuous on $\mathbb{R}^3$.

(i) Prove that $\int_0^1 r^{-1} (\log(3/r))^{-\alpha} \, dr$ exists (is finite) if and only if $\alpha > 1$.

(ii) Let $g(x) := r^{-2} (\log(3/r))^{-1}$ for $x \in \mathbb{R}^3 \setminus \{0\}$ and $r := |x| \le 2$. Show that $g|_B \in L_{3/2}(B)$ and that the Newtonian potential of $g|_B$ exists only in $\mathbb{R}^3 \setminus \{0\}$.

(iii) Let $a_1, a_2, \ldots, a_n, \ldots$, with $a_n = (a_{n1}, a_{n2}, a_{n3})$, be an enumeration of all points in B that have rational co-ordinates. Define $f : B \to \mathbb{R}$ by

$$f(x) := \sum_{n=1}^{\infty} 2^{-n} g(x - a_n), \qquad |x| < 1,$$

where g is as in (ii). Prove that $f \in L_{3/2}(B)$, that the Newtonian potential of f fails to exist on the set $A := (a_n)$, and that $\overline{A} = \overline{B}$.

Exercise A.28 Suppose that $N \ge 3$, that $G = \mathscr{B}(0, b) \setminus \overline{\mathscr{B}(0, a)}$ (with $b > a > 0$) and that $f(x) := r^m$ for some constant $m \in \mathbb{R}$. Then the Newtonian potential of $f : G \to \mathbb{R}$ is

$$u(x) = \begin{cases} \dfrac{b^{m+2} - a^{m+2}}{(m+2)(N-2)}, & 0 \le r < a, \\[2ex] \dfrac{b^{m+2}}{(m+2)(N-2)} - \dfrac{r^{m+2}}{(m+2)(m+N)} - \dfrac{a^{m+N} r^{-N+2}}{(m+N)(N-2)}, & a \le r < b, \\[2ex] \dfrac{b^{m+N} - a^{m+N}}{(m+N)(N-2)} r^{-N+2}, & r \ge b, \end{cases}$$

provided that $m \ne -2$ and $m \ne -N$.

(i) Verify the foregoing claim.

(ii) How can corresponding formulae for $m = -2$ and $m = -N$ be obtained economically?

(iii) For this density function f, many multipole strengths σ_α in Theorem A.12 differ from zero; for example, all those with $\alpha = (2n, 0, \ldots, 0)$ and $n \in \mathbb{N}$. Why are the terms in (A.57a) with $1 \le |\alpha| \le k - 1$ absent from the present representation of u for $r \ge b$?

Exercise A.29 Write $x = (x, y) = (r \cos \theta, r \sin \theta)$ for points of $\mathbb{R}^2$, and contemplate the function u defined by

$$u(x) = \begin{cases} 0, & r = 0, \\[1ex] r^2 \cos 2\theta \left\{ \dfrac{1}{4} + \log \left(\log \dfrac{e}{r} \right) \right\}, & 0 < r < 1, \\[1ex] \frac{1}{4} r^{-2} \cos 2\theta, & r \ge 1. \end{cases}$$

Let $B := \mathscr{B}(0,1)$; define $f := -\Delta u$ on $B \setminus \{0\}$, and define f on ∂B and at $\mathbf{0}$ in such a way that $f \in C(\overline{B})$.

(i) Show that $\partial_1^2 u \notin C(B)$, so that $u \notin C^2(B)$, even though $f \in C(\overline{B})$.

(ii) Prove that u is the Newtonian potential of f.

[The proof of Theorem A.2 implies that, for the present exercise and for $|x_0| > \delta$, $0 < \delta < 1$,

$$u(x_0) = \int_{\delta < r < 1} K_0 f + \int_{r=\delta} \left\{ u \frac{\partial K_0}{\partial r} - K_0 \frac{\partial u}{\partial r} \right\},$$

where $K_0(x) := K(x - x_0)$.]

Exercise A.30 Write $\mathbf{x} = (x,y) = (r\cos\theta, r\sin\theta)$ for points of $\mathbb{R}^2$, and consider the sector

$$S_\beta := \{\, \mathbf{x} \in \mathbb{R}^2 \mid 0 < r < a, \; -\beta < \theta < \beta \,\}, \qquad \beta \in (0, \pi).$$

The Newtonian potential of unit density in S_β is

$$u(\mathbf{x}) = \begin{cases} \dfrac{a^2}{\pi} c_0, & r = 0, \\[2ex] g(r,\theta) + \dfrac{a^2}{\pi} \displaystyle\sum_{n=0}^{\infty} c_n \left(\dfrac{r}{a}\right)^n \cos n\theta, & 0 < r < a, \\[3ex] \dfrac{a^2 \beta}{2\pi} \log \dfrac{a}{r} + \dfrac{a^2}{\pi} \displaystyle\sum_{n=0}^{\infty} (c_n + d_n) \left(\dfrac{a}{r}\right)^n \cos n\theta, & r \geq a, \end{cases}$$

where (if $0 < r < a$, $z = re^{i\theta}$ and Re denotes the real part)

$$g(r,\theta) := \begin{cases} \mathrm{Re}\left\{ \dfrac{\sin 2\beta}{4\pi} z^2 \log \dfrac{a}{z} + \dfrac{\cos 2\beta}{4} z^2 \right\} - \dfrac{1}{4} r^2, & -\beta \leq \theta \leq \beta, \\[3ex] \mathrm{Re}\, \dfrac{\sin 2\beta}{4\pi} z^2 \left[\log \dfrac{a}{z} + i\pi \right], & \beta < \theta < 2\pi - \beta, \end{cases}$$

$$g(r,\theta) = \frac{\sin 2\beta}{4\pi} r^2 \log \frac{a}{r} \cos 2\theta + \frac{r^2}{\pi} \sum_{n=0}^{\infty} d_n \cos n\theta \quad \text{for all } \theta \in \mathbb{R},$$

and where

$$c_0 = \frac{\beta}{2} \log \frac{1}{a} + \frac{\beta}{4}, \quad c_2 = \frac{1}{4} \sin 2\beta - \frac{1}{4}\beta \cos 2\beta, \quad c_n = -\frac{\sin n\beta}{n^2(n-2)}$$

$$\text{if } n = 1 \quad \text{or} \quad n \geq 3,$$

$$d_0 = -\frac{\beta}{4}, \quad d_2 = -\frac{3}{16} \sin 2\beta + \frac{1}{4}\beta \cos 2\beta, \quad d_n = \frac{2\sin n\beta}{n(n^2-4)}$$

$$\text{if } n = 1 \quad \text{or} \quad n \geq 3.$$

(i) Show that $|\partial_1^2 u(\mathbf{x})| \to \infty$ as $\mathbf{x} \to \mathbf{0}$ in $\mathbb{R}^2 \setminus \partial S_\beta$, provided that $\beta \neq \pi/2$.

(ii) Given that in Theorem A.2 the hypotheses $v \in C^2(\mathbb{R}^N \setminus \overline{G})$ and $\Delta v = 0$ in $\mathbb{R}^N \setminus \overline{G}$ are implied by the conditions $v \in C^1(\mathbb{R}^N)$ and

$$\int_{\mathbb{R}^N} \nabla \varphi \cdot \nabla v = 0 \quad \text{for all} \quad \varphi \in C_c^\infty(\mathbb{R}^N \setminus \overline{G}),$$

verify that the foregoing formulae for u do indeed represent the Newtonian potential of unit density in S_β.

Exercise A.31 In order to discuss the Newtonian potential of a density function f on an *unbounded* open set $G \subset \mathbb{R}^N$, we demand that

$$Wf \in L_1(G), \quad \text{where} \quad W(x) := \begin{cases} 1 + |x| & \text{if } N = 1, \\ \log(e + 2|x|) & \text{if } N = 2, \\ 1 & \text{if } N \geq 3. \end{cases} \quad (A.78)$$

As preparation for extending certain theorems, prove that, if (A.78) holds, if α is a multi-index and if $\delta \in (0, 1]$, then

$$\left| \int_{G \setminus \mathscr{B}(x_0, \delta)} (\partial^\alpha K)(x_0 - x) f(x) \, dx \right|$$

$$\leq \begin{cases} \frac{1}{2}(1 + |x_0|) \, \|Wf\| & \text{if } N = 1, |\alpha| = 0, \\[2mm] \frac{1}{2\pi} \left\{ \log \frac{1 + 2|x_0|}{\delta} + 1 \right\} \|Wf\| & \text{if } N = 2, |\alpha| = 0, \\[2mm] C_\alpha \delta^{-N+2-|\alpha|} \|f\| & \text{if } N \geq 3, |\alpha| = 0 \\ & \text{or } N \geq 1, |\alpha| \geq 1, \end{cases} \quad (A.79)$$

where $\|.\| = \|. \mid L_1(G)\|$, the constant C_α is as in (A.31), and G may be unbounded.

Prove also that, if $f \in L_p(G)$ for some $p > 1$, then (A.78) ensures that $f \in L_s(G)$ for $1 \leq s \leq p$.

[Observe that, with the notation (A.22b), we have $\log(1/R) \leq \log(1/\delta)$ if $\delta \leq R \leq 1$; also, $\log R \leq \log 2r_0$ if $R \geq 1$ and $r \leq r_0$; finally, $\log R \leq \log 2r$ if $R \geq 1$ and $r_0 \leq r$.]

Exercise A.32 Use the results of Exercise A.31, and inspection of proofs for a bounded set G, to establish the following results for an *unbounded* open set $G \subset \mathbb{R}^N$.

(i) Theorem A.5 remains valid if the hypothesis $f \in L_1(G)$ is replaced by (A.78).

(ii) Theorems A.6 and A.8, Lemma A.9 and Theorem A.11 remain valid if the hypothesis (A.78) is added.

(iii) Theorem A.16 remains valid if the symbol $C^{0,\mu}(.)$ is replaced by $C_b^{0,\mu}(.)$, and the hypothesis (A.78) is added.

Exercise A.33 This exercise concerns an extension of Theorem A.16 to very smooth density functions f.

(i) Let G be a bounded open subset of $\mathbb{R}^N$, $N \geq 2$; assume that either ∂G is of class C^1 or G is listed in Remark D.4. Let u be the Newtonian potential of a density function $f \in C^\infty(\overline{G})$ and let $x_0 \in G$. Prove that

$$(\partial_i u)(x_0) = -\int_{\partial G} n_i(x) K(x - x_0) f(x) \, dS(x) + \int_G K(x - x_0) (\partial_i f)(x) \, dx,$$
(A.80)

and that, for each multi-index α, in terms of integers $i = i(\alpha, j)$ and multi-indices $\rho = \rho(\alpha, j)$ and $\sigma = \sigma(\alpha, j)$ that need not be specified precisely,

$$\begin{aligned}(\partial^\alpha u)(x_0) &= \sum_{j=1}^{|\alpha|} (-1)^{|\alpha|+j-1} \int_{\partial G} n_i(x) (\partial^\rho K)(x - x_0) (\partial^\sigma f)(x) \, dS(x) \\ &+ \int_G K(x - x_0) (\partial^\alpha f)(x) \, dx,\end{aligned}$$
(A.81)

where $|\rho| + |\sigma| = |\alpha| - 1$ and the sum over j is zero when $|\alpha| = 0$. (If ∂^α is written as $\partial_{i_m} \cdots \partial_{i_2} \partial_{i_1}$, where $m = |\alpha|$, then a precise but labyrinthine form of (A.81) can be found).

(ii) Now let $f \in C^k(\overline{G})$ and let $\partial^\beta f \in C^{0,\mu}(\overline{G})$ if $|\beta| = k$; here $k \in \mathbb{N}_0$, $\mu \in (0,1)$ and μ is the same for all $\partial^\beta f$ of order k. Prove that $\partial^\gamma u \in C^{0,\mu}(\overline{\Omega})$ if $|\gamma| = k+2$ and $\overline{\Omega} \subset G$, and that $u \in C^{k+2}(G)$.

Appendix B. Rudimentary Facts about Harmonic Functions and the Poisson Equation

B.1 Real-analytic functions

The formula

$$\psi(t) := \begin{cases} \exp\left(-\dfrac{1}{1-t^2}\right), & -1 < t < 1, \\ 0, & |t| \geq 1, \end{cases}$$

defines a function $\psi \in C_c^\infty(\mathbb{R})$; it is the case $N = 1$ of the function $\psi : \mathbb{R}^N \to \mathbb{R}$ considered in Exercise 1.15. Such functions have many uses (as test functions and, after scaling, as kernels in the smoothing operation), but we may feel that they are imperfect in some way, that they are man-made, that they are nylon rather than silk. Our instinct is sound, for the flaw of ψ is this: if we form the Taylor expansion about $t = 1$, namely,

$$\psi(1 + x) = \sum_{n=0}^{k-1} \psi^{(n)}(1) \frac{x^n}{n!} + R_k(x),$$

where

$$R_k(x) = \int_0^x \frac{(x-y)^{k-1}}{(k-1)!} \psi^{(k)}(1+y) \, \mathrm{d}y = \psi^{(k)}(1+\theta x) \frac{x^k}{k!}, \quad 0 \leq \theta \leq 1,$$

then there is *no* number $\rho > 0$ such that $R_k(x) \to 0$ for all $x \in (-\rho, \rho)$ as $k \to \infty$. If there were such a number $\rho \in (0, 1)$, say, then we would have

$$\psi\left(1 - \tfrac{1}{2}\rho\right) = \sum_{n=0}^{\infty} \psi^{(n)}(1) \frac{(-\tfrac{1}{2}\rho)^n}{n!},$$

which is a contradiction because $\psi\left(1 - \tfrac{1}{2}\rho\right) > 0$, while $\psi^{(n)}(1) = 0$ for each n.

Real-analytic functions are real-valued functions that lack this flaw;

for a real-analytic function f and for every point p in its domain, the Taylor series of f about p has a positive radius of convergence. Here we make this statement precise for functions defined on open subsets of $\mathbb{R}^N$; then we establish a test for real-analyticity that is often more convenient than the definition.

Notation As always, Ω is an open, non-empty subset of $\mathbb{R}^N$; unless the contrary is stated, N is any positive integer. We use multi-indices α (Definition A.3) with $\alpha! := \alpha_1! \alpha_2! \ldots \alpha_N!$.

Theorem B.1 (Taylor's formula). *If $f \in C^k(\Omega)$ and if the line segment $\{(1-t)p + tx \mid 0 \le t \le 1\}$ is a subset of Ω, then*

$$f(x) = \sum_{|\alpha| \le k-1} (\partial^\alpha f)(p) \frac{(x-p)^\alpha}{\alpha!} + \sum_{|\alpha|=k} (\partial^\alpha f)\big((1-\theta)p + \theta x\big) \frac{(x-p)^\alpha}{\alpha!}, \quad \text{(B.1)}$$

for some number $\theta = \theta(p, x) \in [0, 1]$.

Proof The proof is very similar to several calculations in Appendix A, in particular, to the proof of Theorem A.12, but there the mean-value form of the remainder was not used.

Let $h := x - p$ and $g(t) := f(p + th)$, $0 \le t \le 1$, so that $d/dt = h_1 \partial_1 + \cdots + h_N \partial_N$; the multinomial theorem (Exercise A.22) states that

$$\left(\frac{d}{dt}\right)^m g(t) = \sum_{|\alpha|=m} \frac{m!}{\alpha!} h^\alpha (\partial^\alpha f)(p + th), \quad \text{(B.2)}$$

and repeated integration by parts gives

$$\begin{aligned}
f(p + h) = g(1) &= g(0) + g'(0) + \cdots + \frac{g^{(k-1)}(0)}{(k-1)!} \\
&\quad + \int_0^1 \frac{(1-t)^{k-1}}{(k-1)!} g^{(k)}(t) \, dt \\
&= \sum_{m=0}^{k-1} \frac{g^{(m)}(0)}{m!} + \frac{g^{(k)}(\theta)}{k!}, \quad 0 \le \theta \le 1, \quad \text{(B.3)}
\end{aligned}$$

by the first mean-value theorem for integrals. (The refinement $0 < \theta < 1$ is not needed here.) Inserting (B.2) into (B.3), we obtain

$$f(p + h) = \sum_{m=0}^{k-1} \sum_{|\alpha|=m} \frac{h^\alpha}{\alpha!} (\partial^\alpha f)(p) + \sum_{|\alpha|=k} \frac{h^\alpha}{\alpha!} (\partial^\alpha f)(p + \theta h),$$

which is (B.1) in a slightly different notation. $\square$

Definition B.2 A function $f \in C^\infty(\Omega)$ is *real-analytic* in Ω iff, for each $p \in \Omega$, there is a radius $\rho = \rho(p) > 0$ such that

$$f(x) = \sum_\alpha (\partial^\alpha f)(p) \frac{(x-p)^\alpha}{\alpha!} \quad \text{whenever} \quad x \in \mathscr{B}(p, \rho) \cap \Omega. \tag{B.4}$$

Here $\sum_\alpha$ means summation over all multi-indices of length N, and convergence to $f(x)$ of the series is the main part of the statement. $\square$

Lemma B.3 *For each* $n \in \mathbb{N}$,

$$\mathrm{e}\left(\frac{n}{\mathrm{e}}\right)^n \leq n! \leq \frac{\mathrm{e}^2}{4}\left(\frac{n+1}{\mathrm{e}}\right)^{n+1}. \tag{B.5}$$

Proof For $n = 1$, there is equality on both sides of (B.5). For $n \geq 2$,

$$\log(n!) = \log 2 + \log 3 + \cdots + \log n > \int_1^n \log t \, \mathrm{d}t,$$

because $\log 2 > \log t$ for $1 \leq t < 2, \ldots$, $\log n > \log t$ for $n-1 \leq t < n$. Similarly, for $n \geq 2$,

$$\log(n!) = \log 2 + \log 3 + \cdots + \log n < \int_2^{n+1} \log t \, \mathrm{d}t.$$

Evaluating the integrals and taking exponentials, one obtains (B.5). $\square$

Theorem B.4 *A function* $f \in C^\infty(\Omega)$ *is real-analytic in* Ω *if and only if, for each compact subset* $E \subset \Omega$, *there are positive constants* $A = A(E)$ *and* $B = B(E)$ *such that*

$$|\partial^\alpha f(x)| \leq A\big(B|\alpha|\big)^{|\alpha|} \quad \text{for all} \ \ x \in E \tag{B.6}$$

and for all multi-indices α *of length* N. *(It is to be understood that* $|\alpha|^{|\alpha|} = 1$ *for* $|\alpha| = 0$.)

Proof of sufficiency Let (B.6) hold and let $p \in \Omega$ be given. Referring to (B.1), we write $h := x - p$,

$$R_k(h) := \sum_{|\alpha|=k} (\partial^\alpha f)(p + \theta h) \frac{h^\alpha}{\alpha!} \quad (0 \leq \theta \leq 1), \tag{B.7}$$

and wish to find $\rho = \rho(p) > 0$ such that $R_k(h) \to 0$ as $k \to \infty$ whenever $|h| < \rho$. First, let $\rho \leq \frac{1}{2} d_p$, where $d_p := \mathrm{dist}(p, \partial\Omega)$ if $\Omega \neq \mathbb{R}^N$, or $d_p := 2$

if $\Omega = \mathbb{R}^N$. Then $E := \overline{\mathscr{B}(p, \frac{1}{2}d_p)}$ is suitable for application of (B.6); if A and B are the constants corresponding to this set, we have

$$\left| R_k(h) \right| \leq \sum_{|\alpha|=k} A(Bk)^k \frac{|h_1|^{\alpha_1} \dots |h_N|^{\alpha_N}}{\alpha!}$$

$$= A(Bk)^k \frac{\left(|h_1| + \cdots + |h_N| \right)^k}{k!}$$

by the multinomial theorem (Exercise A.22). In view of (B.5) and the Cauchy–Schwarz inequality,

$$\frac{1}{k!} \leq \frac{1}{e} \left(\frac{e}{k} \right)^k \quad \text{and} \quad \sum_{j=1}^N |h_j| \leq \left\{ \sum_j 1^2 \right\}^{1/2} \left\{ \sum_j h_j^2 \right\}^{1/2} = N^{1/2} |h|,$$

so that

$$\left| R_k(h) \right| \leq \frac{A}{e} (Be N^{1/2} |h|)^k \to 0 \quad \text{as} \quad k \to \infty$$

if $|h| < \rho$ and if we choose ρ to be the smaller of $\frac{1}{2}d_p$ and $1/BeN^{1/2}$. $\quad\square$

The proof of necessity [that real-analyticity in Ω implies (B.6) for every compact set $E \subset \Omega$] is somewhat longer and harder; hints are provided in Exercises B.34 and B.35. These exercises are for enthusiasts; we shall use the sufficiency of (B.6), but we need the necessity only to know that the condition is a good one.

B.2 Smoothness and mean-value properties of harmonic functions

The word *harmonic* has many meanings in mathematics and science, but the phrase *harmonic function* usually means a function that satisfies the Laplace equation in some sense. In this book, the precise meaning must be the following, for reasons to be given presently.

Definition B.5 A function $v : \Omega \to \mathbb{R}$ is *harmonic in* Ω iff it is locally integrable in Ω (integrable on each compact subset of Ω) and

$$\int_\Omega (\triangle \varphi) v = 0 \quad \text{whenever} \quad \varphi \in C_c^\infty(\Omega) \quad \text{and} \quad \varphi \geq 0. \tag{B.8}$$

Henceforth $L_1^{\text{loc}}(\Omega)$ will denote the set of functions that are locally integrable in Ω. $\quad\square$

This definition coincides with a part of Definition 2.20 and states that, according to Definitions 2.10 and A.7, a harmonic function v is a distributional solution of the Laplace equation $\triangle v = 0$ in Ω. In Definition A.7, the condition $\varphi \geq 0$ is omitted; that this has no effect is explained in Remark 8 of §A.2.

The need to adopt Definition B.5 in this book comes from our need to have maximum principles for subharmonic functions that lack second derivatives here and there. That such maximum principles are required in applications is explained before Definition 2.10. Once the word *subharmonic* has been defined in a distributional sense, the corresponding meanings must be given to *superharmonic* and *harmonic*, otherwise our confusion would be total.

Terminology Here as elsewhere, *almost everywhere* or *a.e.* means: except on a set of measure zero. The measure will be Lebesgue measure on $\mathbb{R}^N$ (volume measure) until the contrary is stated.

Theorem B.6 *A function v is harmonic in Ω if and only if it is equal almost everywhere in Ω to a function u satisfying*

$$u \in C^\infty(\Omega) \quad and \quad \triangle u = 0 \quad in \ \Omega. \tag{B.9}$$

Proof (i) 'If'. Let (B.9) hold; then [by (A.47) and (A.48)]

$$0 = \int_\Omega \varphi \triangle u = \int_\Omega (\triangle \varphi) u \quad \text{whenever} \ \varphi \in C_c^\infty(\Omega) \quad \text{and} \quad \varphi \geq 0.$$

Since $v = u$ a.e. in Ω, it follows that $v \in L_1^{\text{loc}}(\Omega)$ and that v satisfies (B.8).

(ii) 'Only if'. Let $v \in L_1^{\text{loc}}(\Omega)$ and let it satisfy (B.8). In this proof a set G will be called *admissible* iff $\overline{G} \subset \Omega$ and G is bounded, open and not empty. It suffices to prove that, for each admissible set G, there is a function $u(.\,;G) \in C^\infty(G)$ such that $u(x;G) = v(x)$ a.e. in G. This is sufficient for the following reasons.

(a) Compatibility: if G and G' are admissible sets that intersect, then $u(x;G) = u(x;G')$ for $x \in G \cap G'$; first, this holds almost everywhere in $G \cap G'$, by equality a.e. to $v(x)$, then it holds everywhere in $G \cap G'$, by continuity of $u(.\,;G)$ and $u(.\,;G')$.

(b) Because of (a), we can define the desired function u by setting $u(x) := u(x;G)$ whenever $x \in G$. [For each point $x \in \Omega$, there is no shortage of admissible sets G containing x.] Then $u \in C^\infty(\Omega)$ because each function $u(.\,;G) \in C^\infty(G)$. That $\triangle u = 0$ in Ω follows from (B.8),

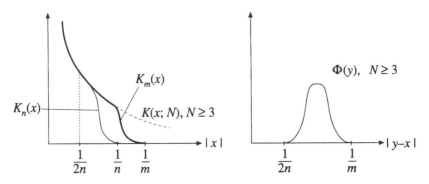

Fig. B.1.

which now implies that

$$\int_\Omega \varphi \triangle u = 0 \qquad \text{whenever} \quad \varphi \in C_c^\infty(\Omega) \quad \text{and} \quad \varphi \ge 0,$$

and from Exercise 1.16.

(c) The boundedness of an admissible set G will be needed, when Ω is unbounded but $\Omega \ne \mathbb{R}^N$, to ensure that $\text{dist}(\overline{G}, \partial\Omega) > 0$ [because then $\overline{G}$ is compact, $\partial\Omega$ is closed and the two are disjoint].

(iii) We prepare to form a special test function φ. First, let $\mu \in C^\infty[0, \infty)$ be non-increasing and such that $\mu(t) = 1$ for $0 \le t \le 1/2$, while $\mu(t) = 0$ for $t \ge 1$. [To construct such a function, apply the smoothing operation of Exercise 1.23, with $\rho = 1/5$, say, to the function that equals 1 on $(-\infty, 3/4)$ and vanishes elsewhere on $\mathbb{R}$.] Let K be the Newtonian kernel, introduced by (A.18), and define, for $m \in \mathbb{N}$ and $x \in \mathbb{R}^N \setminus \{0\}$,

$$K_m(x) := \begin{cases} K(x; N)\,\mu(m|x|) & \text{if } N \ne 2, \\[2mm] \dfrac{1}{2\pi}\left(\log\dfrac{1}{m|x|}\right)\mu(m|x|) & \text{if } N = 2. \end{cases}$$

The function K_m, illustrated in Figure B.1, has two useful offshoots, as follows.

(a) Let

$$h_m(x) := \begin{cases} \triangle K_m(x) & \text{if } x \ne 0, \\ 0 & \text{if } x = 0. \end{cases}$$

We shall see presently that h_m is a smoothing kernel, of smoothing radius $1/m$, that has every property of the smoothing kernel in Exercise

1.23 except (at least in some cases) non-negativity. In other words, $h_m(x) = m^N h_1(mx)$ and

$$h_m \in C_c^\infty(\mathbb{R}^N), \quad \operatorname{supp} h_m \subset \overline{\mathscr{B}(0, 1/m)}, \quad \int_{\mathbb{R}^N} h_m = 1. \tag{B.10}$$

(b) For a given admissible set G, let $\delta := \operatorname{dist}(\overline{G}, \partial\Omega)$ if $\Omega \neq \mathbb{R}^N$, or let $\delta = 1$ if $\Omega = \mathbb{R}^N$. Define (Figure B.1)

$$\left. \begin{array}{l} \Phi(y) := K_m(y - x) - K_n(y - x) \\ \text{for} \quad y \in \mathbb{R}^N \setminus \{x\}, \quad x \in G \quad \text{and} \quad n > m > 2/\delta; \end{array} \right\} \tag{B.11}$$

here x, m and n are parameters. Define $\Phi(x)$ by continuity, so that $\Phi(x) := 0$ if $N \neq 2$ and $\Phi(x) := (1/2\pi)\log(n/m)$ if $N = 2$. The definition implies that $\Phi \leq 0$ if $N = 1$ and that $\Phi \geq 0$ if $N \geq 2$; we shall see that $\Phi \in C_c^\infty(\Omega)$. Since (B.8) extends to functions $\psi \in C_c^\infty(\Omega)$ such that $\psi \leq 0$ [choose $\varphi = -\psi$], we may use Φ as a test function in (B.8).

Regarding the kernel h_m. An easy calculation shows that $h_m(x) = m^N h_1(mx)$. The first two claims in (B.10) are implied by the definitions of K_m and h_m; indeed, $h_m(x) = 0$ also for $|x| \leq 1/2m$, because $\triangle K(x; N) = 0$ if $x \neq 0$. That $\int h_m = 1$ follows from the divergence theorem [or from integration of $(d/dr)(r^{N-1} \, dK_m/dr)$ with respect to r]:

$$\int_{\mathscr{B}(0, 1/m)} \triangle K_m(x) \, dx = \int_{r=1/m} \frac{dK_m}{dr} \, dS - \int_{r=1/2m} \frac{dK_m}{dr} \, dS = 0 + 1$$

by the basic property (A.17) of the Newtonian kernel.

Regarding the function Φ. It is clear that $\Phi \in C_c^\infty(\mathbb{R}^N)$ with $\operatorname{supp} \Phi \subset \overline{\mathscr{B}(x, 1/m)}$. Moreover, $\overline{\mathscr{B}(x, 1/m)} \subset \Omega$ because $x \in G$, so that $\operatorname{dist}(x, \partial\Omega) > \delta > 2/m$ when $\Omega \neq \mathbb{R}^N$; consequently, $\Phi \in C_c^\infty(\Omega)$.

(iv) For a given admissible set G, let δ be the number defined before (B.11), and let

$$B := \bigcup_{x \in G} \mathscr{B}\left(x, \tfrac{1}{2}\delta\right), \quad v_G(y) := \begin{cases} v(y) & \text{if } y \in B, \\ 0 & \text{if } y \in \mathbb{R}^N \setminus B, \end{cases}$$

$$v_{G,m}(x) := \int_{\mathbb{R}^N} h_m(x - y) v_G(y) \, dy$$

for $m \in \mathbb{N}$ and $x \in \mathbb{R}^N$. Then $v_{G,m} \in C^\infty(\mathbb{R}^N)$ [by Exercise 1.23, (i), the proof of which does not need the condition $k_\rho \geq 0$]. Also $v_G \in L_1(\mathbb{R}^N)$, so that $\|v_G - v_{G,m} \mid L_1(\mathbb{R}^N)\| \to 0$ as $m \to \infty$ [by Exercise 1.25 with $p = 1$; the hypotheses there are amply satisfied]. This last implies that a subsequence of $(v_{G,m})$ converges pointwise to v_G almost everywhere in $\mathbb{R}^N$ (Rudin 1970, p.73; Weir 1973, p.171), hence to v a.e. in G.

But, if $x \in G$ and $n > m > 2/\delta$, then $h_n(x-.)$ and $h_m(x-.)$ have supports within $\mathcal{B}(x, \frac{1}{2}\delta)$, which is in Ω and on which $v_G(y) = v(y)$, so that

$$
\begin{aligned}
v_{G,m}(x) - v_{G,n}(x) &= \int_\Omega \{ (\triangle K_m)(x-y) - (\triangle K_n)(x-y) \} v(y) \, \mathrm{d}y \\
&= \int_\Omega (\triangle \Phi)(y) v(y) \, \mathrm{d}y = 0
\end{aligned}
\tag{B.12}
$$

by (B.8) with the test function Φ in (B.11). Thus *all functions $v_{G,m}$ are equal on G for $m > 2/\delta$*; since a 'subsequence converges pointwise' to v a.e. in G, we have $v_{G,m} = v$ a.e. in G for $m > 2/\delta$. Let $u(x; G) := v_{G,m}(x)$ for $x \in G$ and $m > 2/\delta$; then $u(.\,; G) \in C^\infty(G)$ and $u(x; G) = v(x)$ a.e. in G, as desired. $\qquad\square$

Definition B.7 We shall say that u is *smoothly harmonic in Ω* iff $u \in C^\infty(\Omega)$ and $\triangle u = 0$ in Ω. $\qquad\square$

Note that, if v is harmonic and *continuous* in Ω, then v is smoothly harmonic in Ω. For, in Theorem B.6, equality of v and u almost everywhere in Ω becomes equality everywhere in Ω if v is continuous.

Theorem B.8 (the mean-value property of harmonic functions). *If u is smoothly harmonic in a ball $B := \mathcal{B}(c, \rho)$, and $u \in C(\overline{B})$, then the mean values of u over ∂B and over B are both equal to the value of u at the centre of B. That is,*

$$
\frac{1}{|\partial B|} \int_{\partial B} u = u(c), \qquad \frac{1}{|B|} \int_B u = u(c).
\tag{B.13a,b}
$$

Proof   It will be helpful to write integrals over $\partial \mathcal{B}(c, r)$ as integrals over the unit sphere $Y := \partial \mathcal{B}(0, 1)$; we shall do this by means of the transformation $x = c + ry$, in which $r := |x - c|$ and $y \in Y$.

First, let $B_0 := \mathcal{B}(c, r_0)$ with $0 < r_0 < \rho$, and apply the divergence theorem:

$$
\begin{aligned}
0 = \int_{B_0} \triangle u = \int_{\partial B_0} \frac{\partial u}{\partial n} &= \int_Y \left. \frac{\partial u(c+ry)}{\partial r} \right|_{r=r_0} r_0^{N-1} \, \mathrm{d}S(y) \quad \left(\frac{\partial}{\partial n} := n \cdot \nabla \right) \\
&= r_0^{N-1} \frac{\mathrm{d}}{\mathrm{d}r_0} \int_Y u(c + r_0 y) \, \mathrm{d}S(y).
\end{aligned}
$$

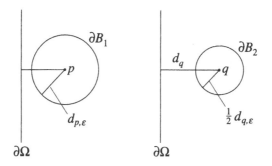

Fig. B.2.

Now discard the factor r_0^{N-1} and integrate the rest over $(0, r_1)$, where $0 < r_1 < \rho$; then

$$0 = \int_Y u(c + r_1 y)\, dS(y) - \int_Y u(c)\, dS(y) = \frac{1}{r_1^{N-1}} \int_{\partial B_1} u - |Y| u(c), \quad (B.14)$$

where $B_1 := \mathscr{B}(c, r_1)$. Since $|\partial B_1| = r_1^{N-1} |Y|$, the result (B.14) is (B.13a) for the sphere ∂B_1. To prove (B.13a) for ∂B, let $r_1 \to \rho$ and use the continuity of u on $\overline{B}$.

To prove (B.13b), multiply (B.14) by r_1^{N-1} and integrate over $(0, \rho)$ with respect to r_1. □

Theorem B.9 *If u is smoothly harmonic in Ω and $\Omega \neq \mathbb{R}^N$, then*

$$|\partial^\alpha u(x)| \leq \left\{ \frac{N|\alpha|}{\operatorname{dist}(x, \partial\Omega)} \right\}^{|\alpha|} \sup_{y \in \Omega} |u(y)| \quad (B.15)$$

for all $x \in \Omega$ and all multi-indices α of length N. (It is to be understood that $|\alpha|^{|\alpha|} = 1$ for $|\alpha| = 0$).

Proof Since $u \in C^\infty(\Omega)$, each derivative $\partial^\alpha u \in C^\infty(\Omega)$; since also ∂^α and Δ commute, $\partial^\alpha u$ is smoothly harmonic in Ω. The proof of (B.15) is by induction and uses the mean-value property of each derivative over balls. We may suppose that $\sup |u(y)| < \infty$ (otherwise, there is nothing to prove); let

$$M := \sup_{y \in \Omega} |u(y)|, \qquad d_x := \operatorname{dist}(x, \partial\Omega), \qquad d_{x,\varepsilon} := d_x(1 - \varepsilon)$$

for some $\varepsilon \in (0, 1)$ that is fixed until we reach step (iv).

(i) Given $p \in \Omega$, we estimate $(\partial_j u)(p)$ as follows for each j. Let $B_1 := \mathscr{B}(p, d_{p,\varepsilon})$ (Figure B.2), apply the mean-value result (B.13b) for balls to $\partial_j u$, and use the divergence theorem:

$$\left|(\partial_j u)(p)\right| = \frac{1}{|B_1|}\left|\int_{B_1}\partial_j u\right| = \frac{1}{|B_1|}\left|\int_{\partial B_1} n_j u\right| \leq \frac{|\partial B_1|}{|B_1|}M = \frac{N}{d_{p,\varepsilon}}M. \quad \text{(B.16)}$$

(ii) Given $q \in \Omega$, we estimate $(\partial_i\partial_j u)(q)$ as follows for each i and j. Let $B_2 := \mathcal{B}(q, \frac{1}{2}d_{q,\varepsilon})$ (Figure B.2), and begin as before:

$$\left|(\partial_i\partial_j u)(q)\right| = \frac{1}{|B_2|}\left|\int_{\partial B_2} n_i\partial_j u\right| \leq \frac{N}{\frac{1}{2}d_{q,\varepsilon}}\sup_{x\in\partial B_2}\left|\partial_j u(x)\right|.$$

Now, for $x \in \partial B_2$ we have $d_x \geq d_q - \frac{1}{2}d_{q,\varepsilon} > \frac{1}{2}d_q$, so that (B.16) yields

$$\left|\partial_j u(x)\right| \leq \frac{N}{d_{x,\varepsilon}}M \leq \frac{N}{\frac{1}{2}d_{q,\varepsilon}}M \quad (x \in \partial B_2),$$

whence

$$\left|(\partial_i\partial_j u)(q)\right| \leq \left\{\frac{2N}{d_{q,\varepsilon}}\right\}^2 M. \quad \text{(B.17)}$$

(iii) We now have

$$\left|\partial^\alpha u(x)\right| \leq \left\{\frac{|\alpha|N}{d_{x,\varepsilon}}\right\}^{|\alpha|} M \qquad \text{for all } x \in \Omega \quad \text{(B.18)}$$

if $|\alpha| = 0, 1$ or 2. Assume that (B.18) holds for $|\alpha| = k-1$; to extend it to $|\alpha| = k$ at a given point $c \in \Omega$, choose $B_k := \mathcal{B}\left(c,(1/k)d_{c,\varepsilon}\right)$ and proceed as in step (ii). Then

$$\left|(\partial_j\partial^\alpha u)(c)\right| \leq \frac{N}{(1/k)d_{c,\varepsilon}}\sup_{x\in\partial B_k}\left|\partial^\alpha u(x)\right| \qquad (|\alpha| = k-1);$$

for $x \in \partial B_k$,

$$d_x \geq d_c - \frac{1}{k}d_{c,\varepsilon} > \frac{k-1}{k}d_c \implies \left|\partial^\alpha u(x)\right| \leq \left\{\frac{kN}{d_{c,\varepsilon}}\right\}^{k-1} M,$$

whence

$$\left|(\partial_j\partial^\alpha u)(c)\right| \leq \left\{\frac{kN}{d_{c,\varepsilon}}\right\}^k M.$$

Thus (B.18) is extended by induction to every multi-index α of length N.

(iv) Since (B.18) holds for every $\varepsilon \in (0,1)$, it holds also for $\varepsilon = 0$. [Otherwise, for some point $y \in \Omega$ and some multi-index β,

$$\left|(\partial^\beta u)(y)\right| > \left\{\frac{|\beta|N}{d_y}\right\}^{|\beta|} M, \qquad \text{whence} \qquad \left|(\partial^\beta u)(y)\right| = \left\{\frac{|\beta|N}{d_{y,\delta}}\right\}^{|\beta|} M$$

for some number $\delta > 0$, and this contradicts (B.18) if $\varepsilon = \delta/2$.] $\qquad\square$

Theorem B.9 is remarkable in that, outside the class of holomorphic (or complex analytic) functions, it is rare for the supremum of the modulus of a function to control the magnitudes of all derivatives. For example, the real-analytic function defined by

$$u(x) = \cos \lambda x, \qquad x \in (-b, b) \subset \mathbb{R}, \quad \lambda = \text{const.} > 0,$$

satisfies (B.15), for $|\alpha| = 2m$, $m \in \mathbb{N}$, and $x = 0$, only if $\lambda \leq 2m/b$. Perhaps better evidence of the strength of Theorem B.9 is that it yields rather easily Theorems B.10 and B.11, both of which are reminiscent of properties of holomorphic functions.

Theorem B.10 *If u is smoothly harmonic in Ω, then it is real-analytic in Ω.*

Proof By Theorem B.4 it is sufficient to prove that, for each compact set $E \subset \Omega$, there are positive constants $A = A(E)$ and $B = B(E)$ such that

$$|\partial^\alpha u(x)| \leq A(B|\alpha|)^{|\alpha|} \qquad \text{for all } x \in E \tag{B.19}$$

and for all multi-indices α of length N. Let E be given; observing that $\text{dist}(E, \partial\Omega) > 0$ when $\Omega \neq \mathbb{R}^N$ [because E is compact, $\partial\Omega$ is closed and the two are disjoint], choose a bounded open set G such that $E \subset G$ and $\overline{G} \subset \Omega$, and define

$$A := \sup_{y \in G} |u(y)| = \max_{y \in \overline{G}} |u(y)|.$$

By Theorem B.9, applied to the set G rather than to Ω,

$$|\partial^\alpha u(x)| \leq A \left\{ \frac{N|\alpha|}{\text{dist}(x, \partial G)} \right\}^{|\alpha|} \leq A \left\{ \frac{N|\alpha|}{\text{dist}(E, \partial G)} \right\}^{|\alpha|} \qquad \text{for all } x \in E,$$

where $\text{dist}(E, \partial G) > 0$ for the same reason that $\text{dist}(E, \partial\Omega) > 0$ when $\Omega \neq \mathbb{R}^N$. Therefore (B.19) holds if we choose $B := N/\text{dist}(E, \partial G)$. $\square$

Theorem B.11 (a Liouville theorem). *If u is smoothly harmonic in $\mathbb{R}^N$ and $u(x) = o(r)$ as $r := |x| \to \infty$, then u is a constant.*

Proof We shall prove that $(\partial_j u)(x_0) = 0$ for each $x_0 \in \mathbb{R}^N$ and each $j \in \{1, \dots, N\}$ by applying Theorem B.9 with $|\alpha| = 1$ and with $\Omega = \mathcal{B}(0, R_m)$. Here (R_m) is an increasing sequence such that $R_1 > 2|x_0|$ for given $x_0 \in \mathbb{R}^N$ and such that $R_m \to \infty$ as $m \to \infty$. Then

$$|(\partial_j u)(x_0)| \leq \frac{N}{R_m - |x_0|} \max_{|x| \leq R_m} |u(x)| < \frac{2N}{R_m} |u(y^m)|, \tag{B.20}$$

where $|u(y^m)|$ is the maximum of $|u(x)|$ for $|x| \leq R_m$. If $|u(y^{m+1})| > |u(y^m)|$, then $|y^{m+1}| > |y^m|$; if $|u(y^{m+1})| = |u(y^m)|$, then we either have, or may take, $y^{m+1} = y^m$. Thus the sequence $(|y^m|)$ is non-decreasing. The theorem now follows from (B.20) because $|u(y^m)|/R_m \to 0$ as $m \to \infty$. [If $|y^m| \to \infty$, then $u(y^m) = o(|y^m|)$ and $|y^m| \leq R_m$; if $|y^m|$ is bounded, then so is $u(y^m)$.]   □

B.3 The Kelvin transformation

This section concerns a conformal map of $\mathbb{R}^N \setminus \{0\}$, and a related transformation of functions, under which smoothly harmonic functions remain smoothly harmonic.

Definition B.12 Given a sphere $\partial \mathcal{B}(0, a)$ in $\mathbb{R}^N$, we call

$$Sx := \frac{a^2}{r^2} x \qquad (x \in \mathbb{R}^N \setminus \{0\}, \ r := |x|) \tag{B.21}$$

the *reflection* or *inverse point* of x relative to $\partial \mathcal{B}(0, a)$. The *Kelvin transform* Tf relative to $\partial \mathcal{B}(0, a)$, of a function $f : \Omega \to \mathbb{R}$, is then defined by

$$(Tf)(x) := \left(\frac{a}{r}\right)^{N-2} f(Sx) \qquad (Sx \in \Omega), \tag{B.22}$$

provided that $\Omega \subset \mathbb{R}^N \setminus \{0\}$.   □

To calculate with S and T, we use points $\xi \in \mathbb{R}^N$ with $\rho := |\xi|$. If $\xi = Sx$, then $\rho r = a^2$ and

$$S\xi = \frac{a^2}{\rho^2} \xi = \frac{r^2}{a^2}\left(\frac{a^2}{r^2} x\right) = x,$$

so that S is its own inverse operator: $S^{-1} = S$. If $\xi = Sx$ and $\varphi = Tf$, then

$$\varphi(x) = \left(\frac{a}{r}\right)^{N-2} f(Sx) \ \Rightarrow \ f(\xi) = \left(\frac{a}{\rho}\right)^{N-2} \varphi(S\xi) = (T\varphi)(\xi),$$

so that T is also its own inverse operator: $T^{-1} = T$.

As regards the geometry of the transformation $\xi = Sx$, it is obvious that $\partial \mathcal{B}(0, a)$, the sphere of inversion, is mapped onto itself, and that the sets $\mathcal{B}(0, a) \setminus \{0\}$ and $\mathbb{R}^N \setminus \overline{\mathcal{B}(0, a)}$ are mapped onto each other. The Fréchet or total or linear derivative of the transformation is

$$S'(x) = \left(\frac{\partial \xi_i}{\partial x_j}(x)\right) = \frac{a^2}{r^2} M(x), \qquad \text{where} \ \ M_{ij}(x) := \delta_{ij} - \frac{2x_i x_j}{r^2}. \tag{B.23}$$

This shows that the *matrix $M(x)$ is orthogonal as well as symmetric*, because one checks easily that $M(x)M(x) = I$, the identity matrix. The orthogonality of $M(x)$ implies that, for all h and k in $\mathbb{R}^N \setminus \{0\}$, the angle between $S'(x)h$ and $S'(x)k$ equals that between h and k. Consequently, the map S is *conformal* in the sense that, if two smooth arcs intersect at a point, then the angle between them at that point is conserved by the mapping. [For, let $x = f(t)$ and $x = g(t)$ be descriptions of two smooth arcs that intersect at $x_0 = f(t_0) = g(t_0)$. The orthogonality of $M(x_0)$ ensures that the angle between $S'(x_0)f'(t_0)$ and $S'(x_0)g'(t_0)$ equals that between $f'(t_0)$ and $g'(t_0)$; the latter angle is that between tangents at x_0 to the original arcs $x = f(t)$ and $x = g(t)$, while the former angle is that between tangents at Sx_0 to the transformed arcs $\xi = S(f(t))$ and $\xi = S(g(t))$.]

Further properties of the transformation S require a small definition.

Definition B.13 By a *half-space in $\mathbb{R}^N$* we mean a set $\{ x \in \mathbb{R}^N \mid x \cdot k > \mu \}$, where $k \in \mathbb{R}^N \setminus \{0\}$ and $\mu \in \mathbb{R}$. $\square$

Exercise B.14 Let S be the reflection operator in (B.21), and let G be a ball or a half-space in $\mathbb{R}^N$.

(i) Prove that $S(G)$ is a ball if $0 \notin \overline{G}$, that $S(G)$ is a half-space if $0 \in \partial G$, and that $S(G \setminus \{0\})$ is the complement of a closed ball if $0 \in G$.

(ii) Prove that, for $N \geq 2$, $S(G) = G$ if and only if ∂G intersects $\partial \mathscr{B}(0,a)$ orthogonally.

[Orthogonality means here that normals n_G to ∂G and n_B to $\partial \mathscr{B}(0,a)$ satisfy $n_G(y) \cdot n_B(y) = 0$ at a point y of intersection.] $\square$

The next theorem shows that, if u is smoothly harmonic in Ω, then its Kelvin transform Tu is smoothly harmonic in $S(\Omega)$; Definition 2.20 and Exercise B.37 deal with an extension to merely subharmonic functions. It is this property of the Kelvin transform that makes it useful. Indeed, almost everything that can be done with holomorphic (or complex analytic) functions in the complex plane $\mathbb{C}$ by means of a Möbius (or 'bilinear') transformation, can be done with smoothly harmonic functions in $\mathbb{R}^N$ by a composition of translation ($u \mapsto u(.+c)$, where $c \in \mathbb{R}^N$), dilation ($u \mapsto u(\lambda .)$, where $\lambda \in \mathbb{R} \setminus \{0\}$) and the Kelvin transformation.

The proof of Theorem B.15 is not the shortest route to the formula (B.24), but includes details that are useful in applications of the transformation.

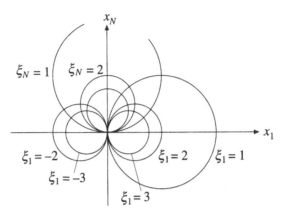

Fig. B.3.

Theorem B.15 *Let $f \in C^2(\Omega)$, where $\Omega \subset \mathbb{R}^N \setminus \{0\}$, and let Tf be its Kelvin transform, as in (B.22). Then, for $x \in S(\Omega)$,*

$$\triangle(Tf)(x) = \left(\frac{a}{r}\right)^{N+2} (\triangle f)(Sx). \tag{B.24}$$

Proof (i) Let $\xi = Sx$, $x \neq 0$, so that $x = S\xi$, $\xi \neq 0$. Consider two copies of $\mathbb{R}^N \setminus \{0\}$: the first has Cartesian co-ordinates $\xi_1, \ldots, \xi_N$; the second (Figure B.3) is the image of the first by S and has Cartesian co-ordinates $x_1, \ldots, x_N$, while $\xi_1, \ldots, \xi_N$ are *curvilinear co-ordinates* there.

Let $\partial x / \partial \xi_k := (\partial_k S)(\xi)$. In the second copy of $\mathbb{R}^N \setminus \{0\}$, the vector $\partial x / \partial \xi_k$ at any given point is tangential to a curve along which ξ_k increases while the other ξ_j are fixed. The vectors $\partial x / \partial \xi_1, \ldots, \partial x / \partial \xi_N$ at the given point are mutually orthogonal because they are images, under the conformal map S, of vectors along the co-ordinate lines in the first copy of $\mathbb{R}^N \setminus \{0\}$. To verify this orthogonality of $\partial x / \partial \xi_1, \ldots, \partial x / \partial \xi_N$, we infer from (B.23) that

$$\frac{\partial x}{\partial \xi_j} \cdot \frac{\partial x}{\partial \xi_k} = \frac{a^4}{\rho^4} \sum_{p=1}^{N} M_{pj}(\xi) M_{pk}(\xi) = \frac{a^4}{\rho^4} \delta_{jk}. \tag{B.25}$$

Thus $\xi_1, \ldots \xi_N$ are *orthogonal* curvilinear co-ordinates in the second copy of $\mathbb{R}^N \setminus \{0\}$.

(ii) We use arc-length functions h_j defined by

$$h_j(\xi) := \left| \frac{\partial x}{\partial \xi_j}(\xi) \right| = \frac{a^2}{\rho^2},$$

and basis vectors $b^j := (1/h_j)\partial x/\partial \xi_j$, for $j = 1,\ldots,N$. The rules (Kellogg 1929, pp. 181 and 183; Spiegel 1959, pp. 148 and 151)

$$\left(\frac{\partial}{\partial x_1},\ldots,\frac{\partial}{\partial x_N}\right) = \frac{b^1}{h_1}\frac{\partial}{\partial \xi_1} + \cdots + \frac{b^N}{h_N}\frac{\partial}{\partial \xi_N}, \tag{B.26a}$$

$$\begin{aligned}
\frac{\partial^2}{\partial x_1^2} + \cdots + \frac{\partial^2}{\partial x_N^2} &= \frac{1}{h_1 \ldots h_N}\left\{\frac{\partial}{\partial \xi_1}\left(\frac{h_2 \ldots h_N}{h_1}\frac{\partial}{\partial \xi_1}\right) + \cdots\right.\\
&\quad \left.+ \frac{\partial}{\partial \xi_N}\left(\frac{h_1 \ldots h_{N-1}}{h_N}\frac{\partial}{\partial \xi_N}\right)\right\}
\end{aligned} \tag{B.26b}$$

simplify here because all the h_j are equal.

(iii) Let $g(x) := f(Sx)$ and recall that $\triangle r^{-N+2} = 0$ in $\mathbb{R}^N \setminus \{0\}$ for all $N \in \mathbb{N}$; then, by the definition (B.22) of Tf and the Leibniz rule for repeated differentiation of products,

$$\triangle(Tf)(x) = 2\left\{\nabla\left(\frac{a}{r}\right)^{N-2}\right\} \cdot \nabla g(x) + \left(\frac{a}{r}\right)^{N-2}\triangle g(x),$$

where $a/r = \rho/a$ and $g(x) = f(Sx) = f(\xi)$. Application of (B.26a, b) now gives

$$\begin{aligned}
&\triangle(Tf)(x)\\
&= 2(N-2)a^{-N-2}\rho^N \sum_{j=1}^{N} \xi_j\frac{\partial f(\xi)}{\partial \xi_j}\\
&\quad + a^{-N-2}\rho^{3N-2} \sum_{j=1}^{N}\left\{(-2N+4)\rho^{-2N+2}\xi_j\frac{\partial f(\xi)}{\partial \xi_j} + \rho^{-2N+4}\frac{\partial^2 f(\xi)}{\partial \xi_j^2}\right\}\\
&= \left(\frac{\rho}{a}\right)^{N+2}(\triangle f)(\xi),
\end{aligned}$$

and this is the desired formula (B.24). $\qquad\qquad\qquad\qquad\qquad\qquad\square$

B.4 On the Dirichlet and Neumann problems

In Appendix A we encountered the Poisson equation, $-\triangle u = f$ in G, as an equation satisfied by the Newtonian potential u of a smooth density function f on G. Here we regard the Poisson equation as one to be solved when a boundary condition for u is specified.

Let $f : \Omega \to \mathbb{R}$ and $g : \partial\Omega \to \mathbb{R}$ be given functions. The *Dirichlet problem for* $-\triangle$ *in* Ω is to find $u \in C(\overline{\Omega}) \cap C^2(\Omega)$ such that

$$-\triangle u = f \quad \text{in } \Omega, \qquad u\big|_{\partial\Omega} = g. \tag{B.27}$$

The *Neumann problem for* $-\triangle$ *in* Ω is to find $u \in C^1(\overline{\Omega}) \cap C^2(\Omega)$ such that

$$-\triangle u = f \quad \text{in} \quad \Omega, \qquad \frac{\partial u}{\partial n}\bigg|_{\partial \Omega} = g, \qquad \text{(B.28)}$$

where n denotes the unit normal outward from Ω and $\partial/\partial n := n \cdot \nabla$.

These statements omit a great deal. Only pointwise solutions (C^2-solutions) have been mentioned. We have not said what smoothness the data f, g and $\partial\Omega$ must have in order that the solution u have the continuity properties that we have demanded. When Ω is bounded, we can hope to solve the Neumann problem only if the *compatibility condition*

$$-\int_{\Omega} f = \int_{\partial \Omega} g \qquad \text{(B.29)}$$

holds, because both sides of (B.29) equal $\int_{\partial\Omega} \partial u/\partial n$ for a smooth solution of (B.28) with smooth data. In fact, if Ω is not connected, (B.29) must hold for each component (each maximal connected subset) of Ω. When Ω is unbounded, *growth conditions* (or decay conditions), specifying how large $|u(x)|$ and perhaps $|\nabla u(x)|$ are allowed to become as $|x| \to \infty$, must be added to the statements of the Dirichlet and Neumann problems. Some of these gaps will be filled as we proceed.

Reduction to the Laplace equation or to zero boundary data (i) Often one can remove the term f from the Poisson equation in (B.27) and (B.28) by the substitution $u = U_f + v$, where U_f is the Newtonian potential of f. For example, if Ω is bounded and $f \in C^{0,\mu}(\overline{\Omega})$ for some $\mu \in (0, 1)$ [Definition A.10, with the suffix b omitted from $C_b^{0,\mu}$ because Ω is bounded], then Theorems A.11 and A.16 ensure that $U_f \in C^1(\overline{\Omega}) \cap C^2(\Omega)$ and that $-\triangle U_f = f$ pointwise in Ω. The Dirichlet problem is then to find $v \in C(\overline{\Omega}) \cap C^2(\Omega)$ such that

$$\triangle v = 0 \quad \text{in} \quad \Omega, \qquad v\big|_{\partial\Omega} = g - U_f\big|_{\partial\Omega}; \qquad \text{(B.30)}$$

the Neumann problem is to find $v \in C^1(\overline{\Omega}) \cap C^2(\Omega)$ such that

$$\triangle v = 0 \quad \text{in} \quad \Omega, \qquad \frac{\partial v}{\partial n}\bigg|_{\partial\Omega} = g - \frac{\partial}{\partial n} U_f\bigg|_{\partial\Omega}. \qquad \text{(B.31)}$$

In both cases, v (if it exists) will be not merely in $C^2(\Omega)$ but in $C^\infty(\Omega)$ (Theorem B.6), indeed, it will be real-analytic (Theorem B.10). Note that, if the compatibility condition (B.29) holds for f and g in the Neumann case, then it holds also for 0 and $g - \partial U_f/\partial n$.

(ii) Alternatively, if one can find a sufficiently smooth function $h : \overline{\Omega} \to \mathbb{R}$ such that $h = g$ on $\partial\Omega$ in the Dirichlet case, or $\partial h/\partial n = g$ on $\partial\Omega$ in the Neumann case, then the substitution $u = h + w$ yields a problem for w with boundary condition $w = 0$ on $\partial\Omega$ or $\partial w/\partial n = 0$ on $\partial\Omega$. However, for a Poisson equation with forcing function in $C^{0,\mu}(\overline{\Omega})$, it is necessary that $\triangle h \in C^{0,\mu}(\overline{\Omega})$; to find such a function h may not be easy if g and $\partial\Omega$ are less than beautiful.

We begin work on the Dirichlet and Neumann problems by proving uniqueness of solutions in some cases. Then *any* method of solution becomes acceptable in those cases, however contrived or squalid the method may seem, because what it produces is *the* solution.

Theorem B.16 *The Dirichlet problem for* $-\triangle$ *in* Ω *has at most one solution if either*

(a) Ω *is bounded, or*

(b) Ω *is a half-space* (Definition B.13) *and we add the growth condition*: $u(x) = o(r)$ *for* $x \in \overline{\Omega}$ *and* $r := |x| \to \infty$.

Proof Let $v := u_1 - u_2$ be the difference of two solutions; then $v \in C(\overline{\Omega}) \cap C^2(\Omega)$, we have $\triangle v = 0$ in Ω, and $v = 0$ on $\partial\Omega$. The additional smoothness implied by Theorems B.6 and B.10 is not needed here.

(a) If Ω is bounded, then Theorem 2.5 (our first and simplest maximum principle) states that $v \leq 0$ on $\overline{\Omega}$, because v is a C^2-subsolution relative to $\triangle$ and Ω, and that $v \geq 0$ on $\overline{\Omega}$, because $-v$ is also a C^2-subsolution. Thus $v = 0$ on $\overline{\Omega}$, as desired.

(b) If Ω is a half-space, we so choose co-ordinates that $\Omega = \{ x \in \mathbb{R}^N \mid x_N > 0 \}$, and note that $v(x) = o(r)$ as $r \to \infty$. For $N = 1$, the equation $\triangle v = 0$ implies that $v(x) = c_0 + c_1 x$, where c_0 and c_1 are constants; then $c_0 = 0$ because $v(0) = 0$, and $c_1 = 0$ because $v(x) = o(x)$ as $x \to \infty$. For $N \geq 2$, we apply Theorem 2.30 (a relatively advanced maximum principle). The hypotheses of that theorem are satisfied by both v and $-v$ (with room to spare in both smoothness and growth condition); hence $\sup_\Omega v = 0$ and $\inf_\Omega v = 0$, so that $v = 0$ on $\overline{\Omega}$ once again. $\qquad\square$

For the Neumann problem stated in (B.28), uniqueness in the strictest sense is impossible: if we add to a solution u any function k that is constant on each component [on each maximal connected subset] of $\overline{\Omega}$, then $u + k$ is again a solution. However, in many applications only uniqueness of ∇u is required, and we have already encountered an example of such uniqueness in Theorem A.1 and the remark following it.

Theorem B.17 *The Neumann problem for* $-\triangle$ *in* Ω *admits at most one function* ∇u *for solutions* u *if*
   (a) Ω *is bounded and* $\partial\Omega$ *is of class* C^1, *or*
   (b) Ω *is bounded and is listed in Remark D.4, or*
   (c) Ω *is a half-space and we add the growth condition:* $u(x) = o(r)$ *for* $x \in \overline{\Omega}$ *and* $r := |x| \to \infty$.

Proof The difference $v := u_1 - u_2$ of two solutions is now in $C^1(\overline{\Omega}) \cap C^2(\Omega)$ and satisfies $\triangle v = 0$ in Ω, $\partial v/\partial n = 0$ on $\partial\Omega$.

(a),(b) If Ω is bounded, and either $\partial\Omega$ is of class C^1 or Ω is listed in Remark D.4, then we have both a divergence theorem and approximations Ω_m to Ω as in Theorem D.9. Applying the divergence theorem to the vector field $v\nabla v$ and one of these sets Ω_m [which is legitimate because $v \in C^2(\overline{\Omega}_m)$], and observing that $\nabla \cdot (v\nabla v) = |\nabla v|^2$ because $\triangle v = 0$ in Ω [*a fortiori* in Ω_m], we obtain

$$\int_{\Omega_m} |\nabla v|^2 = \int_{\partial\Omega_m} v\frac{\partial v}{\partial n}.$$

Let $m \to \infty$; by Theorem D.9 and because $v \in C^1(\overline{\Omega})$, $\partial v/\partial n = 0$ on $\partial\Omega$,

$$\int_{\Omega} |\nabla v|^2 = 0,$$

and this implies that $|\nabla v| = 0$ on $\overline{\Omega}$, because $v \in C^1(\overline{\Omega})$.

(c) If Ω is a half-space, we choose co-ordinates again that make $\Omega = \{ x \in \mathbb{R}^N \mid x_N > 0 \}$, and notice again that $v(x) = o(r)$ as $r \to \infty$.

Let $\Omega_m := \{ x \in \mathbb{R}^N \mid x_N > 1/m \}$ for $m \in \mathbb{N}$. Integration by parts along co-ordinate lines, very much as in Remark 7 of §A.2, shows that

$$\int_{\Omega_m} (\triangle\varphi)v = -\int_{\partial\Omega_m} \left\{ (\partial_N\varphi)v - \varphi\partial_N v \right\} + \int_{\Omega_m} \varphi\triangle v \quad \text{if } \varphi \in C_c^\infty(\mathbb{R}^N);$$

the last integral vanishes because $\triangle v = 0$ in Ω. Let $m \to \infty$ and recall that $v \in C^1(\overline{\Omega})$ and $\partial_N v = 0$ on $\partial\Omega$; then

$$\int_{\Omega} (\triangle\varphi)v = -\int_{\partial\Omega} (\partial_N\varphi)v \qquad \text{if } \varphi \in C_c^\infty(\mathbb{R}^N). \qquad (B.32)$$

Now extend v to $\mathbb{R}^N$ as an even function of x_N by setting $v(x', -x_N) := v(x)$, where $x' := (x_1, \ldots, x_{N-1})$ and $x_N > 0$. Let $G := \{ x \in \mathbb{R}^N \mid x_N < 0 \}$; repeating for G the steps that led to (B.32), we obtain

$$\int_{G} (\triangle\varphi)v = \int_{\partial G} (\partial_N\varphi)v \qquad \text{if } \varphi \in C_c^\infty(\mathbb{R}^N). \qquad (B.33)$$

Also, $v \in C(\mathbb{R}^N)$. (In fact, $v \in C^1(\mathbb{R}^N)$, but we do not need this.) Add (B.32) and (B.33) for the same function φ; the boundary terms cancel and the result shows, in view of Definition B.5, that v is harmonic in $\mathbb{R}^N$. Since also v is continuous, it is smoothly harmonic in $\mathbb{R}^N$, by the remark following Definition B.7, and the extended function v is still $o(r)$ as $r \to \infty$. Therefore Theorem B.11 states that v is a constant. □

Regarding existence of solutions, we shall consider only some aspects of the method of Green functions, which is perhaps the most classical and constructive of the various approaches that are now available. Our limited treatment will yield results only for balls and half-spaces, with a slender hint of how the method proceeds for other sets. The main ingredients of the method are the notion of a fundamental solution, and the representation formula (B.35) below.

Definition B.18 Let K be the Newtonian kernel introduced as $K(.\,;N)$ in (A.18). A function F defined by

$$F(x_0, x) := K(x_0 - x) + q(x_0, x) \qquad \text{for } x_0 \in \Omega, \ x \in \overline{\Omega} \ \text{ and } \ x_0 \neq x$$

is a *fundamental solution of* $-\Delta$ *in* Ω iff, for each fixed $x_0 \in \Omega$,

$$\left. \begin{array}{l} q(x_0, .\,) \in C^1(\overline{\Omega}) \cap C^2(\Omega), \\ \Delta q(x_0, x) = 0 \quad \text{for all } x \in \Omega, \end{array} \right\} \tag{B.34}$$

where Δ is with respect to x. The function q will be called the *non-singular part of* F. □

Theorems B.6 and B.10 imply once again that $q(x_0, .\,)$ is real-analytic in Ω; we have written $C^2(\Omega)$ as a demand for a pointwise solution, not as a result. Obviously q is not determined uniquely by (B.34); presently we shall add one of several boundary conditions.

Exercise B.19 In both the following situations F is a fundamental solution of $-\Delta$ in Ω, and $u \in C^1(\overline{\Omega}) \cap C^2(\Omega)$ with $\Delta u \in L_1(\Omega)$.

(i) Assume that Ω is bounded and that either $\partial\Omega$ is of class C^1 or Ω is listed in Remark D.4. Prove that

$$\begin{aligned} u(x_0) \ = \ & -\int_\Omega F(x_0, x)\,(\Delta u)(x)\,\mathrm{d}x \\ & + \int_{\partial\Omega} \{F(x_0, x)\frac{\partial u(x)}{\partial n} - \frac{\partial F(x_0, x)}{\partial n}u(x)\}\,\mathrm{d}S(x), \end{aligned} \tag{B.35}$$

where $x_0 \in \Omega$ and $\partial/\partial n$ is with respect to x.

(ii) Assume that Ω is a half-space and that (in addition to the foregoing hypotheses about F and u) one or other of the following growth conditions holds. For $x \in \overline{\Omega}$ and $r := |x| \to \infty$, for multi-indices α (Definition A.3) of order $|\alpha| = 0$ or 1, for each fixed $x_0 \in \Omega$ and for some constant $\delta > 0$, either

$$\left(\frac{\partial}{\partial x}\right)^\alpha F(x_0, x) = O\left(r^{-N+1-|\alpha|}\right), \qquad \partial^\alpha u(x) = O\left(r^{1-\delta-|\alpha|}\right), \quad \text{(B.36a,b)}$$

or

$$\left(\frac{\partial}{\partial x}\right)^\alpha F(x_0, x) = \begin{cases} O\left(r^{-N+2-|\alpha|}\right) & \text{if } N \neq 2, \\ O\left(r^{-|\alpha|} \log r\right) & \text{if } N = 2, \end{cases} \qquad \partial^\alpha u(x) = O\left(r^{-\delta-|\alpha|}\right).$$
$$\text{(B.37a,b)}$$

Prove that (B.35) still holds.

[For both (i) and (ii), adapt the proof of Theorem A.2, which contains all essential steps.] $\qquad\qquad\qquad\qquad\qquad\qquad\qquad\qquad\qquad\qquad\qquad$ $\square$

Now we specify boundary conditions for $q(x_0, .)$ and hence for $F(x_0, .)$. In order to have a fundamental solution that is useful for the *Dirichlet problem*, we demand that

$$q(x_0, x) = -K(x_0 - x) \qquad \text{if } x_0 \in \Omega \text{ and } x \in \partial\Omega. \quad \text{(B.38)}$$

In other words, $F(x_0, .)$ is to vanish on $\partial\Omega$; then on the right-hand side of (B.35) only the terms involving $\triangle u$ and $u\big|_{\partial\Omega}$ remain, and these functions are prescribed in the Dirichlet problem. By Theorem B.16, there is at most one solution of (B.34) and (B.38) if Ω is a bounded set or a half-space, provided that for a half-space we add the growth condition: $q(x_0, x) = o(r)$ as $r := |x| \to \infty$ with x_0 fixed in Ω.

For the *Neumann problem*, if Ω is *bounded* and either $\partial\Omega$ is of class C^1 or Ω is listed in Remark D.4, a suitable boundary condition is

$$\frac{\partial q(x_0, x)}{\partial n} = -\frac{\partial K(x - x_0)}{\partial n} - \frac{1}{|\partial\Omega|} \quad \text{if } x_0 \in \Omega \text{ and } x \in \partial\Omega; \quad \text{(B.39)}$$

here $\partial/\partial n$ is with respect to x and $|\partial\Omega|$ is the surface area of $\partial\Omega$. We must not demand that $\partial F(x_0, x)/\partial n = 0$, because (A.20) and (B.34) imply that

$$\int_{\partial\Omega} \frac{\partial K(x - x_0)}{\partial n} = -1 \quad \text{and} \quad \int_{\partial\Omega} \frac{\partial q(x_0, x)}{\partial n} = 0$$

for fixed $x_0 \in \Omega$.

For the *Neumann problem* with Ω a *half-space*, we do demand that

$$\frac{\partial q(x_0, x)}{\partial n} = -\frac{\partial K(x - x_0)}{\partial n} \quad \text{if } x_0 \in \Omega \text{ and } x \in \partial\Omega; \quad \text{(B.40)}$$

in this case, whatever emerges from the point source at x_0 can escape to infinity.

Exercise B.20 Let Ω be bounded and either have $\partial\Omega$ of class C^1 or be listed in Remark D.4.

(a) Assume that the solution $q(x_0,.)$ of (B.34) and (B.38) exists, and denote the corresponding fundamental solution by G. Show that, if the solution u of the Dirichlet problem (B.27) exists and belongs to $C^1(\overline{\Omega}) \cap C^2(\Omega)$ with $\triangle u \in L_1(\Omega)$, then it is given by

$$u(x_0) = \int_\Omega G(x_0, x) f(x) \, dx - \int_{\partial\Omega} \frac{\partial G(x_0, x)}{\partial n} g(x) \, dS(x), \qquad (B.41)$$

where $x_0 \in \Omega$ and $\partial/\partial n$ is with respect to x.

(b) Assume that a solution $q(x_0,.)$ of (B.34) and (B.39) exists, and denote the corresponding fundamental solution by H. Show that, if a solution u of the Neumann problem (B.28) exists and belongs to $C^1(\overline{\Omega}) \cap C^2(\Omega)$ with $\triangle u \in L_1(\Omega)$, then it is given by

$$u(x_0) = \int_\Omega H(x_0, x) f(x) \, dx + \int_{\partial\Omega} H(x_0, x) g(x) \, dS(x) + k, \qquad (B.42)$$

where $x_0 \in \Omega$ and the constant k is the mean value of u over $\partial\Omega$. □

Remarks 1. A fundamental solution that satisfies a useful boundary condition, and leads to a representation formula like (B.41) or (B.42), is called a *Green function*. This phrase can be lengthened; for example, the function G in Exercise B.20, (a), is (if it exists) the *Green function of the Dirichlet problem for* $-\triangle$ in Ω. Green functions exist for differential operators other than $\triangle$ and for boundary conditions other than those of the Dirichlet and Neumann problems.

2. At this stage of the theory, the formula (B.41) rests on two *assumptions*: that the non-singular part q of the Green function G, and the solution u of the Dirichlet problem, both exist. This state of affairs can be improved.

(a) For a few simple sets Ω, such as a ball in $\mathbb{R}^N$, there is an explicit formula for the Green function G. (In the case of the half-space, this formula is an obvious one, but at the moment we are considering bounded sets Ω.) For these particular sets Ω, the first assumption is not needed, and we can dispose of the second by a change of direction. Instead of proceeding from the Dirichlet problem (B.27) to the representation formula (B.41) by means of assumptions about the solution u, we can show (when G is known explicitly) that the function u *defined by* (B.41) satisfies equations (B.27) for suitable data f and g. This will be done for a ball in §B.5.

(b) For an arbitrary set Ω, one cannot expect an explicit formula for the Green function (however smooth the boundary $\partial\Omega$ may be). Rather, a more abstract argument is needed to prove existence of solutions of the Dirichlet problem. This material is outside the range of this book, but we remark that one can cast the Dirichlet problem into a form to which the Fredholm alternatives apply. This means, in effect, that uniqueness of solutions implies their existence, so that Theorem B.16 is again a corner stone. For a pleasant boundary $\partial\Omega$, the existence theory for the Dirichlet problem implies existence of $q(x_0, .)$ and hence of the Green function; then the formula (B.41) has the advantage of representing the solution u for *all* admissible data f and g.

Similar remarks apply to the Neumann problem.

3. For the Neumann problem, we observed before Theorem B.17 that a constant can always be added, on each component of $\overline{\Omega}$, to a solution u. For the function $q(x_0, .)$ this arbitrary constant becomes an arbitrary function of x_0 that can be added to $q(x_0, .)$ and hence to $F(x_0, .)$. To remedy this partly, we demand that, *for a Green function F of the Neumann problem for* $-\Delta$ *in a bounded set* Ω (with $\partial\Omega$ of class C^1 or with Ω listed in Remark D.4),

$$\int_{\partial\Omega} F(x_0, x) \, dS(x) = c \qquad \text{for all } x_0 \in \Omega, \qquad (\text{B.43})$$

where c is independent of x_0 but is otherwise arbitrary.

Given a fundamental solution F_* suitable for the present Neumann problem in that its non-singular part satisfies (B.39), and such that $\int_{\partial\Omega} F_*(x_0, .) = \gamma_*(x_0)$ for all $x_0 \in \Omega$, we must now define

$$F(x_0, x) = F_*(x_0, x) + \frac{c - \gamma_*(x_0)}{|\partial\Omega|},$$

and here only c is arbitrary.

4. In our discussion up to now of fundamental solutions F, the field point x_0 has been *fixed*, although an arbitrary point of Ω. However, when we come to use the representation formulae (B.41) and (B.42), the dependence on x_0 of the Green functions there (that is, the behaviour of $G(x_0, x)$ and $H(x_0, x)$ as x_0 varies) will be of importance. This dependence on the first variable is given to us cheaply by the next theorem, which

ensures that results postulated or established for $F(p,.)$, with p fixed in Ω, apply also to $F(.,p)$.

Theorem B.21 *Let Ω be bounded and either have $\partial\Omega$ of class C^1 or be listed in Remark D.4. Assume the existence of Green functions of the Dirichlet and Neumann problems for $-\triangle$ in Ω. In the Dirichlet case, the non-singular part is to satisfy* (B.34) *and* (B.38); *in the Neumann case, conditions* (B.34), (B.39) *and* (B.43) *are to hold.*

Then, after extension by continuity, both Green functions have the symmetry property

$$F(y,z) = F(z,y) \qquad if \ \ (y,z) \in (\Omega \times \overline{\Omega}) \cup (\overline{\Omega} \times \Omega) \quad and \ \ y \neq z.$$

Proof (i) We need prove only that $F(y,z) = F(z,y)$ for $(y,z) \in \Omega \times \Omega$ and $y \neq z$. For, suppose that this has been shown and that $y \in \partial\Omega$, $z \in \Omega$. There is a sequence (y^n) in Ω such that, for each n, $y^n \neq z$ and such that $y^n \to y$ as $n \to \infty$. Then $F(y^n, z) = F(z, y^n)$ for each n, and $F(z, y^n) \to F(z, y)$ by the continuity of $F(z,.)$ on $\overline{\Omega} \setminus \{z\}$. Extending $F(.,z)$ by continuity, we have $F(y,z) := \lim_{y^n \to y} F(y^n, z) = F(z, y)$.

(ii) To prove symmetry for $(y,z) \in \Omega \times \Omega$ and $y \neq z$, it will be sufficient to consider the Neumann case; the proof for the Dirichlet case is similar but a little easier. We apply the Green identity (A.26),

$$\int_A \{v\triangle w - w\triangle v\} = \int_{\partial A} \left\{ v\frac{\partial w}{\partial n} - w\frac{\partial v}{\partial n} \right\},$$

to $v = F(y,.)$, $w = F(z,.)$ and $A = \Omega_m \setminus \{ \overline{\mathscr{B}(y,\varepsilon)} \cup \overline{\mathscr{B}(z,\varepsilon)} \}$, where Ω_m is the mth approximation to Ω in Theorem D.9; once again this approximation is required in order that $v \in C^2(\overline{A})$ and $w \in C^2(\overline{A})$. Given distinct points y and z in Ω, we can choose $m \in \mathbb{N}$ so large, and $\varepsilon > 0$ so small, that $\overline{\mathscr{B}(y,\varepsilon)}$ and $\overline{\mathscr{B}(z,\varepsilon)}$ are within Ω_m and are disjoint. The integral over A vanishes because $\triangle v = \triangle w = 0$ in A. Dealing with the integrals over $\partial\mathscr{B}(y,\varepsilon)$ and $\partial\mathscr{B}(z,\varepsilon)$ as we dealt with the integral over $\partial\mathscr{B}(x_0,\varepsilon)$ in the proof of Theorem A.2, we find that

$$0 = F(y,z) - F(z,y) + \int_{\partial\Omega_m} \left\{ F(y,x)\frac{\partial F(z,x)}{\partial n} - F(z,x)\frac{\partial F(y,x)}{\partial n} \right\} dS(x),$$

where $\partial/\partial n$ is with respect to x. Suppose that our choice of m was $m \geq M$. Since $F(y,.)$ and $F(z,.)$ are in $C^1(\overline{\Omega} \setminus \Omega_M)$, it follows from Theorem D.9 that the integral over $\partial\Omega_m$ tends to the corresponding integral over $\partial\Omega$

as $m \to \infty$. By (B.39),

$$F(y,z) - F(z,y) = \frac{1}{|\partial\Omega|} \int_{\partial\Omega} \{ F(y,x) - F(z,x) \} \, dS(x), \qquad (B.44)$$

and this integral over $\partial\Omega$ vanishes for all y and z in Ω if and only if (B.43) holds. Thus $F(y,z) = F(z,y)$. $\square$

We turn to particular cases: the Green functions of the Dirichlet and Neumann problems for half-space and ball. The ordering of this material is: notation, formulae for the half-space (presented first because of their simplicity), formulae for the ball, explanations and comments.

Notation Let $D := \left\{ x \in \mathbb{R}^N \mid x_N > 0 \right\}$ and $B := \mathscr{B}(0,a)$ in $\mathbb{R}^N$, for all $N \in \mathbb{N}$ in both cases. The reflection in ∂D of any point $x \in \mathbb{R}^N$ will be written $x_* := (x', -x_N)$, where $x' := (x_1, \ldots, x_{N-1})$. The reflection in ∂B, or inverse point, of any point $x \in \mathbb{R}^N \setminus \{0\}$ will be written $Sx := (a^2/r^2)x$, where $r := |x|$, as in Definition B.12. The Newtonian kernel, introduced as $K(.\,;N)$ in (A.18), continues to be denoted by K. Field points will often be denoted by y instead of x_0.

Statement B.22 The following formulae define Green functions on $\overline{\Omega} \times \overline{\Omega} \setminus \left\{ (y,x) \mid x = y \right\}$, where either $\Omega = D$ or $\Omega = B$. The non-singular parts are defined on

$$\overline{\Omega} \times \overline{\Omega} \setminus \left\{ (y,x) \mid x = y \in \partial\Omega \right\};$$

this domain is bigger than that demanded in Definition B.18, and bigger than that established by extension in Theorem B.21. The reader should verify that, if $x \in \overline{D}$, $y \in \overline{D}$ and $x = y_*$, then $x = y \in \partial D$; also that, if $x \in \overline{B}$, $y \in \overline{B}$ and $x = Sy$, then $x = y \in \partial B$.

(i) The Green function of the Dirichlet problem for $-\Delta$ in D is

$$G_D(y,x) = K(y-x) - K(y_* - x), \qquad (B.45)$$

where

$$|y_* - x| = \left\{ |y' - x'|^2 + (y_N + x_N)^2 \right\}^{1/2} = |y - x_*|. \qquad (B.46)$$

(ii) A Green function of the Neumann problem for $-\Delta$ in D is

$$H_D(y,x) = K(y-x) + K(y_* - x). \qquad (B.47)$$

(iii) The Green function of the Dirichlet problem for $-\Delta$ in B is

$$G_B(y,x) = K(y-x) - K\left(\frac{|y|}{a}(Sy - x) \right), \qquad (B.48)$$

where

$$\frac{|y|}{a}|Sy - x| = \frac{1}{a}\left\{ a^4 - 2a^2 y \cdot x + |y|^2 |x|^2 \right\}^{1/2} = \frac{|x|}{a}|y - Sx|. \quad (B.49)$$

It is to be understood that limiting values are taken for $y = 0$ or $x = 0$, both here and, later, in formulae for derivatives of G_B.

(iv) A Green function of the Neumann problem for $-\Delta$ in B is

$$H_B(y, x) = \begin{cases} K(y - x) & \text{if } N = 1, \\[2ex] K(y - x) + K\left(\dfrac{|y|}{a}(Sy - x)\right) & \text{if } N = 2, \\[2ex] K(y - x) + K\left(\dfrac{|y|}{a}(Sy - x)\right) + \dfrac{E(y, x)}{\sigma_N a^{N-2}} & \text{if } N \geq 3, \end{cases}$$

$$(B.50a,b,c)$$

where $\sigma_N := |\partial \mathscr{B}_N(0, 1)|$ and $E(y, x)$ is defined as follows. Let $r := |x|$, $s := |y|$ and $rs \cos \lambda := x \cdot y$; then

$$E(y, x) = \int_{a^2/s}^{\infty} \left\{ \rho^{N-3}(\rho^2 - 2\rho r \cos \lambda + r^2)^{-\frac{1}{2}(N-2)} - \frac{1}{\rho} \right\} d\rho, \quad (B.51a)$$

$$= \int_{1}^{\infty} \left\{ t^{N-3}\left(t^2 - 2t\frac{rs \cos \lambda}{a^2} + \frac{r^2 s^2}{a^4} \right)^{-\frac{1}{2}(N-2)} - \frac{1}{t} \right\} dt, \quad (B.51b)$$

$$= \int_{a^2/r}^{\infty} \left\{ \rho^{N-3}(\rho^2 - 2\rho s \cos \lambda + s^2)^{-\frac{1}{2}(N-2)} - \frac{1}{\rho} \right\} d\rho. \quad (B.51c)$$

Again it is to be understood that limiting values are taken for $y = 0$ or $x = 0$. $\qquad \square$

Remarks 5. Construction of these Green functions. (a) Definition B.18 obliges us always to begin with the potential $K(y-.)$ [equally, $K(.,-y)$] of a unit source at y. The artifice of making $G_D(y, x)$ an odd function of x_N (hence vanishing for $x_N = 0$) by adding a negative unit source at the reflected point y_*, and of making $H_D(y, x)$ an even function of x_N (hence satisfying $\partial H_D(y, x)/\partial x_N = 0$ for $x_N = 0$) by adding a positive unit source at y_*, is an application of a standard device in mathematics and physics. A shadow of this method can be found in virtually all formulae for Green functions.

(b) In the case of G_B, there is an obvious resemblance of Sy to y_*, but the factor $|y|/a$, accompanying $Sy - x$ in the formula (B.48), requires

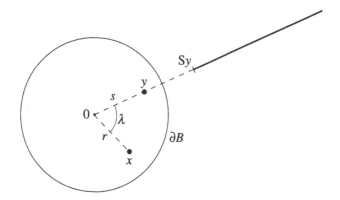

Fig. B.4.

explanation. The situation is this: with y a fixed parameter and $y \in B$, we seek the function $-q(y,.)$ that is smoothly harmonic in B and equals $K(y-.)$ on ∂B. These conditions suggest the Kelvin transform, relative to ∂B, of $K(y-.)\big|_{\mathbb{R}^N \setminus B}$. [Here it is the dot that is restricted to $\mathbb{R}^N \setminus B$, and with respect to which the transform is taken.] By Definition B.12 and Theorem B.15, this Kelvin transform is defined on $\overline{B} \setminus \{0\}$, is smoothly harmonic in $B \setminus \{0\}$ and equals $K(y-.)$ on ∂B. If $N \neq 2$, we extend it to $\overline{B}$ by continuity to obtain $-q(y,.)$. If $N = 2$, this Kelvin transform tends to minus infinity at the origin and is not symmetrical in . and y, but addition of the function $x \mapsto (1/2\pi) \log(a/r)$ corrects both faults.

(c) For the Green function H_B of the Neumann problem, addition of the point singularity $K\big((|y|/a)(Sy-.)\big)$ is completely successful only for $N = 2$. When $N = 1$, this term is not needed for the boundary condition $\partial H_B/\partial n = -1/|\partial B|$; when $N \geq 3$, it is not enough. Equation (B.51a) shows that the field $-\nabla E(y,.)$ of the extra term (for $N \geq 3$ and $y \neq 0$) is that of sources distributed on the radial line outward from Sy (Figure B.4); their strength (charge per length or mass per length) is proportional to ρ^{N-3} at the point distant ρ from the origin. The term $-1/\rho$ in (B.51a) is required for convergence of the integral defining the potential $E(y,.)$, but makes no contribution when we differentiate with respect to x.

6. *Verification of basic properties.* To check that the Green functions in Statement B.22 have non-singular parts $q(y,.)$ that are smoothly harmonic in Ω, and satisfy appropriate auxiliary conditions, is a matter of inspection and direct calculation that is left mainly to the reader. Here are a few details that may be helpful.

(a) That the non-singular part of $G_B(y,.)$ is smoothly harmonic in B, for fixed parameter $y \in B$ (hence for fixed $Sy \notin \bar{B}$), is shown by the formulae

$$K\left(\frac{|y|}{a}(Sy - x)\right) = \begin{cases} \left(\dfrac{|y|}{a}\right)^{-N+2} K(x - Sy) & \text{if } N \neq 2,\ y \neq 0, \\[2mm] \kappa_N a^{-N+2} & \text{if } N \neq 2,\ y = 0, \\[2mm] K(x - Sy) + \dfrac{1}{2\pi} \log \dfrac{a}{|y|} & \text{if } N = 2,\ y \neq 0, \\[2mm] \dfrac{1}{2\pi} \log \dfrac{1}{a} & \text{if } N = 2,\ y = 0, \end{cases}$$

(B.52)

where the values for $y = 0$ are limits of those for $y \neq 0$ [because $|Sy| = a^2/|y|$ when $y \neq 0$].

(b) That the function $E(y,.)$, occurring in $H_B(y,.)$ for $N \geq 3$, is smoothly harmonic in B, for fixed parameter $y \in B$, is established as follows. For $y \neq 0$, we use (B.51a); differentiation with respect to x under the integral sign is legitimate and is justified very much as in the proof of Theorem A.5. For $y = 0$, we observe from (B.51b) that $E(0, x) = 0$ for all $x \in \bar{B}$.

(c) To verify that H_B, for $N \geq 3$, satisfies the boundary condition (B.39), we use the notation introduced before (B.51), write

$$R := \left\{ r^2 - 2rs \cos \lambda + s^2 \right\}^{1/2},$$

$$A := \left\{ a^2 - 2rs \cos \lambda + \frac{r^2 s^2}{a^2} \right\}^{1/2},$$

and use (B.51c) for the function E. Then (for $r \neq 0$)

$$\frac{\partial H_B(y, x)}{\partial r} = -\frac{1}{\sigma_N} \left\{ R^{-N}(r - s \cos \lambda) + A^{-N}\left(\frac{rs^2}{a^2} - s \cos \lambda\right) \right.$$
$$\left. - \frac{1}{r}\left(A^{-N+2} - a^{-N+2}\right) \right\};$$

setting $r = a$ and observing a pleasing cancellation, we obtain

$$\left.\frac{\partial H_B(y, x)}{\partial r}\right|_{r=a} = -\frac{1}{\sigma_N a^{N-1}}.$$

(d) The function H_B satisfies the normalization condition (B.43) because it has the symmetry property $H_B(y, z) = H_B(z, y)$; the equivalence of these two conditions was established by (B.44).

(e) For the half-space D, the Green function G_D satisfies the growth condition (B.36a), while H_D satisfies (B.37a). This shows that the hypotheses leading to the representation formula (B.35) for the half-space were not artificial.

7. *On uniqueness of the Green functions.* (a) As was remarked after (B.38), Theorem B.16 implies uniqueness of G_D and G_B, provided that, in the case of D, the non-singular part $q_D(x_0, x) = o(r)$ as $r := |x| \to \infty$. This growth condition is not satisfied for $N = 1$. However, if the condition is changed to $G_D(x_0, x) = o(r)$ as $r \to \infty$, then this is satisfied even for $N = 1$, and the proof of Theorem B.16 still applies to the difference of two Green functions, because this difference has no singularity at $x = x_0 \in \Omega$.

(b) The function H_D is unique apart from an additive constant, provided that for $N = 1$ we demand symmetry: $H_D(x, y) = H_D(y, x)$, and provided that for $N \geq 2$ we impose the growth condition: $H_D(x_0, x) = o(r)$ or $q_D(x_0, x) = o(r)$ as $r \to \infty$. For $N = 1$, this uniqueness follows from elementary consideration of what may be added to the function H_D in (B.47); for $N \geq 2$, it follows from Theorem B.17.

(c) One can concoct a partial uniqueness result for the function H_B, but this is hardly worthwhile, because there are alternatives to the conditions (B.39) and (B.43), and because it is the uniqueness statement in Theorem B.17 that is the important one.

B.5 The solution of the Dirichlet problem for a ball

According to the representation formula (B.41), the solution of the Dirichlet problem for the ball $B := \mathcal{B}(0, a)$ is

$$u(x_0) = \int_B G(x_0, x) f(x) \, dx - \int_{\partial B} \frac{\partial G(x_0, x)}{\partial n} g(x) \, dS(x), \quad x_0 \in \Omega, \quad \text{(B.53)}$$

where $G = G_B$ is now given explicitly by (B.48). In this section we discard the strong assumptions on u under which (B.41) was derived. Rather, we impose various conditions on the data f and g, and show that (B.53) yields various kinds of solution of the Dirichlet problem.

Notation Throughout this §B.5 we write $r := |x|$, $r_0 := |x_0|$, $\sigma_N := |\partial \mathcal{B}_N(0, 1)|$ and [as in §B.3] $Sx := (a^2/r^2)x$, $Sx_0 := (a^2/r_0^2)x_0$.

It is helpful to decompose the right-hand member of (B.53) into three parts by the definitions

$$u_f(x_0) := \int_B G(x_0, x) f(x) \, dx = v(x_0) - w(x_0), \tag{B.54}$$

where

$$v(x_0) := \int_B K(x_0 - x) f(x) \, dx, \quad x_0 \in \mathbb{R}^N, \tag{B.55}$$

$$w(x_0) := \int_B K\left(\frac{r}{a}(x_0 - Sx)\right) f(x) \, dx, \quad x_0 \in \overline{B}, \tag{B.56}$$

and

$$u_g(x_0) := \begin{cases} \displaystyle\int_{\partial B} P(x_0, x) g(x) \, dS(x) & \text{if } x_0 \in B, \tag{B.57a} \\[6mm] g(x_0) & \text{if } x_0 \in \partial B, \tag{B.57b} \end{cases}$$

where

$$P(x_0, x) := -\frac{\partial G(x_0, x)}{\partial n}\bigg|_{r=a} = \frac{a^2 - r_0^2}{\sigma_N a |x_0 - x|^N} \quad (x_0 \in B, \ x \in \partial B). \tag{B.58}$$

Here u_f may be called the *Green volume potential* of the ball B and the density function f; the function u_g is the *Poisson integral* for B of the boundary-value function g; and P is the *Poisson kernel* of B.

The two parts of the volume potential are the Newtonian potential v of f and a function $-w$ that has two useful representations. The first form of w results from changing the variable of integration to $\xi := Sx = (a^2/r^2)x$; the Jacobian determinant of the transformation is $\det(\partial x_i/\partial \xi_j) = -(a/|\xi|)^{2N}$ by (B.23). Then, with the notation

$$\rho := |\xi|, \qquad f_*(\xi) := f\left(\frac{a^2}{\rho^2}\xi\right)\left(\frac{a}{\rho}\right)^{N+2}, \tag{B.59}$$

and apart from an added constant when $N = 2$, $w(x_0)$ is the Newtonian potential at $x_0 \in \overline{B}$ of the density function f_*, which has support outside B. That is,
for $N \neq 2$,

$$w(x_0) = \int_{\rho > a} K(x_0 - \xi) f_*(\xi) \, d\xi \quad (x_0 \in \overline{B}), \tag{B.60a}$$

for $N = 2$,

$$w(x_0) = \int_{\rho>a} K(x_0 - \xi) f_*(\xi) \, d\xi + C \quad (x_0 \in \overline{B}), \quad \left.\begin{array}{c}\\\\\\\end{array}\right\} \quad \text{(B.60b)}$$
$$\text{where} \quad C = \frac{1}{2\pi} \int_{\rho>a} \log \frac{\rho}{a} \, f_*(\xi) \, d\xi.$$

The second form of w results from the identity (B.49), now written as $(r/a)|x_0 - Sx| = (r_0/a)|Sx_0 - x|$. This implies that, apart from an added harmonic function when $N = 2$, $w(x_0)$ is $(a/r_0)^{N-2}$ times the Newtonian potential, at the field point $\xi_0 := Sx_0$ outside B, of the original density function f. That is,

$$w(x_0) = \int_B K\left(\frac{a}{\rho_0}(\xi_0 - x)\right) f(x) \, dx \quad (\rho_0 := |\xi_0| \geq a)$$

$$= \begin{cases} \left(\dfrac{\rho_0}{a}\right)^{N-2} v(\xi_0) & \text{if } N \neq 2, \quad \text{(B.61a)} \\[2ex] v(\xi_0) + \dfrac{1}{2\pi} \log \dfrac{\rho_0}{a} \displaystyle\int_B f(x) \, dx & \text{if } N = 2. \quad \text{(B.61b)} \end{cases}$$

The case $N = 1$ of these formulae is discussed in Remark 1.5 and Exercise B.32; *it is to be understood in the remainder of this §B.5 that the dimension $N \geq 2$.*

Theorem B.23 *Let u denote the Green volume potential called u_f in* (B.54).

(i) *If $f \in L_p(B)$ for some $p > N/2$, then $u \in C(\overline{B})$, $u|_{\partial B} = 0$ and u is a distributional solution (Definition A.7) of $-\triangle u = f$ in B; it is the only function with these three properties.*

(ii) *If $f \in L_p(B)$ for some $p > N$, then $u \in C^1(\overline{B})$, $u|_{\partial B} = 0$ and u is a generalized solution (Definition A.7) of $-\triangle u = f$ in B; again it is the only function with these three properties.*

Proof We use the decomposition $u = v - w$ described above, and results in Appendix A.

(i) (a) First, $v \in C(\overline{B})$ when $f \in L_p(B)$ with $p > N/2$ because $v \in C(\mathbb{R}^N)$ by Theorem A.6. The same is true for w, by the extension of Theorem A.6 in Exercise A.32, if the function f_* introduced in (B.59) has two properties: $f_* \in L_p(\mathbb{R}^N \setminus \overline{B})$ with $p > N/2$, and $Wf_* \in L_1(\mathbb{R}^N \setminus \overline{B})$, where W is the weight function in (A.78). For the first of these we use the same p as in $f \in L_p(B)$; then

$$
\int_{\rho > a} \left| f_*(\xi) \right|^p \, d\xi = \int_{\rho > a} \left| f\left(\frac{a^2}{\rho^2} \xi \right) \right|^p \left(\frac{a}{\rho} \right)^{p(N+2)} \, d\xi
$$

$$
\leq \int_{\rho > a} \left| f\left(\frac{a^2}{\rho^2} \xi \right) \right|^p \left(\frac{a}{\rho} \right)^{2N} \, d\xi = \int_B |f(x)|^p \, dx,
$$

because $p(N+2) > \frac{1}{2} N^2 + N \geq 2N$. For the L_1 property of Wf_*, first let $N = 2$. Then $f_*(\xi) = f(x)(r/a)^4$ and $d\xi = (a/r)^4 \, dx$, so that

$$
\int_{\rho > a} \log(e + 2|\xi|) \, |f_*(\xi)| \, d\xi = \int_B \log\left(e + \frac{2a^2}{|x|} \right) |f(x)| \, dx
$$

$$
\leq \| \ell \mid L_q(B) \| \, \| f \mid L_p(B) \|,
$$

where $\ell(x) := \log(e + 2a^2/|x|)$, $1/p + 1/q = 1$ and $p > 1$ implies that $q < \infty$, hence that $\ell \in L_q(B)$. For $N \geq 3$, the L_1 property of $Wf_* = f_*$ is established again by means of the Hölder inequality. Thus $w \in C(\overline{B})$.

(b) That $u|_{\partial B} = 0$ follows from (B.61), which shows that $w|_{\partial B} = v|_{\partial B}$.

(c) Theorem A.8 establishes v as a distributional solution of $-\triangle v = f$ in B, while $\triangle w = 0$ in B pointwise by (B.60), which shows B to be outside the support of f_*, and by the extension of Theorem A.5 in Exercise A.32.

(d) Let u_0 be the difference of two functions u having the three properties stated in (i) of the theorem. Then $u_0 \in C(\overline{B})$, $u_0|_{\partial B} = 0$ and both u_0 and $-u_0$ are distributional subsolutions (Definition 2.10) relative to $\triangle$ and B. It follows from Theorem 2.11 that $u_0 = 0$ on $\overline{B}$.

(ii) The proof for $f \in L_p(B)$ with $p > N$ is entirely analogous to the foregoing proof for $p > N/2$. In place of Theorems A.6 and A.8 we now use Theorem A.11. $\qquad \square$

It is to be expected that, if $f \in C^{0,\mu}(\overline{B})$ for some $\mu \in (0, 1)$, then u_f is the pointwise solution of the Dirichlet problem for $-\wedge$ in B with boundary condition $u = 0$ on ∂B. To prove this, we need a long (but straightforward) calculation like that in Lemma A.14. Reverting to the symbol q for the non-singular part of a fundamental solution, we now write

$$
q(x_0, x) := -K\left(\frac{r}{a}(x_0 - Sx) \right), \tag{B.62a}
$$

$$
q_{ij}(x_0, x) := \frac{\partial^2 q(x_0, x)}{\partial x_{0i} \, \partial x_{0j}} = -\left(\frac{a}{\rho} \right)^{-N+2} \left(\partial_i \partial_j K \right) (x_0 - \xi), \tag{B.62b}
$$

where again $\xi := Sx$ and $\rho = |\xi|$.

Lemma B.24 *Let q_{ij} be as in* (B.62) *and let $H := \mathscr{B}(x_0, 2|h|)$ with $x_0 \in B$ and $0 < |h| \leq \frac{1}{6}a$. Then*

$$\sum_{i,j=1}^{N} \left\{ \int_{B \setminus H} q_{ij}(x_0, x) \, dx \right\}^2 \leq \text{const.,}$$

where the constant depends only on N.

Proof To a large extent the proof follows that of Lemma A.14. Although $q_{ij}(x_0, x)$ has no singularity when $x_0 \in B$ and $x \in \overline{B}$, it is unbounded because $|x_0 - \xi|$ can be arbitrarily small when x_0 and x are near ∂B and near each other.

(i) In contrast to the function $K_{ij}(x_0-.)$ in the earlier proof, $q_{ij}(x_0, .)$ is integrable on B and on $H \cap B$, so that the integrals over these two sets can be calculated separately. Moreover, $q_{ij}(x_0, .) \in C^\infty(\overline{B})$ and is smoothly harmonic in B [because $q(x_0, .)$ has these properties by (B.49) and (B.52), and because the operators $\partial^2/\partial x_{0i} \partial x_{0j}$ and $\triangle$ commute]. Therefore we can use the mean-value property of harmonic functions, Theorem B.8, to integrate $q_{ij}(x_0, .)$ over any ball contained in B. In particular, for all $i, j \in \{1, \ldots, N\}$,

$$\int_B q_{ij}(x_0, x) \, dx = |B| q_{ij}(x_0, 0) = 0, \tag{B.63}$$

because (B.62b) shows that $q_{ij}(x_0, x) = O(\rho^{-2})$ as $x \to 0$ and $\rho \to \infty$.

(ii) Under a rotation of the co-ordinate frame, again written $x = T(x_0)\tilde{x}$ and $x_0 = T(x_0)\tilde{x}_0$, where $T(x_0)$ is an orthogonal $N \times N$ matrix depending on x_0, we also have $\xi = T(x_0)\tilde{\xi}$, where $\tilde{\xi} = S\tilde{x} = \widetilde{(Sx)}$, and q_{ij} transforms exactly as K_{ij} did in the proof of Lemma A.14. Again we so choose $T(x_0)$ that $\tilde{x}_0 = (|x_0|, 0, \ldots, 0)$, and *use these special co-ordinates $\tilde{x}_j$* in the rest of this proof, but *omit the tilde* for ease of writing. For the same reasons as before,

$$\int_{B \setminus H} q_{ij}(x_0, x) \, dx = 0 \qquad \text{if } i \neq j,$$

$$\int_{B \setminus H} q_{11}(x_0, x) \, dx = -\sum_{j=2}^{N} \int_{B \setminus H} q_{jj}(x_0, x) \, dx,$$

so that we need bound only the integrals of q_{jj} for $j = 2, \ldots, N$, and these integrals are all equal.

(iii) For $\overline{H} \subset B$, we evaluate the integral of $q_{jj}(x_0, .)$, $j \neq 1$, over H as

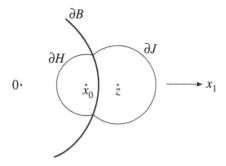

Fig. B.5.

follows. Set $|x_0| =: ta$, $0 \le t < 1$, and apply the mean-value property of harmonic functions once more; there results

$$\left| \int_H q_{jj}(x_0, x) \, dx \right| = |H| \, |q_{jj}(x_0, x_0)|$$

$$= \frac{\sigma_N (2|h|)^N}{N} \frac{a^{-N} t^2 (1 - t^2)^{-N}}{\sigma_N} \qquad (j \ne 1).$$

Since $\overline{H} \subset B$, we have $2|h| < a - |x_0| = a(1 - t)$ and $0 \le t < 1$, so that

$$\left| \int_H q_{jj}(x_0, x) \, dx \right| < \frac{t^2 (1 + t)^{-N}}{N} < \frac{1}{N} \qquad (\overline{H} \subset B, \ j \ne 1). \qquad (B.64)$$

(iv) It remains to bound the integral of $q_{jj}(x_0, .)$ over $H \cap B$ when $\overline{H}$ intersects ∂B and $j \ne 1$. Changing the variable of integration from x to $\xi := Sx$, we find from the solution of Exercise B.14 that

$$S(H) = \mathscr{B}(z, \gamma), \qquad \text{where} \ \ z := \frac{a^2}{r_0^2 - 4|h|^2} x_0, \ \ \gamma := \frac{a^2}{r_0^2 - 4|h|^2} 2|h|;$$
$$(B.65)$$

also, $r_0 - 2|h| \ge \frac{1}{3} a$ because $\overline{H}$ intersects ∂B and $|h| \le \frac{1}{6} a$. The Jacobian determinant of the transformation is $\det(\partial x_i / \partial \xi_j) = -(a/\rho)^{2N}$. Let $J := S(H)$; the configuration is shown in Figure B.5. Accordingly,

$$A_j(x_0, h) := \int_{H \cap B} q_{jj}(x_0, x) \, dx = - \int_{J \setminus \overline{B}} (\partial_j^2 K)(\xi - x_0) \left(\frac{a}{\rho} \right)^{N+2} d\xi$$

$$= \sum_{k=1}^{3} A_{jk}(x_0, h) \qquad (B.66)$$

if we define

$$A_{j,1}(x_0, h) := -\int_{\partial J \setminus B} n_j(\xi) \, (\partial_j K)(\xi - x_0) \left(\frac{a}{\rho}\right)^{N+2} dS(\xi),$$

$$A_{j,2}(x_0, h) := -\int_{\partial B \cap J} n_j(\xi) \, (\partial_j K)(\xi - x_0) \, dS(\xi),$$

$$A_{j,3}(x_0, h) := \int_{J \setminus \overline{B}} (\partial_j K)(\xi - x_0) \left\{ \frac{\partial}{\partial \xi_j} \left(\frac{a}{\rho}\right)^{N+2} \right\} d\xi.$$

Here $n(\xi)$ is the unit normal outward from $J \setminus \overline{B}$; use of the divergence theorem is legitimate because $J \setminus \overline{B}$ is a set listed in Remark D.4.

For the contribution $A_{j,1}$ of $\partial J \setminus B$, we set

$$\xi_1 = z_1 - \gamma \cos \varphi,$$

$$(\xi_2^2 + \cdots + \xi_N^2)^{1/2} = \gamma \sin \varphi \qquad (\xi \in \partial J, \ 0 \le \varphi \le 2\pi).$$

Then, very much as in the proof Lemma A.14, step (iv),

$$|\xi - x_0|^2 = \gamma^2 \left\{ (\zeta - 1)^2 + 4\zeta \sin^2 \frac{\varphi}{2} \right\}, \qquad \text{where } \zeta := \frac{|z| - |x_0|}{\gamma} > 0,$$

and, for $j \ne 1$ and some angle $\mu \in [0, \pi/2)$ that depends on $|x_0|$ and $|h|$,

$$|A_{j,1}(x_0, h)| = \frac{\sigma_{N-1}}{(N-1)\sigma_N} \int_\mu^\pi \left[\frac{4 \cos^2 \frac{\varphi}{2} \sin^2 \frac{\varphi}{2}}{(\zeta - 1)^2 + 4\zeta \sin^2 \frac{\varphi}{2}} \right]^{N/2} \left(\frac{a}{\rho}\right)^{N+2} d\varphi$$

$$< \frac{\sigma_{N-1} 2^{N+1}}{(N-1)\sigma_N} \int_0^{\pi/2} (\sin \varphi)^N \, d\varphi \qquad (j \ne 1), \qquad (\text{B.67})$$

where we have used first that $a/\rho \le 1$ and then the step from (A.69a) to (A.69b); this step remains valid if $\zeta > 1$.

In the contribution $A_{j,2}$ of $\partial B \cap J$, the integrand, apart from sign, is precisely that considered on $\partial B \setminus H$ in the earlier proof; therefore, a coarse bound is

$$|A_{j,2}(x_0, h)| < \frac{\sigma_{N-1} 2^{N+1}}{(N-1)\sigma_N} \int_0^{\pi/2} (\sin \theta)^N \, d\theta \qquad (j \ne 1). \qquad (\text{B.68})$$

For the contribution $A_{j,3}$ of $J \setminus \overline{B}$, we use once more the cylindrical symmetry about the x_1-axis:

$$|A_{j,3}(x_0, h)| = \int_{J \setminus \overline{B}} \frac{\xi_j}{\sigma_N |\xi - x_0|^N} \frac{(N+2) a^{N+2} \xi_j}{\rho^{N+4}} \, d\xi \qquad (j \ne 1)$$

$$= \frac{(N+2) a^{N+2}}{(N-1)\sigma_N} \int_{J \setminus \overline{B}} \frac{\xi_2^2 + \cdots + \xi_N^2}{|\xi - x_0|^N} \frac{1}{\rho^{N+4}} \, d\xi$$

$$< \frac{(N+2)a^{-2}}{(N-1)\sigma_N} \int_{J\backslash \overline{B}} \frac{1}{|\xi - x_0|^{N-2}} \, d\xi \, .$$

One verifies from (B.65) that $J \setminus \overline{B} \subset \mathscr{B}(x_0, 7a/3)$ for all $x_0 \in B$ and $|h| \le \frac{1}{6}a$, so that

$$|A_{j,3}(x_0, h)| < \frac{(N+2)a^{-2}}{(N-1)\sigma_N} \int_0^{7a/3} R^{-N+2}\sigma_N R^{N-1} \, dR$$

$$= \frac{49}{18}\frac{N+2}{N-1} \quad (j \ne 1). \tag{B.69}$$

In view of remarks in (ii), the Lemma now follows from taking the bigger of two constants: that implied by (B.63) and (B.64) if $\overline{H} \subset B$, and that implied by (B.63) and (B.66) to (B.69) if $\overline{H}$ intersects ∂B. $\qquad\square$

We observe in passing that, with q as in (B.62a),

$$\int_B K(x_0 - x) \, dx = -\frac{r_0^2}{2N} + c_N(a) \quad (x_0 \in \overline{B}), \tag{B.70}$$

$$\int_B q(x_0, x) \, dx = \frac{a^2}{2N} - c_N(a) \quad (x_0 \in \overline{B}), \tag{B.71}$$

$$\int_B G(x_0, x) \, dx = \frac{a^2 - r_0^2}{2N} \quad (x_0 \in \overline{B}); \tag{B.72}$$

here (B.70) is a part of (A.23a) and $c_N(a)$ is as in (A.23b). The identity (B.71) can be derived by means of the mean-value property of harmonic functions, Theorem B.8; then addition of (B.70) and (B.71) yields (B.72). Alternatively, we may observe that (B.72) is a particular case of the representation formula (B.41), because $u(x) = (a^2 - r^2)/2N$ is the (unique, pointwise) solution of $-\Delta u = 1$ in B, $u = 0$ on ∂B; in the present case, the formula extends to $\overline{B}$ by continuity. Note that (B.71) implies (B.63).

Evidently the Green function G and the Green volume potential u_f have much in common with the Newtonian kernel K and the Newtonian potential v of f. Two small lemmas now precede a theorem in this direction.

Lemma B.25 (i) *If* $x_0 \in B$ *and* $x \in B \setminus \{0\}$, *then*

$$a \, |x_0 - Sx| > |Sx| \, |x_0 - x| \, . \tag{B.73}$$

(ii) *If* $x_0 \in B$, $x \in B \setminus \{x_0\}$ *and limiting values are taken for* $x_0 = 0$ *or* $x = 0$, *then*

$$\left| \left(\frac{\partial}{\partial x_0} \right)^\beta q(x_0, x) \right| \le C_\beta |x_0 - x|^{-N+2-|\beta|} \quad (N \ge 2), \tag{B.74}$$

where q is the function in (B.62a), C_β is the constant in (A.31), and $|\beta| \geq 1$ if $N = 2$.

Proof (i) For $x_0 = 0$, the inequality (B.73) reduces to $a > |x|$, which is true. Therefore assume that $x_0 \neq 0$ and let λ be the angle between the vectors x_0 and x. Continuing to write $\xi := Sx$, $\rho := |\xi|$ and hence $r = a^2/\rho$, we wish to prove that

$$a^2 \left(r_0^2 - 2r_0\rho \cos\lambda + \rho^2 \right) > \rho^2 \left(r_0^2 - 2r_0\frac{a^2}{\rho}\cos\lambda + \frac{a^4}{\rho^2} \right),$$

equivalently, that $\rho^2 \left(a^2 - r_0^2 \right) > a^2 \left(a^2 - r_0^2 \right)$, which is true.

(ii) First,

$$
\begin{aligned}
-\left(\frac{\partial}{\partial x_0}\right)^\beta q(x_0, x) &= \left(\frac{a}{\rho}\right)^{|\beta|} \left(\partial^\beta K\right) \left(\frac{a}{\rho}(x_0 - \xi)\right) \\
&= \left(\frac{a}{\rho}\right)^{-N+2} \left(\partial^\beta K\right) (x_0 - \xi)
\end{aligned}
$$

because $\partial^\beta K$ is algebraically homogeneous of degree $-N + 2 - |\beta|$, as is displayed in the proof of Lemma A.4. Next, (A.31) and then (B.73) imply that

$$
\begin{aligned}
\left| \left(\frac{\partial}{\partial x_0}\right)^\beta q(x_0, x) \right| &\leq \left(\frac{a}{\rho}\right)^{-N+2} C_\beta |x_0 - \xi|^{-N+2-|\beta|} \\
&\leq C_\beta |x_0 - x|^{-N+2} |x_0 - \xi|^{-|\beta|} \\
&\leq C^\beta |x_0 - x|^{-N+2-|\beta|}
\end{aligned}
$$

because (B.73), in which $a < |Sx|$, certainly implies that $|x_0 - \xi| > |x_0 - x|$. □

Lemma B.26 *In the Green volume potential u_f, defined by (B.54), let $f \in C^{0,\mu}(\overline{B})$ for some $\mu \in (0, 1)$. Then*

$$\left(\partial_i \partial_j u_f\right)(x_0) = \int_B \frac{\partial^2 G(x_0, x)}{\partial x_{0i} \partial x_{0j}} \{ f(x) - f(x_0) \} \, dx - \delta_{ij} \frac{f(x_0)}{N}, \qquad (B.75)$$

whenever $i, j \in \{1, \ldots, N\}$ and $x_0 \in B$.

Proof Let $(.)_{ij}$ denote $\partial^2(.)/\partial x_{0i}\partial x_{0j}$, except in the Kronecker delta δ_{ij}. We use the decomposition $u_f = v - w$ and the formulae (to be explained presently)

$$v_{ij}(x_0) = \int_B K_{ij}(x_0 - x)\{f(x) - f(x_0)\}\, dx - \delta_{ij}\frac{f(x_0)}{N},$$

$$-w_{ij}(x_0) = \int_B q_{ij}(x_0, x)\, f(x)\, dx,$$

$$0 = -\int_B q_{ij}(x_0, x)\, f(x_0)\, dx,$$

where $x_0 \in B$. The first is the result of Lemma A.13. The second comes from the form (B.60) of w and from the extension of Theorem A.5 in Exercise A.32. The third is (B.63) multiplied by $-f(x_0)$; alternatively, it is an implication of (B.71). The sum of the three formulae gives (B.75).

$\square$

Theorem B.27 *Let u denote the Green volume potential called u_f in (B.54). If $f \in C^{0,\mu}(\overline{B})$ for some $\mu \in (0,1)$, then $u \in C^2(\overline{B})$, $u\big|_{\partial B} = 0$ and u is a C^2-solution of $-\triangle u = f$ in B. Indeed, each second derivative $\partial_i\partial_j u\big|_B$ has an extension to $\overline{B}$ that is in $C^{0,\mu}(\overline{B})$.*

Proof Apart from the result $u\big|_{\partial B} = 0$, which follows from (B.61), the proof is an exact parallel of the proof of Theorem A.15. The properties enjoyed there by the Newtonian kernel K and Newtonian potential v of f are now shared by the Green function G and Green volume potential u; this is the content of Lemmas B.24 to B.26. In other words, to establish the uniform Hölder continuity of second derivatives of u, and to prove that $-\triangle u = f$ in B pointwise, we merely replace $K(x_0-.)$ by $G(x_0,.)$ in the proof of Theorem A.15, and replace the condition $0 < |h| \leq \frac{1}{4}a$ there by $0 < |h| \leq \frac{1}{6}a$. $\square$

We turn now to the Poisson integral u_g, defined by (B.57) and expected to solve the problem $\triangle u = 0$ in B, $u = g$ on ∂B. The Poisson kernel P, introduced in (B.58), has two essential properties: first, as (B.58) shows, $P(x_0, x) \to 0$ as x_0 tends to any point of ∂B other than x [so that $r_0 \to a$ with $|x_0 - x| > 0$]; second, P satisfies the following analogue of the identity (B.72) satisfied by G.

Lemma B.28

$$\int_{\partial B} P(x_0, x)\, dS(x) = 1 \qquad \text{for all} \quad x_0 \in B. \tag{B.76}$$

First proof Recall from (B.58) that $P(x_0, x) = -\partial G(x_0, x)/\partial n$, where

$x \in \partial B$. Therefore (B.76) is a particular case of the representation formula (B.41), because $u(x) = 1$ is the (unique, pointwise) solution of $\triangle u = 0$ in B, $u = 1$ on ∂B. □

Second proof By the basic property (A.20) of the Newtonian kernel,

$$- \int_{\partial B} \frac{\partial K(x - x_0)}{\partial n} \, dS(x) = 1.$$

Since $q(x_0,.) \in C^\infty(\overline{B})$ and $(\triangle q)(x_0,.) = 0$ in B [by (B.52) and the accompanying remark],

$$- \int_{\partial B} \frac{\partial q(x_0, x)}{\partial n} \, dS(x) = - \int_B (\triangle q)(x_0, x) \, dx = 0.$$

Finally, $P(x_0, x) = -\partial\{K(x - x_0) + q(x_0, x)\}/\partial n$ evaluated at $x \in \partial B$. □

To define the space $L_\infty(\partial B)$ of essentially bounded functions on ∂B, we merely replace the open set Ω by the sphere ∂B in the definition of $L_\infty(\Omega)$ in Chapter 0, (xii), it being understood that sets of measure zero are now those having zero surface area on ∂B.

Theorem B.29 *Let u denote the function called u_g in (B.57); let A be a closed (possibly empty) subset of ∂B. If $g \in L_\infty(\partial B) \cap C(\partial B \setminus A)$, with the understanding that $|g(x)| \le \|g \mid L_\infty(\partial B)\|$ at every $x \in \partial B$, then*

(a) $u \in C(\overline{B} \setminus A) \cap C^\infty(B)$,
(b) $|u(x)| \le \|g \mid L_\infty(\partial B)\|$ for all $x \in \overline{B}$,
(c) $\triangle u = 0$ in B pointwise.

Proof (i) Note first that, for fixed $x \in \partial B$, the function $P(.,x)$ is smoothly harmonic in B, because this is true for $G(.,x)$ and because the operators $\triangle_0 := \sum \partial^2/\partial x_{0j}^2$ and $-\partial/\partial n$ commute for $x_0 \ne x$. The proof that $u \in C^\infty(B)$ and that $\triangle u = 0$ in B is now very like that of Theorem A.5. We omit the details because, for given $x_0 \in B$, we have $\delta_0 := \text{dist}(x_0, \partial B) > 0$.

(ii) The inequality $|u(x)| \le \|g\|$, in which $\|.\|$ denotes the norm of $L_\infty(\partial B)$, holds at every $x \in \partial B$ by hypothesis. To prove it for $x_0 \in B$, we use the Hölder inequality with exponents 1 and ∞, the fact that $P(x_0, x) > 0$ for $x_0 \in B$ and $x \in \partial B$, and the result (B.76); accordingly,

$$\left| \int_{\partial B} P(x_0, x) \, g(x) \, dS(x) \right| \le \|g\| \int_{\partial B} P(x_0, x) \, dS(x) = \|g\|.$$

(iii) It remains to prove that $u \in C(\overline{B} \setminus A)$. Since $u \in C^\infty(B)$ and

$u \in C\left(\partial B \setminus A\right)$, we need prove only this: for given $p \in \partial B \setminus A$ and $\varepsilon > 0$, there is a number $\delta = \delta(p, \varepsilon) > 0$ such that

$$|u(x_0) - u(p)| < \varepsilon \quad \text{whenever} \quad x_0 \in B \quad \text{and} \quad |x_0 - p| < \delta.$$

In the following argument, (B.76) and the positivity of P are used repeatedly; we also use an abbreviated (but self-explanatory) notation for subsets of ∂B. Observe that, for every $\rho > 0$,

$$\begin{aligned}
&|u(x_0) - u(p)| \\
&= \left| \int_{\partial B} P(x_0, x)\{g(x) - g(p)\} \, dS(x) \right| \\
&\leq \int_{|x-p|<\rho} P(x_0, x)\,|g(x) - g(p)| \, dS(x) + \int_{|x-p|\geq\rho} P(x_0, x)\,2\|g\| \, dS(x) \\
&\leq \sup_{|x-p|<\rho} |g(x) - g(p)| + \left\{ \sup_{|x-p|\geq\rho} P(x_0, x) \right\} 2\|g\| \, |\partial B|. \quad \text{(B.77)}
\end{aligned}$$

Since g is continuous at p, and $\partial B \setminus A$ is open relative to the metric space ∂B, we can choose $\rho = \rho(p, \varepsilon)$ to be positive and so small that

$$\sup_{|x-p|<\rho} |g(x) - g(p)| < \tfrac{1}{2}\varepsilon.$$

With ρ now fixed, we prepare to choose δ. If $|x - p| \geq \rho$ and $|x_0 - p| < \delta \leq \tfrac{1}{2}\rho$, then $|x - x_0| > \tfrac{1}{2}\rho$ and $a - r_0 < \delta$, so that

$$P(x_0, x) = \frac{a^2 - r_0^2}{\sigma_N a} |x - x_0|^{-N} < \frac{2\delta}{\sigma_N} \left(\frac{1}{2}\rho \right)^{-N}.$$

Therefore we can choose δ to be in $(0, \tfrac{1}{2}\rho]$ and to be so small that the last term of (B.77) is less than $\tfrac{1}{2}\varepsilon$. $\qquad \square$

If in Theorem B.29 the set A is empty, then $g \in C(\partial B)$, mention of $L_\infty(\partial B)$ is not necessary, and the theorem shows that $u_g \in C(\overline{B}) \cap C^\infty(B)$. In that case, Theorem B.29 establishes existence of the solution of the particular Dirichlet problem $\triangle u = 0$ in B, $u = g$ on ∂B, and Theorem B.16 establishes uniqueness. (In addition, the solution is real-analytic in B, by Theorem B.10.) If A is not empty and g is discontinuous on A, then Theorem B.29 may still be useful (there is an application of the theorem in Appendix C), but the question of uniqueness depends on details of A and g.

The next theorem states that uniform Hölder continuity of g on ∂B implies uniform Hölder continuity of u_g on $\overline{B}$ with the same exponent. This requires a proof longer than that of Theorem B.29 because Hölder continuity with exponent λ at every point of a compact set does not imply uniform Hölder continuity with exponent λ (Exercise B.38).

By $C_b^{0,\lambda}(\partial B)$, usually abbreviated to $C^{0,\lambda}(\partial B)$, we mean the normed linear space that results from Definition A.10 if both Ω and $\overline{\Omega}$ are replaced by ∂B there.

Theorem B.30 *Again let u denote the function called u_g in* (B.57). *If $g \in C^{0,\lambda}(\partial B)$ for some $\lambda \in (0,1)$, then $u \in C^{0,\lambda}(\overline{B})$.*

Proof (i) Theorem B.29 ensures that $u \in C(\overline{B})$, the set A being empty; therefore, it remains to establish uniform Hölder continuity in B, with exponent λ. This is immediate on any closed ball $\overline{\mathscr{B}(0,\rho)}$ with $\rho < a$ because $u \in C^\infty(B)$: if $x,y \in \overline{\mathscr{B}(0,\rho)}$, we bound $|u(x)-u(y)|$ by integrating ∇u along the line segment from x to y. In the remainder of this proof we consider field points

$$x_0 \text{ and } x_0 + h \text{ in } B \setminus \mathscr{B}\left(0, \tfrac{2}{3}a\right), \quad \text{with } 0 < |h| \leq \tfrac{1}{6}a; \qquad (B.78)$$

the point $p := (a/r_0)x_0 \in \partial B$ will be the centre of a small ball $\mathscr{B}(p, 3|h|)$; and the norm $\|g\| := \|g \mid C^{0,\lambda}(\partial B)\|$ is to be so defined that

$$|g(x) - g(p)| \leq \|g\| \, |x - p|^\lambda \quad \text{for all } x \in \partial B. \qquad (B.79)$$

The identity (B.76) implies that

$$u(y) = \int_{\partial B} P(y,x)\{g(x) - g(p)\} \, \mathrm{d}S(x) + g(p) \qquad \text{for all } y \in B,$$

whence, with the abbreviation $\mathrm{d}S := \mathrm{d}S(x)$,

$$\begin{aligned} u(x_0 + h) - u(x_0) &= \int_{\partial B} \{ P(x_0 + h, x) - P(x_0, x) \}\{ g(x) - g(p) \} \, \mathrm{d}S \\ &= I_1(x_0, h) + I_2(x_0, h) \end{aligned} \qquad (B.80)$$

if we define $S_1 := \partial B \cap \mathscr{B}(p, 3|h|)$, $S_2 := \partial B \setminus S_1$ and

$$I_j(x_0, h) := \int_{S_j} \{ P(x_0 + h, x) - P(x_0, x) \}\{ g(x) - g(p) \} \, \mathrm{d}S \qquad (j = 1, 2). \qquad (B.81)$$

(ii) It is easy to bound I_1: using first (B.79), then the positivity of P and (B.76), we obtain

$$\begin{aligned} |I_1(x_0, h)| &\leq \|g\| \, (3|h|)^\lambda \int_{S_1} |P(x_0 + h, x) - P(x_0, x)| \, \mathrm{d}S \\ &\leq \|g\| \, (3|h|)^\lambda \int_{\partial B} \{ P(x_0 + h, x) + P(x_0, x) \} \, \mathrm{d}S \\ &= 2\|g\| \, (3|h|)^\lambda. \end{aligned} \qquad (B.82)$$

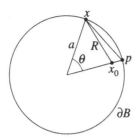

Fig. B.6.

(iii) In order to bound I_2, we let θ denote the angle between the radiusvectors to $x \in \partial B$ and $p \in \partial B$ (see Figure B.6); then

$$|x - p|^2 \;=\; 4a^2 \sin^2 \frac{\theta}{2} \qquad (x \in \partial B, \;\; 0 \le \theta \le \pi),$$

$$R^2 \;:=\; |x - x_0|^2 = (a - r_0)^2 + 4ar_0 \sin^2 \frac{\theta}{2}.$$

Since $r_0 \ge (2/3)a$ by (B.78), we have $|x - p|^2/R^2 \le a/r_0 \le 3/2$; then, for $0 \le t \le 1$ and $|x - p| \ge 3|h|$,

$$\begin{aligned}
\left| x_0 + th - x \right| \ge R - |h| \;&\ge\; R - \tfrac{1}{3}|x - p| \\
&\ge\; R - \tfrac{1}{3} \left(\tfrac{3}{2} \right)^{1/2} R \;>\; \tfrac{1}{2} R. \qquad \text{(B.83)}
\end{aligned}$$

The inequality (B.83) allows us to estimate $P(x_0 + h, x) - P(x_0, x)$, for x outside $\mathscr{B}(p, 3|h|)$, by the method used repeatedly in Appendix A, from Theorem A.5 onwards, for difference kernels with x outside a small ball. Let Γ denote constants independent of x_0, x, h, g and a; then, for $|x - p| \ge 3|h|$,

$$\begin{aligned}
& \left| P(x_0 + h, x) - P(x_0, x) \right| \\[4pt]
&= \frac{1}{\sigma_N a} \left| \frac{|x_0|^2 - |x_0 + h|^2}{|x_0 + h - x|^N} + (a^2 - |x_0|^2) \left(\frac{1}{|x_0 + h - x|^N} - \frac{1}{|x_0 - x|^N} \right) \right| \\[4pt]
&\le \Gamma |h| \, R^{-N} + \Gamma(a - r_0) \, |h| \, R^{-N-1} \\[4pt]
&\le \Gamma |h| \, R^{-N}, \qquad\qquad\qquad\qquad\qquad\qquad\qquad\qquad\qquad\qquad \text{(B.84)}
\end{aligned}$$

since $a - r_0 < R$. Accordingly,

$$\begin{aligned}
\left| I_2(x_0, h) \right| \;&\le\; \Gamma \int_{S_2} |h| \, R^{-N} \, \|g\| \, |x - p|^\lambda \, \mathrm{d}S \\[4pt]
&= \; \Gamma \, \|g\| \, |h| \, a^{\lambda + N - 1} \int_{\gamma(h)}^{\pi} \frac{(\sin(\theta/2))^\lambda \, (\sin \theta)^{N-2}}{\left\{ (a - r_0)^2 + 4ar_0 \sin^2(\theta/2) \right\}^{N/2}} \, \mathrm{d}\theta,
\end{aligned}$$

where $\gamma(h) := 2\sin^{-1}(3|h|/2a)$ with $\sin^{-1}$ taking values in $[-\pi/2, \pi/2]$. Discarding the $(a-r_0)^2$ in the denominator of the integrand, and recalling that $r_0 \geq 2a/3$, we obtain

$$|I_2(x_0, h)| \leq \Gamma \|g\| |h| a^{\lambda-1} \int_{3|h|/a}^{\pi} \theta^{\lambda-2} \, d\theta \leq \Gamma \|g\| |h|^{\lambda}. \qquad (B.85)$$

The uniform Hölder continuity of u in B, with exponent λ, now follows from (B.80), (B.82) and (B.85).   □

B.6 Exercises

Exercise B.31 Use the Taylor formula (1.6) to solve the Neumann problem

$$-u'' = f \quad \text{in} \ (a,b), \qquad u'(a) = \alpha, \qquad u'(b) = \beta, \qquad (B.86)$$

where $f \in C[a,b]$ and $-\int_a^b f = \beta - \alpha$; this last is the compatibility condition (B.29) for the present case. Verify that your solution satisfies (B.86), and adapt it to the problem

$$-v'' = f \quad \text{in} \ (a,\infty), \qquad v'(a) = \alpha,$$

where $f \in C[a,\infty)$.

Exercise B.32 Consider the Dirichlet problem for a ball in one dimension $(N = 1)$:

$$-u'' = f \quad \text{in} \ (-a,a), \qquad u(-a) = c_1, \qquad u(a) = c_2. \qquad (B.87)$$

(i) Deferring conditions on f, derive the formula $u = u_f + u_c$, where

$$u_f(x) := \frac{a-x}{2a} \int_{-a}^{x} (a+t) f(t) \, dt + \frac{a+x}{2a} \int_{x}^{a} (a-t) f(t) \, dt \quad (-a \leq x \leq a),$$

$$u_c(x) := \frac{1}{2a} \big\{ c_1(a-x) + c_2(a+x) \big\} \qquad (-a \leq x \leq a),$$

from the proof of Remark 1.5, or else from (B.41) and (B.48), or, ideally, from both.

(ii) Verify that, if $f \in C[-a,a]$, then $u_f \in C^2[-a,a]$ and $u_f + u_c$ satisfies (B.87).

(iii) Prove that, if $f \in L_1(-a,a)$, then $u_f \in C^1[-a,a]$, u_f satisfies the differential equation almost everywhere in $(-a, a)$, and $u_f + u_c$ satisfies the boundary conditions. In what respects is this result better than the result in Theorem B.23, (ii), for $N \geq 2$?

[If $g \in L_1(\alpha, \beta)$, then the function $x \mapsto \int_\alpha^x g$ is continuous on $[\alpha, \beta]$, and $(d/dx) \int_\alpha^x g = g(x)$ almost everywhere in (α, β). But for the proof in (iii) that $u_f \in C^1[-a, a]$, almost everywhere is not enough; one method is to show that

$$u'_f(x) = \left(\frac{1}{2a} \int_{-a}^a F \right) - F(x), \qquad \text{where} \quad F(x) := \int_{-a}^x f. \Big]$$

Exercise B.33 Consider the Dirichlet problem for a half-line:

$$-u'' = f \text{ in } (0, \infty), \qquad u(0) = c, \qquad u(x) = o(x) \text{ as } x \to \infty. \quad \text{(B.88)}$$

(i) Again deferring conditions on f, derive the formula $u = u_f + c$, where

$$u_f(x) := \int_0^x t f(t) \, dt + x \int_x^\infty f(t) \, dt \qquad (x \geq 0),$$

from Exercise B.19, (ii) [that is, from (B.35)] and from (B.45).

(ii) Verify that, if $f \in C[0, \infty)$ and $f(t) = O\left(t^{-1-\delta}\right)$ as $t \to \infty$, for some constant $\delta > 0$, then $u_f \in C^2[0, \infty)$ and $u_f + c$ satisfies (B.88).

(iii) Prove that, if $f \in L_1(0, \infty)$, then $u_f \in C^1[0, \infty)$, u_f satisfies the differential equation almost everywhere in $(0, \infty)$, and $u_f + c$ satisfies the second and third conditions in (B.88).

[The hint about g in Exercise B.32 extends from (α, β) to (α, ∞) and here

$$u_f(x) = \int_0^x F, \qquad \text{where} \quad F(x) := \int_x^\infty f. \Big]$$

Exercise B.34 This exercise and the next concern the necessity of condition (B.6) in Theorem B.4. Let $Q(p, \sigma) := (p - \sigma, p + \sigma)^N$ [an open cube in $\mathbb{R}^N$ with centre p and edges of length 2σ], and let $\sigma_\varepsilon := \sigma(1 - \varepsilon)$.

(i) Given that

$$f(x) = \sum_\alpha c_\alpha (x - p)^\alpha \qquad \text{if } x \in Q(p, \sigma),$$

where $c_\alpha := \left(\partial^\alpha f\right)(p)/\alpha!$ and summation is over all multi-indices α of length N, prove that for every $\varepsilon \in (0, 1)$ there is a number M_ε such that

$$\left| c_\alpha \right| \leq M_\varepsilon \sigma_\varepsilon^{-|\alpha|} \qquad \text{for all } \alpha,$$

and hence that, for each multi-index β of length N,

$$\left| \partial^\beta f(x) \right| \leq M_\varepsilon \sum_{\alpha \geq \beta} \sigma_\varepsilon^{-|\alpha|} \frac{\alpha!}{(\alpha - \beta)!} \sigma_{2\varepsilon}^{|\alpha - \beta|} \quad \text{if } \varepsilon \in (0, \tfrac{1}{2}) \text{ and } x \in \overline{Q(p, \sigma_{2\varepsilon})}.$$

Here $\alpha \geq \beta$ means that $\alpha_j \geq \beta_j$ for each j.

(ii) Noting that, for $-1 < t < 1$,

$$\sum_{n=k}^{\infty} \frac{n!}{(n-k)!} t^{n-k} = \left(\frac{\mathrm{d}}{\mathrm{d}t} \right)^k \sum_{n=0}^{\infty} t^n = k!\,(1-t)^{-1-k},$$

prove that

$$\sum_{\alpha \geq \beta} \sigma_\varepsilon^{-|\alpha|} \frac{\alpha!}{(\alpha - \beta)!} \sigma_{2\varepsilon}^{|\alpha - \beta|} = \beta! \left(\frac{1 - \varepsilon}{\varepsilon} \right)^N (\varepsilon \sigma)^{-|\beta|} \qquad \left(0 < \varepsilon < \tfrac{1}{2} \right),$$

and hence that, for each multi-index β of length N,

$$\left| \partial^\beta f(x) \right| \leq M_{1/4} 3^N \beta! \left(\frac{4}{\sigma} \right)^{|\beta|} \qquad \text{if } x \in \overline{Q(p, \tfrac{1}{2}\sigma)}.$$

Exercise B.35 (i) Prove that $\beta! \leq |\beta|!$ for all multi-indices β. [Induction on N is possible.]

(ii) Use the result of Exercise B.34 to prove that, if $f : \Omega \to \mathbb{R}$ is real-analytic, then for each compact subset $E \subset \Omega$ there are constants $A = A(E)$ and $B = B(E)$ such that

$$\left| \partial^\beta f(x) \right| \leq A \left(B\, |\beta| \right)^{|\beta|} \qquad \text{for all } x \in E$$

and for all multi-indices β of length N.

Exercise B.36 Under a co-ordinate transformation $(r, \eta_1, \ldots, \eta_{N-1}) \mapsto x$, in which $r = |x|$ and $\eta := (\eta_1, \ldots, \eta_{N-1})$ labels points of the unit sphere $\partial \mathscr{B}_N(0, 1)$, $N \geq 2$, we have

$$\triangle = r^{-N+1} \frac{\partial}{\partial r} \left(r^{N-1} \frac{\partial}{\partial r} \right) + \frac{1}{r^2} \triangle_\eta,$$

where $\triangle_\eta$ is independent of r. (Exercises 1.18, 1.20 and D.18 give examples of such transformations.) Use this form of $\triangle$ to prove Theorem B.15 for $N \geq 2$.

Exercise B.37 Let the operators S and T of the Kelvin transformation be as in Definition B.12, and let $\Omega \subset \mathbb{R}^N \setminus \{0\}$. Show that, if u is subharmonic in Ω (Definition 2.20), then Tu is subharmonic in $S(\Omega)$.

Exercise B.38 *Hölder continuity on a compact set need not be uniform.*
Define

$$u(x) := \begin{cases} x\sin(1/x) & \text{if } 0 < x \le 1, \\ 0 & \text{if } x = 0, \end{cases} \qquad v(x) := \begin{cases} x\sin(e^{1/x}) & \text{if } 0 < x \le 1, \\ 0 & \text{if } x = 0. \end{cases}$$

Recall Definition A.10 and the remarks preceding it.

(i) Show that u and v are Lipschitz continuous (Hölder continuous with exponent $\lambda = 1$) at each point of $[0, 1]$.

(ii) Prove that there are constants A and B, independent of x and h, such that

$$|u(x+h) - u(x)| \le A h^{1/2}, \qquad |v(x+h) - v(x)| \le B/\log\frac{1}{h}$$

whenever $x \in [0, 1)$, $x + h \in (0, 1]$ and $0 < h \le \frac{1}{2}$.
[For all such x and h we have both

$$|u(x+h) - u(x)| \le 2x + h \quad \text{and} \quad |u(x+h) - u(x)| \le h/x + h \quad \text{if } x > 0;$$

the former is better for $x < (h/2)^{1/2}$ and the latter for $x > (h/2)^{1/2}$. In the case of v, the value of x at which two such bounds are equal must be *estimated*.]

(iii) Prove that in (ii) the bound $A h^{1/2}$ cannot be replaced by one that is $o\left(h^{1/2}\right)$ as $h \downarrow 0$, and that $B/\log(1/h)$ cannot be replaced by a bound that is $o\left(1/\log(1/h)\right)$ as $h \downarrow 0$.
[For u, contemplate points $x_n := (2n\pi + \pi/2)^{-1}$ and $x_n + h_n := (2n\pi)^{-1}$ with $n \in \mathbb{N}$.]

Exercise B.39 Show that, for the Green function H_B of the Neumann problem for a ball, the formula given in (B.50c) for $N \ge 3$ is valid for all $N \in \mathbb{N}$.
[For $N = 1$, this function differs by a constant from that in (B.50a).]

Exercise B.40 Let $B := \mathscr{B}_N(0, a)$, let $r := |x|$ and write $\partial/\partial r := (x/r) \cdot \nabla$ if $x \neq 0$. Consider for $N \ge 2$ the Neumann problem of finding u such that

$$\left. \begin{aligned} u \in C\left(\overline{B}\right) \cap C^2(B), \quad & \partial u/\partial r \in C\left(\overline{B} \setminus \{0\}\right), \\ \Delta u = 0 \quad \text{in } B, \quad & \left.\frac{\partial u}{\partial r}\right|_{\partial B} = g, \end{aligned} \right\} \tag{B.89}$$

where $g \in C(\partial B)$ and $\int_{\partial B} g = 0$ (the compatibility condition).

The solution of Exercise 2.42 shows (even though less smoothness is demanded here) that any two solutions of (B.89) differ only by a constant.

Let H_B be the Green function in (B.50), and define

$$u(x_0) := \int_{\partial B} H_B(x_0, \xi)\, g(\xi)\, dS(\xi) \qquad \text{for } x_0 \in \overline{B}. \tag{B.90}$$

Given the result that this function $u \in C(\partial B)$, prove that it is a solution of (B.89).

[One can verify the boundary condition, and prove that $\partial u/\partial r \in C(\overline{B} \setminus \{0\})$, by means of the identity

$$r_0 \frac{\partial H_B}{\partial r_0}(x_0, \xi) = a\, P_B(x_0, \xi) - \frac{1}{\sigma_N\, a^{N-2}} \qquad \text{for } x_0 \in B \setminus \{0\}, \quad \xi \in \partial B, \tag{B.91}$$

and by use of Theorem B.29; here P_B denotes the Poisson kernel called P in (B.58).

The proof that $u \in C(\partial B)$ is not difficult but not short.]

Exercise B.41 Show by an example that a function u satisfying (B.89) need not belong to $C^1(\overline{B})$.

[One possibility is to choose $N = 2$, to write (x, y) for points of $\mathbb{R}^2$ and $z = x + iy$ for points of $\mathbb{C}$, and to contemplate the real part of

$$w(z) := 3ia\, \zeta\, \log(\log \zeta), \qquad \text{where} \quad \zeta := \frac{z+a}{3a}$$

and a convenient branch of $\arg(\log \zeta)$ is chosen.]

Exercise B.42 (i) Show that the function u in (B.90) may be written $u = v + w$ if, for $N \geq 2$ and $x_0 \in \overline{B}$,

$$v(x_0) := 2 \int_{\partial B} K(x_0 - \xi)\, g(\xi)\, dS(\xi), \tag{B.92}$$

$$\left. w(x_0) := \frac{a}{\sigma_N} \int_{r>a} \left\{ \frac{r^{N-3}}{|x_0 - x|^{N-2}} - \frac{1}{r} \right\} \frac{g(\xi)}{r^{N-1}}\, dx \atop \text{where} \quad r := |x|, \quad \xi := \frac{a}{r} x. \right\} \tag{B.93}$$

Note that $w = 0$ (the zero function) if $N = 2$.

(ii) By a *lateral* derivative of u at x we mean any directional derivative

$$m(x) \cdot \nabla u(x) \qquad \text{with} \quad x \in B \setminus \{0\}, \quad m(x) \neq 0, \quad x \cdot m(x) = 0.$$

Show that every lateral derivative of u at x is a linear combination of $N - 1$ lateral derivatives of form

$$u_{ij}(x) := x_j\big(\partial_i u\big)(x) - x_i\big(\partial_j u\big)(x)$$

if $x_i \neq 0$, i is fixed and $j \in \{1, 2, \ldots, N\} \setminus \{i\}$.

Caution This $(.)_{ij}$ must not be confused with that in (A.65) and (B.62b).

(iii) Prove that lateral derivatives of the v in (B.92) and w in (B.93) are given by

$$v_{ij}(x_0) = 2 \int_{\partial B} K_{ij}(x_0, \xi) \{ g(\xi) - g(\xi_0) \} \, dS(\xi), \qquad (B.94)$$

$$w_{ij}(x_0) = a(N-2) \int_{r>a} K_{ij}(x_0, x) \frac{g(\xi) - g(\xi_0)}{r^2} \, dx, \qquad (B.95)$$

where $x_0 \in B \setminus \{0\}$, $\xi_0 := (a/r_0)x_0$, $i \ne j$ and, for $x \in \mathbb{R}^N \setminus \{x_0\}$,

$$K_{ij}(x_0, x) := x_{0j}(\partial_i K)(x_0 - x) - x_{0i}(\partial_j K)(x_0 - x). \qquad (B.96)$$

Exercise B.43 This exercise concerns the following result. If in the Neumann problem (B.89) we have $g \in C^{0,\mu}(\partial B)$ for some $\mu \in (0,1)$, and $\int_{\partial B} g = 0$, then not only does the solution u in (B.90) belong to $C^1(\overline{B})$ [cf. Exercise B.41], but also $\nabla u \in C^{0,\mu}(\overline{B}, \mathbb{R}^N)$.

Give the main parts of the proof by showing that the functions $x \mapsto r\partial u(x)/\partial r$ and v_{ij}, all defined to be zero at the origin, belong to $C^{0,\mu}(\overline{B})$. Here $v_{ij}(x_0)$ is as in (B.94) for $x_0 \in B \setminus \{0\}$.

[The uniform Hölder continuity of $r\partial u/\partial r$ follows from (B.91) and Theorem B.30. For v_{ij}, the proof of Theorem B.30 is relevant, but perhaps the key step is a decomposition of $v_{ij}(x_0 + h) - v_{ij}(x_0)$ that resembles (A.71), despite the different meaning here of $(.)_{ij}$.

The proof for w_{ij}, which is not demanded here, is similar to that for v_{ij} but easier.]

Exercise B.44 Consider for $N \ge 2$ the Green volume potential

$$u_f(x_0) := \int_D G_D(x_0, x) f(x) \, dx, \qquad x_0 \in \overline{D}, \qquad (B.97)$$

of the half-space $D := \{ x \in \mathbb{R}^N \mid x_N > 0 \}$; here, G_D is the Green function in (B.45). Assume that f is measurable and that, for some $\delta \in (0,1)$,

$$|||f||| := \operatorname{ess\,sup}_{x \in D}(1 + r)^{1+\delta} |f(x)| < \infty \qquad (r := |x|). \qquad (B.98)$$

Prove that, for all $x_0 \in \overline{D}$,

$$|(\partial^\alpha u_f)(x_0)| \le \Gamma |||f||| (1 + r_0)^{1-\delta-|\alpha|} \qquad \text{if } |\alpha| = 0 \text{ or } 1, \qquad (B.99)$$

where the constants Γ depend only on N, δ and α.

Note that (B.98) fails to imply that $f \in L_1(D)$ [since $N \ge 2$], hence fails to imply the condition (A.78) assumed for f in Exercise A.31, but that (B.98) is enough to imply the growth conditions assumed for u in Theorem B.16 and in (B.36b) of Exercise B.19.

[It may be helpful to separate the cases $r_0 < 1$ and $r_0 \geq 1$; and, for $r_0 \geq 1$, to integrate separately over the subsets of D in which (a) $R := |x - x_0| < \frac{1}{2}r_0$, (b) $R \geq \frac{1}{2}r_0$ and $r < 2r_0$, (c) $r \geq 2r_0$.]

Exercise B.45 For the half-space D and for $N \geq 2$, the Poisson kernel $-\partial G_D/\partial n$ is

$$P_D(x_0, \xi) := -2(\partial_N K)(x_0 - \xi) = \frac{2x_{0N}}{\sigma_N |x_0 - \xi|^N} \qquad \text{with } x_0 \in D, \ \xi \in \partial D,$$

and we shall also need the Marcel Riesz kernels

$$M_j(x_0, \xi) \ := \ -2(\partial_j K)(x_0 - \xi) = \frac{2(x_{0j} - \xi_j)}{\sigma_N |x_0 - \xi|^N}$$

$$\text{with } j \in \{1, \ldots, N-1\}, \ x_0 \in \overline{D} \setminus \{\xi\}, \ \xi \in \partial D.$$

The corresponding Poisson and Riesz integrals are, respectively,

$$u_g(x_0) := \begin{cases} \displaystyle\int_{\partial D} P_D(x_0, \xi) \, g(\xi) \, dS(\xi) & \text{if } x_0 \in D, \\ g(x_0) & \text{if } x_0 \in \partial D, \end{cases}$$

and

$$v_{g,j}(x_0) := \begin{cases} \displaystyle\int_{\partial D} M_j(x_0, \xi) \, g(\xi) \, dS(\xi) & \text{if } x_0 \in D, \\ \displaystyle\int_{\partial D} M_j(x_0, \xi) \, \{ g(\xi) - g(x_0)m(|\xi - x_0|) \} \, dS(\xi) \\ \hspace{5cm} \text{if } x_0 \in \partial D, \end{cases}$$

where m is a 'mollifier': $m \in C^\infty[0, \infty)$, $m(t) = 1$ for $0 \leq t \leq 1$, $m(t) = 0$ for $t \geq 2$, and m is non-increasing.

Still for $N \geq 2$, prove (or merely accept, in order to do Exercises B.46 and B.47) that the map $g \mapsto u_g$ is a bounded linear operator from $C_b^{0,\mu}(\partial D)$ into $C_b^{0,\mu}(\overline{D})$, and that $g \mapsto v_{g,j}$ is a bounded linear operator from $C_b^{0,\mu}(\partial D) \cap L_p(\partial D)$ into $C_b^{0,\mu}(\overline{D})$; here $\mu \in (0,1)$, $p \in [1, \infty)$ and ∂D may be identified with $\mathbb{R}^{N-1}$.

[For u_g, the proof resembles that of Theorem B.30. Let $\xi_0 := (x_0', 0)$; then

$$v_{g,j}(x_0) \ = \ \int_{\partial D} M_j(x_0, \xi) \, \{ g(\xi) - g(\xi_0)m(|\xi - \xi_0|) \} \, dS(\xi)$$

$$\text{for all } x_0 \in \overline{D}.$$

The uniform Hölder continuity of $v_{g,j}$ follows once again from a decomposition like (A.71).]

Exercise B.46 Use the results of Exercise B.45 to prove an analogue, for the Neumann problem in the half-space D with $N \geq 3$, of the theorem in Exercise B.43. More precisely, defining

$$u(x_0) := \int_{\partial D} H_D(x_0, \xi)\, g(\xi)\, dS(\xi) \qquad \text{for } x_0 \in \overline{D},\ N \geq 3,$$

where H_D is the Green function in (B.47) and $g \in C_b^{0,\mu}(\partial D) \cap L_p(\partial D)$ for some $\mu \in (0,1)$ and $p \in [1, N-1)$, prove that $u \in C^1(\overline{D}) \cap C^\infty(D)$, that $\nabla u \in C_b^{0,\mu}(\overline{D}, \mathbb{R}^N)$ and that

$$\triangle u = 0 \quad \text{in}\quad D, \qquad -\partial_N u\big|_{\partial D} = g.$$

Exercise B.47 Here we are concerned with the smoothness on $\overline{D}$ of the Poisson integral u_ψ, defined in Exercise B.45, when $\psi \in C_c^\infty(\partial D)$. The compact support of ψ could be the result of applying a partition of unity to a more general function in $C^\infty(\partial D)$.

Derive the formulae

$$\left(\partial_N^{2k} u_\psi\right)(x_0) = \int_{\partial D} P_D(x_0, \xi)\left((-\triangle')^k \psi\right)(\xi)\, dS(\xi),$$

$$\left(\partial_N^{2k+1} u_\psi\right)(x_0) = 2\int_{\partial D} (\nabla' K)(x_0 - \xi) \cdot \left(\nabla'(-\triangle')^k \psi\right)(\xi)\, dS(\xi),$$

in which $k \in \mathbb{N}_0$, $x_0 \in D$, $\triangle' := \partial_1^2 + \cdots + \partial_{N-1}^2$ and $\nabla' := (\partial_1, \ldots, \partial_{N-1})$. For $x_0 \in D$, write any derivative $(\partial^\alpha u_\psi)(x_0)$ as an integral with kernel P_D or kernel $2\nabla' K$. Use the results of Exercise B.45 to show that $u_\psi \in C^\infty(\overline{D})$.

Appendix C. Construction of the Primary Function of Siegel Type

In this appendix we use the Poisson integral for a ball, which is the subject of Theorems B.29 and B.30, to prove Theorem 2.27, which describes the primary function of Siegel type. The dimension $N \geq 2$. There is a clash of notation between §B.5 and §2.5; the primary function of Siegel type, denoted by g in Theorem 2.27, will be called v in this appendix.

We recall from (B.57) and (B.58) that the Poisson integral for the ball $B := \mathscr{B}(0, a)$ may be written

$$
u(x) := \begin{cases}
\displaystyle\int_{\partial B} P(x, \xi)\, g(\xi)\, dS(\xi) & \text{if } x \in B, & \text{(C1.a)} \\[2mm]
g(x) & \text{if } x \in \partial B, & \text{(C1.b)}
\end{cases}
$$

the Poisson kernel P being defined by

$$
P(x, \xi) := \frac{a^2 - r^2}{\sigma_N\, a\, |x - \xi|^N} \qquad (r := |x|,\ x \in B,\ \xi \in \partial B). \tag{C.2}
$$

The symbol ξ, used consistently in Appendix B for the inverse point $(a^2/r^2)x$ of x, is now merely a variable of integration.

Theorem B.29 shows that, if $g \in C(\partial B)$, then $u \in C(\overline{B}) \cap C^\infty(B)$ and the formula (C.1) describes the unique pointwise solution of the particular Dirichlet problem $\Delta u = 0$ in B, $u = g$ on ∂B. The theorem also establishes properties of the function u in (C.1) when g is discontinuous but bounded. In the present application, however, we shall have

$$
g(x) = \frac{a}{x_N} \text{ on } \partial B \setminus E, \qquad \text{where } E := \partial B \cap \{ x \in \mathbb{R}^N \mid x_N = 0 \}, \tag{C.3}
$$

so that g is neither continuous nor bounded on ∂B; something beyond the results of Theorem B.29 will be needed. Our first task is to examine the particular form of Poisson's integral for B when g is an odd function of its Nth argument.

Notation For upper and lower hemispheres we write

$$S_+ := \{ x \in \mathbb{R}^N \mid |x| = a, \ x_N > 0 \},$$
$$S_- := \{ x \in \mathbb{R}^N \mid |x| = a, \ x_N < 0 \}.$$

As elsewhere, $x' = (x_1, \ldots, x_{N-1})$ and $x_* := (x', -x_N)$ for any point $x \in \mathbb{R}^N$, so that

$$|x_* - \xi|^2 = |x' - \xi'|^2 + (x_N + \xi_N)^2 = |x - \xi_*|^2.$$

The dimension $N \geq 2$.

If g is an odd function of its Nth argument, so that $g(x_*) = -g(x)$, then the substitution $\xi = \eta_*$ yields

$$\int_{S_-} P(x, \xi) g(\xi) \, dS(\xi) = -\int_{S_+} P(x, \eta_*) g(\eta) \, dS(\eta);$$

consequently, (C.1) becomes

$$u(x) := \begin{cases} \displaystyle\int_{S_+} Q(x, \xi) g(\xi) \, dS(\xi) & \text{if } x \in B, & \text{(C.4a)} \\[2ex] g(x) & \text{if } x \in \partial B, & \text{(C.4b)} \end{cases}$$

where

$$Q(x, \xi) := P(x, \xi) - P(x, \xi_*)$$
$$= \frac{a^2 - r^2}{\sigma_N a} \left\{ \frac{1}{|x - \xi|^N} - \frac{1}{|x - \xi_*|^N} \right\}, \quad x \in B, \xi \in \overline{S}_+. \quad \text{(C.5)}$$

Lemma C.1 *The kernel $Q(\cdot, \zeta)$ is an odd function of its Nth argument. If $x_N \geq 0$ (and $x \in B, \xi \in \overline{S}_+$), then*

$$0 \leq Q(x, \xi) \leq \frac{2N(a^2 - r^2) x_N \xi_N}{\sigma_N a \, |x - \xi|^{N+2}}, \quad \text{(C.6)}$$

and

$$\left| \left(\frac{\partial}{\partial x} \right)^\gamma Q(x, \xi) \right| \leq \begin{cases} M_\gamma \, a^2 \xi_N \, |x - \xi|^{-N-3} & \text{if } |\gamma| = 1, & \text{(C.7a)} \\[2ex] M_\gamma a^{|\gamma|} \xi_N \, |x - \xi|^{-N-2|\gamma|} & \text{if } |\gamma| \geq 2; & \text{(C.7b)} \end{cases}$$

the constant M_γ depends only on the multi-index γ (*the length of which specifies* N).

Proof (i) Replacing x_N by $-x_N$ changes $|x - \xi|$ to $|x - \xi_*|$ and changes $|x - \xi_*|$ to $|x - \xi|$; then $Q(x, \xi)$ changes sign, by (C.5).

(ii) To derive (C.6) we write $R := |x - \xi|$ and $R_* := |x - \xi_*|$. In the remainder of this proof we assume that $x_N \geq 0$; then $0 < R \leq R_*$ and

$$R_* - R = \frac{R_*^2 - R^2}{R_* + R} = \frac{4x_N \xi_N}{R_* + R} \leq \frac{2x_N \xi_N}{R}.$$

Accordingly, for $m \in \mathbb{N}$,

$$
\begin{aligned}
0 \;\leq\; R^{-m} - R_*^{-m} \;&=\; \frac{R_*^m - R^m}{(R_* R)^m} \\
&=\; (R_* - R)\left(\frac{1}{R_* R^m} + \frac{1}{R_*^2 R^{m-1}} + \cdots + \frac{1}{R_*^m R}\right) \\
&\leq\; \frac{2m\, x_N \xi_N}{R^{m+2}}.
\end{aligned}
\tag{C.8}
$$

The definition (C.5) of Q, and the inequality (C.8) with $m = N$, imply (C.6).

(iii) To derive (C.7), we begin with an application of the Leibniz rule (Exercise A.23) to obtain

$$
\left(\frac{\partial}{\partial x}\right)^\gamma Q(x, \xi) \;=\; \frac{1}{\sigma_N a} \sum_{\beta \leq \gamma} \frac{\gamma!}{(\gamma - \beta)!\,\beta!} \left\{\left(\frac{\partial}{\partial x}\right)^{\gamma - \beta} (a^2 - r^2)\right\} \times
$$
$$
\left\{\left(\frac{\partial}{\partial x}\right)^\beta \left(R^{-N} - R_*^{-N}\right)\right\},
\tag{C.9}
$$

and observe [since $|\gamma - \beta| = |\gamma| - |\beta|$ when $\beta \leq \gamma$] that

$$
\left|\left(\frac{\partial}{\partial x}\right)^{\gamma - \beta} (a^2 - r^2)\right| \leq 2a^{2 - |\gamma| + |\beta|}.
\tag{C.10}
$$

This last is an overestimate if $\beta = \gamma$ or $|\gamma - \beta| > 2$, but a harmless one.

(iv) For derivatives of $R^{-N} - R_*^{-N}$, we use a slight variant of (A.32):

$$
\partial^\beta r^{-N} = r^{-N - 2|\beta|} p_\beta(x) \qquad (x \neq 0),
$$

where p_β is a homogeneous polynomial in $x_1, \ldots, x_N$ of degree $|\beta|$; say

$$p_\beta(x) = \sum_{|\alpha|=|\beta|} C_{\beta,\alpha} x^\alpha.$$

It follows that

$$\left(\frac{\partial}{\partial x}\right)^\beta \left(R^{-N} - R_*^{-N}\right) = \left(R^{-N-2|\beta|} - R_*^{-N-2|\beta|}\right) p_\beta(x - \xi)$$

$$+ R_*^{-N-2|\beta|} \left\{ p_\beta(x - \xi) - p_\beta(x - \xi_*) \right\}. \quad \text{(C.11)}$$

In this proof, *const.* will denote positive numbers independent of x, ξ and a. By (C.8), and because $|p_\beta(x - \xi)| \leq \text{const.} R^{|\beta|}$,

$$\left(R^{-N-2|\beta|} - R_*^{-N-2|\beta|}\right) |p_\beta(x - \xi)| \leq \text{const.} x_N \xi_N R^{-N-|\beta|-2}.$$

Next, with the notation $\alpha' := (\alpha_1, \ldots, \alpha_{N-1})$,

$$R_*^{-N-2|\beta|} |p_\beta(x - \xi) - p_\beta(x - \xi_*)|$$

$$= R_*^{-N-2|\beta|} \left| \sum_{|\alpha|=|\beta|} C_{\beta,\alpha}(x' - \xi')^{\alpha'} \left\{ (x_N - \xi_N)^{\alpha_N} - (x_N + \xi_N)^{\alpha_N} \right\} \right|$$

$$\leq \text{const.} a^{|\beta|-1} \xi_N R^{-N-2|\beta|};$$

we have ignored the favourable inequality $\left|(x' - \xi')^{\alpha'})\right| \leq R^{|\alpha'|}$ because there are terms with $C_{\beta,\alpha} \neq 0$, $|\alpha'| = 0$ and $\alpha_N = |\beta|$. Insertion of the last two estimates into (C.11) yields

$$\left| \left(\frac{\partial}{\partial x}\right)^\beta (R^{-N} - R_*^{-N}) \right| \leq \text{const.} a \xi_N \left(R^{|\beta|-2} + a^{|\beta|-2}\right) R^{-N-2|\beta|}, \quad \text{(C.12)}$$

where $R^{|\beta|-2} + a^{|\beta|-2} < \text{const.} a^{|\beta|-2}$ if $|\beta| \geq 2$, because $R < 2a$; the term $R^{|\beta|-2}$ will be retained only for $|\beta| \leq 1$.

(v) Finally, we return to (C.9) and use (C.10), (C.12) and the remark following (C.12); there results

$$\left| \left(\frac{\partial}{\partial x}\right)^\gamma Q(x, \xi) \right|$$

$$\leq \text{const.} \xi_N \left\{ a^{2-|\gamma|} R^{-N-2} \sum_{|\beta| \leq 1} \left(\frac{2a}{R}\right)^{|\beta|} + a^{-|\gamma|} R^{-N} \sum_{\beta \leq \gamma} \left(\frac{2a}{R}\right)^{2|\beta|} \right\}.$$

Since $1 < 2a/R < \infty$, the terms with $|\beta| = 1$ dominate the first sum, and that with $\beta = \gamma$ dominates the second; therefore

$$\left| \left(\frac{\partial}{\partial x} \right)^{\gamma} Q(x,\xi) \right| \leq \text{const.}\, \xi_N \left\{ a^{3-|\gamma|} R^{-N-3} + a^{|\gamma|} R^{-N-2|\gamma|} \right\}. \quad (C.13)$$

For $|\gamma| = 1$, the first term on the right of (C.13) dominates [because $1 < 2a/R < \infty$]; for $|\gamma| \geq 2$, the second term does. $\qquad\square$

The primary function of Siegel type is now defined by

$$v(x) = v(x;a) := \begin{cases} \displaystyle\int_{S_+} Q(x,\xi)\, \frac{a}{\xi_N}\, dS(\xi) & \text{if } x \in B, \qquad\quad\ (C.14a) \\[12pt] a/x_N & \text{if } x \in \partial B \setminus E, \quad (C.14b) \end{cases}$$

where E denotes the equator of B, as in (C.3). Recall that the *signum function* sgn is defined by $\operatorname{sgn} t = -1, 0, 1$ for $t < 0$, $t = 0$, $t > 0$ respectively.

Theorem C.2 *The function* $v = v(\,.\,;a)$ *defined by* (C.14) *belongs to* $C(\overline{B} \setminus E) \cap C^{\infty}(B)$. *Also,*

$$\triangle v = 0 \quad \text{in } B;$$

$$\text{for } x \in \overline{B} \setminus E, \quad v(x', -x_N) = -v(x', x_N) \quad \text{and} \quad \operatorname{sgn} v(x) = \operatorname{sgn} x_N;$$

$$v(x;a) \text{ depends only on } x/a;$$

and

$$|v(x)| \leq \text{const.}\, |x_N|/a \qquad \text{if } |x| \leq a/2,$$

where the constant depends only on N.

Proof (i) To prove that $v \in C^{\infty}(B)$ and $\triangle v = 0$ in B, we note first that $Q(\,.\,,\xi) \in C^{\infty}(B)$, by Lemma C.1, and that $\triangle Q(\,.\,,\xi) = 0$ in B, because this is true for $P(\,.\,,\xi)$. In addition, for $x \in B$ and $\xi \in S_+$ the integrand $Q(x,\xi)/\xi_N$ is bounded in terms of $\operatorname{dist}(x, \partial B)$, because of the factor ξ_N in the bound (C.6) for $Q(x,\xi)$. Similarly, (C.7) shows that $(\partial/\partial x)^{\gamma} Q(x,\xi)/\xi_N$ is bounded in terms of $\operatorname{dist}(x, \partial B)$ for each multi-index γ. Then an argument like the proof of Theorem A.5 shows that $v \in C^{\infty}(B)$ and $\triangle v = 0$ in B.

(ii) On its domain $\overline{B} \setminus E$, v is an odd function of its Nth argument, and $\operatorname{sgn} v(x) = \operatorname{sgn} x_N$. This is true in B by the definition (C.14a) and because $Q(\,.\,,\xi)$ has these two properties; it is true on $\partial B \setminus E$ by the definition (C.14b).

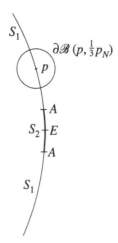

Fig. C.1.

(iii) Now we prove that $v \in C(\overline{B} \setminus E)$. Since $v \in C(\partial B \setminus E)$ by definition, and $v \in C^\infty(B)$ by (i) above, it remains only to prove that $v(x) \to v(p)$ whenever $x \in B$, $p \in \partial B \setminus E$ and $x \to p$. Let $p \in \partial B \setminus E$ by given; in view of (ii), we may assume that $p_N > 0$. We shall prove that

$$v \text{ is continuous on } \mathscr{B}\left(p, \tfrac{1}{3} p_N\right) \cap \overline{B}, \qquad (C.15)$$

which is more than enough. Partition ∂B and v by defining (Figure C.1)

$$S_1 := \partial B \cap \left\{ x \mid |x_N| > \tfrac{1}{3} p_N \right\}, \qquad S_2 := \partial B \setminus S_1,$$

$$A := \partial B \cap \left\{ x \mid |x_N| = \tfrac{1}{3} p_N \right\},$$

$$g_1(x) := \begin{cases} a/x_N & \text{on } S_1, \\ 0 & \text{on } S_2, \end{cases} \qquad g_2(x) := \begin{cases} 0 & \text{on } S_1 \cup E, \\ a/x_N & \text{on } S_2 \setminus E, \end{cases}$$

and, for $j = 1, 2$,

$$v_j(x) := \begin{cases} \displaystyle\int_{S_+} Q(x, \xi)\, g_j(\xi)\, dS(\xi) & \text{if } x \in B, \\ g_j(x) & \text{if } x \in \partial B. \end{cases}$$

Then $v = v_1 + v_2$ on $\overline{B} \setminus E$.

Since $g_1 \in L_\infty(\partial B) \cap C(\partial B \setminus A)$, Theorem B.29 ensures that $v_1 \in C(\overline{B} \setminus A)$; certainly v_1 satisfies (C.15). Now consider v_2 on $\overline{B} \setminus S_2$, observing that S_2 is closed. Adapting remarks in (i), we see that $(\partial/\partial x)^\nu Q(x, \xi) g_2(\xi)$

is bounded in terms of $\text{dist}(x, S_2)$ for each γ, and that $v_2 \in C^\infty(\overline{B} \setminus S_2)$. (Lemma C.1 extends to points $x \in \partial B$, provided that $|x - \xi| > 0$.) In particular, (C.7a) shows that for points x and $x + h$ in $\overline{B}$, such that $|x_N| \geq \frac{2}{3} p_N$ and $|x_N + h_N| \geq \frac{2}{3} p_N$, we have

$$|v_2(x + h) - v_2(x)| \leq \text{const.}\, a^{N+2} \left(\tfrac{1}{3} p_N\right)^{-N-3} |h|,$$

with a constant that depends only on N. Thus v_2 satisfies (C.15) with room to spare.

(iv) To prove that $v(x; a)$ depends only on x/a, we write $Q(x, \xi; a)$ for the function defined by (C.5) and set $x =: a\hat{x}$, $\xi =: a\hat{\xi}$ in the definition (C.14a); there results

$$v(x; a) = \int Q(\hat{x}, \hat{\xi}; 1)\, (1/\hat{\xi}_N)\, dS(\hat{\xi}) \quad \text{if } x \in B,$$

the integral being over the unit hemisphere $\left\{ \hat{\xi} \ \middle| \ |\hat{\xi}| = 1, \hat{\xi}_N > 0 \right\}$. This proves the claim for $x \in B$; the result is immediate from (C.14b) if $x \in \partial B \setminus E$.

(v) It remains to bound $|v(x)|$ for $|x| \leq a/2$. Now, $|x - \xi| \geq a/2$ when $|x| \leq a/2$ and $\xi \in S_+$; then the definition of v and the bound (C.6) for Q imply that

$$|v(x)| \leq \int_{S_+} \frac{2Na^2|x_N|\xi_N}{\sigma_N a(a/2)^{N+2}} \frac{a}{\xi_N} \, dS(\xi) = 2^{N+2}N \frac{|x_N|}{a}.$$

$\square$

Remark C.3 *For $N = 2$, let $(x, y) \in \mathbb{R}^2$ and $z = x + iy \in \mathbb{C}$. Then, on $\overline{B} \setminus E \subset \mathbb{R}^2$,*

$$v(x, y; a) = \text{Im} \left(\frac{a}{a - z} - \frac{a}{a + z} \right) = \frac{ay}{(a - x)^2 + y^2} + \frac{ay}{(a + x)^2 + y^2}. \tag{C.16}$$

Proof The proof is an exercise in contour integration for which we choose $a = 1$ (as we may, because $v(x, y; a)$ depends only on x/a and y/a) and use (C.1) rather than its special form (C.4) for anti-symmetric functions g. The Poisson kernel may be written

$$P(z, \zeta) := \frac{1 - r^2}{2\pi|z - \zeta|^2}, \qquad \text{where } |z| = r < 1, \ \zeta = e^{it},$$

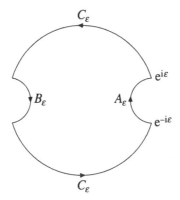

Fig. C.2.

and, if we define

$$I_\varepsilon(z) := \int_{J_\varepsilon} \frac{1}{|z - \zeta|^2} \frac{dt}{\sin t}, \quad J_\varepsilon := (-\pi + \varepsilon, -\varepsilon) \cup (\varepsilon, \pi - \varepsilon), \qquad \text{(C.17)}$$

then

$$v(x, y; 1) = \frac{1 - r^2}{2\pi} \lim_{\varepsilon \downarrow 0} I_\varepsilon(z) \qquad \text{(C.18)}$$

for $|z| = |x + iy| < 1$. The result for $|z| = 1$, $z \neq \pm 1$, is known *a priori* and need not be calculated.

Now, since $|z - \zeta|^2 = (\zeta - z)(1/\zeta - \bar{z})$, where $\bar{z} := x - iy$, and $\sin t = -(i/2)(\zeta - 1/\zeta)$, it follows from (C.17) that

$$I_\varepsilon(z) = \int_{C_\varepsilon} F(z, \zeta) \, d\zeta, \qquad \text{where } C_\varepsilon := \left\{ e^{it} \mid t \in J_\varepsilon \right\}$$

and

$$F(z, \zeta) := \frac{-2\zeta}{(\bar{z}\zeta - 1)(\zeta - z)(\zeta^2 - 1)}.$$

Let A_ε and B_ε be circular arcs in the closed unit disc, centred at $\zeta = 1$ and -1, respectively; A_ε proceeds from $e^{-i\varepsilon}$ to $e^{i\varepsilon}$ and B_ε from $-e^{-i\varepsilon}$ to $-e^{i\varepsilon}$ (Figure C.2). Choose ε to be positive and so small that z belongs to the indented open disc bounded by $A_\varepsilon \cup B_\varepsilon \cup C_\varepsilon$. The only singularity of $F(z, .)$ in this indented disc is the pole at $\zeta = z$; consequently,

$$I_\varepsilon(z) + \int_{A_\varepsilon} F(z, \zeta) \, d\zeta + \int_{B_\varepsilon} F(z, \zeta) \, d\zeta = \frac{-4\pi iz}{(\bar{z}z - 1)(z^2 - 1)}.$$

Calculation of $-\pi\mathrm{i}$ times the residues of $F(z,.)$ at $\zeta = 1$ and $\zeta = -1$ yields

$$\int_{A_\varepsilon} F(z,\zeta)\,\mathrm{d}\zeta = -\frac{\pi\mathrm{i}}{(1 - \bar{z})(1 - z)} + O(\varepsilon),$$

$$\int_{B_\varepsilon} F(z,\zeta)\,\mathrm{d}\zeta = \frac{\pi\mathrm{i}}{(1 + \bar{z})(1 + z)} + O(\varepsilon).$$

In view of (C.18), the rest is manipulation.   □

Appendix D. On the Divergence Theorem and Related Matters

D.1 A first divergence theorem

Our task in this appendix is to extend the fundamental theorem of the calculus to functions defined on subsets of $\mathbb{R}^N$. If the result is to have a modest generality and to be of some use, then this task cannot be short and easy, for several reasons. First, it is not obvious how to pass from the merely local description of $\partial\Omega$ in Chapter 0, (viii), to the evaluation of integrals over $\partial\Omega$ as a whole; Definition D.1 and Lemma D.2 are preparations for this step. Second, we must attend to the smoothness both of the function being integrated and of the boundary $\partial\Omega$. Third, conditions that allow a straightforward proof of the divergence theorem (such as those in Theorem D.3) are often too restrictive for applications. Although we take only two primitive steps towards relaxing the conditions in Theorem D.3, those steps require a certain length if they are to be elementary and transparent.

Notation Throughout this appendix, $\mathbb{R}^N$ is to have dimension $N \geq 2$. The notation of *Cartesian products* will be taken beyond the definition of $A \times B$ in Chapter 0, (i); for example, $(-\beta, \beta)^{N-1}$ denotes the cube $Q'(0, \beta)$ described rather fully in Definition D.1, and $\{a\} \times [0, a]^{N-1}$ denotes a face of the cube $(0, a)^N$, more precisely, the intersection of the hyperplane $\{x \in \mathbb{R}^N \mid x_1 = a\}$ and the closed cube $[0, a]^N$.

Definition D.1 Let $\partial\Omega$ be [at least] of class C and let $p \in \partial\Omega$. Referring to Chapter 0, (viii) and Figure 0.3, we note that Ω is defined globally in terms of co-ordinates x_j, while y_j are 'local' co-ordinates such that $\partial\Omega \cap U(p)$ (where $U(p)$ is open in $\mathbb{R}^N$ and contains p) has a representation

$$y_N = h(y'), \qquad \text{where} \quad h = h(., p), \quad y' = (y_1, \ldots, y_{N-1}). \tag{D.1}$$

279

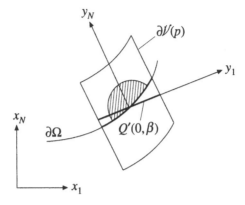

Fig. D.1.

(a) The co-ordinate transformation is

$$y = Y_p(x) := A(p)(x - p), \qquad (D.2)$$

where $A(p)$ is an orthogonal $N \times N$ matrix depending on p.

(b) Denote a cube about the origin in $\mathbb{R}^{N-1}$, with edges parallel to the co-ordinate axes and of length $2\beta > 0$, by

$$Q'(0, \beta) := \left\{ y' \in \mathbb{R}^{N-1} \mid -\beta < y_j < \beta \quad \text{for } j = 1, \ldots, N - 1 \right\}.$$

Define (Figure D.1)

$$\mathscr{V}(p) := \left\{ y \in \mathbb{R}^N \mid y' \in Q'(0, \beta), \quad -\beta < y_N - h(y') < \beta \right\}$$

and its x-image $\mathscr{U}(p) := Y_p^{-1}(\mathscr{V}(p))$, where $\beta = \beta(p)$ and β is so small that $\mathscr{U}(p) \subset U(p)$. The sets $\mathscr{V}(p)$ and $\mathscr{U}(p)$ will be called *quasi-cubical neighbourhoods* of the points 0 and p, respectively (because they are easily transformed to cubes in $\mathbb{R}^N$). $\qquad \square$

The y_N-axis will point *into* Ω throughout this appendix; then the definition implies that

$$x \in \partial\Omega \cap \mathscr{U}(p) \quad \text{iff} \quad y = Y_p(x) \in \mathscr{V}(p) \quad \text{and} \quad y_N = h(y'), \qquad (D3.a)$$

$$x \in \Omega \cap \mathscr{U}(p) \quad \text{iff} \quad y = Y_p(x) \in \mathscr{V}(p) \quad \text{and} \quad y_N > h(y'). \qquad (D3.b)$$

The functions $\psi_1, \ldots, \psi_M$ in the following lemma are an example of a *partition of unity on* $\overline{\Omega}$. They are a key to our proof of the divergence theorem because we use them to reduce the proof to one for functions of small support in convenient neighbourhoods. [The *support* of a function is defined in Chapter 0, (iv); essentially, it is the smallest closed set outside which the function equals zero.] The hatched set in Figure D.1 represents the support of a function ψ_m when $p = p^m$ in the figure.

Lemma D.2 *Let Ω be bounded and $\partial\Omega$ of class C. Then for some integer M there exist functions $\psi_1, \psi_2, \ldots, \psi_M$ as follows.*

(a) $\sum_{m=1}^{M} \psi_m(x) = 1$ *whenever $x \in \overline{\Omega}$.*

(b) $\psi_m \in C^\infty(\overline{\Omega})$ *and $0 \le \psi_m(x) \le 1$ for each $m \in \{1, \ldots, M\}$ and all $x \in \overline{\Omega}$.*

(c) *The functions ψ_m are of two kinds: there is an integer $K < M$ such that, if $m \in \{1, \ldots, K-1\}$, then $\psi_m \in C_c^\infty(\Omega)$, while, if $m \in \{K, \ldots, M\}$, then $\mathrm{supp}\,\psi_m$ intersects $\partial\Omega$. In the second case $\mathrm{supp}\,\psi_m \subset \overline{\Omega} \cap \mathcal{U}(p^m)$, where $\mathcal{U}(p^m)$ is a quasi-cubical neighbourhood of a point $p^m \in \partial\Omega$.*

Proof (i) Assign to every point $c \in \overline{\Omega}$ a ball $\mathscr{B}(c, \rho)$ as follows. If $c \in \Omega$, choose $\rho = \rho(c)$ so small that $\overline{\mathscr{B}(c, \rho)} \subset \Omega$; if $c \in \partial\Omega$, choose $\rho = \rho(c)$ so small that $\overline{\mathscr{B}(c, \rho)} \subset \mathcal{U}(c)$, where $\mathcal{U}(c)$ is a quasi-cubical neighbourhood of c. The family $\{\mathscr{B}(c, \rho) \mid c \in \overline{\Omega}\}$ is then an open cover of $\overline{\Omega}$ [in other words, $\overline{\Omega}$ is a subset of the union of all these open balls]. Since $\overline{\Omega}$ is compact, there is a finite subfamily, say $\{\mathscr{B}(c^m, \rho_m) \mid m = 1, \ldots, M\}$, that covers $\overline{\Omega}$, and we so label this subfamily that $c^m \in \Omega$ if $m < K$, while $c^m \in \partial\Omega$ if $m \ge K$.

(ii) Define for each $m \in \{1, \ldots, M\}$ (see Figure D.2)

$$\varphi_m(x) := \begin{cases} \exp\left(-\dfrac{1}{1-s^2}\right) & \text{if } s := \dfrac{|x - c^m|}{\rho_m} < 1, \\ 0 & \text{if } s \ge 1, \end{cases} \tag{D.4}$$

$$\psi_m(x) := \varphi_m(x) \Big/ \sum_{k=1}^{M} \varphi_k(x) \qquad \text{if } x \in \overline{\Omega}, \tag{D.5}$$

and let $p^m := c^m$ when $c^m \in \partial\Omega$ (when $m \ge K$). We observe that $\sum_{k=1}^{M} \varphi_k(x) > 0$ when $x \in \overline{\Omega}$, because then $x \in \mathscr{B}(c^k, \rho_k)$ for at least one k, and that $\varphi_m \in C_c^\infty(\mathbb{R}^N)$ for each m, by Exercise 1.15. The properties (a) to (c) of the functions ψ_m now follow directly from the definitions (D.4) and (D.5). $\qquad\square$

Notation We shall use the abbreviations

$$\int_\Omega g := \int_\Omega g(x)\,dx \quad \text{and} \quad \int_{\partial\Omega} g := \int_{\partial\Omega} g(x)\,dS(x),$$

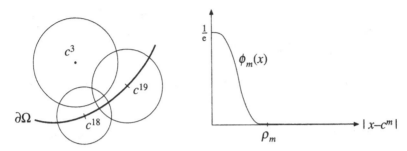

Fig. D.2.

where $dx := dx_1 \, dx_2 \ldots dx_N$ and $dS(x)$ is as in Chapter 0, (viii), unless a transformation of co-ordinates or detailed evaluation of an integral should demand a fuller display.

Theorem D.3 (a first divergence theorem). *Let Ω be bounded, with $\partial\Omega$ of class C^1, and let $n = (n_1, \ldots, n_N)$ denote the outward unit normal on $\partial\Omega$. Then*

$$\int_{\Omega} \partial_j f = \int_{\partial\Omega} n_j f \qquad (D.6)$$

whenever $f \in C^1(\overline{\Omega})$ and $j \in \{1, \ldots, N\}$.

Proof (i) Let $\psi_1, \ldots, \psi_M$ be as in Lemma D.2, and define $f_m := f\psi_m$ for each $m \in \{1, \ldots, M\}$. By (a) of Lemma D.2,

$$\sum_{m=1}^{M} f_m(x) = f(x) \sum_{m=1}^{M} \psi_m(x) = f(x) \qquad \text{for } x \in \overline{\Omega}, \qquad (D.7)$$

and each $f_m \in C^1(\overline{\Omega})$ by (b) of the lemma. (If f should happen to be smoother than C^1, then so is f_m.) Since

$$\sum_{m=1}^{M} \int_{\Omega} \partial_j f_m = \int_{\Omega} \partial_j \left(\sum_{m=1}^{M} f_m \right) = \int_{\Omega} \partial_j f \quad \text{and} \quad \sum_{m=1}^{M} \int_{\partial\Omega} n_j f_m = \int_{\partial\Omega} n_j f,$$

it is sufficient to prove (D.6) for each f_m; the result for f then follows by summation over m.

Let $B_m := \mathscr{B}(c^m, \rho_m)$, where c^m and ρ_m are as in (D.4). There are two cases: if $c^m \in \Omega$ (if $m < K$), then $\operatorname{supp} f_m \subset \overline{B}_m \subset \Omega$; if $c^m \in \partial\Omega$ (if $m \geq K$), then $p^m := c^m$ and $\operatorname{supp} f_m \subset \overline{\Omega} \cap \overline{B}_m$, where $\overline{B}_m \subset \mathscr{U}(p^m)$.

(ii) Let $c^m \in \Omega$ (let $m < K$). We shall prove that

$$\int_\Omega \partial_j f_m = 0 = \int_{\partial\Omega} n_j f_m \qquad (c^m \in \Omega). \tag{D.8}$$

The right-hand equality is immediate because $f_m = 0$ on $\partial\Omega$. To prove the left-hand equality, we define f_m to be zero not merely on $\overline{\Omega} \setminus B_m$ but on $\mathbb{R}^N \setminus B_m$; then

$$\int_\Omega \partial_j f_m = \int_{Q_m} \partial_j f_m$$

for any cube Q_m (in $\mathbb{R}^N$) that contains B_m. Suppose that $j = 1$ and that $Q_m =: (a,b) \times Q''$, where Q'' is a cube in $\mathbb{R}^{N-1}$, and let $x'' := (x_2, \ldots, x_N)$. By Fubini's theorem,

$$\begin{aligned}
\int_{Q_m} \partial_1 f_m &= \int_{Q''} \left\{ \int_a^b (\partial_1 f_m)(x_1, x'')\, dx_1 \right\} dx'' \\
&= \int_{Q''} \{ f_m(b, x'') - f_m(a, x'') \}\, dx'' = 0
\end{aligned}$$

because $f_m = 0$ on ∂Q_m. The argument is similar for $j \in \{2, \ldots, N\}$: the Fubini theorem allows us to replace the volume integral by a repeated integral, and we integrate $\partial_j f_m$ first with respect to x_j.

(iii) Let $p^m = c^m \in \partial\Omega$ (let $m \geq K$). With m fixed, we abbreviate p^m to p, adopt the notation in Definition D.1, and let $\tilde{f}_m(y) := f_m(x)$ under the transformation $y = Y_p(x)$. Define

$$V_+ := \left\{ y \in \mathcal{V}(p) \mid y_N > h(y') \right\}, \qquad \Gamma := \left\{ y \in \mathcal{V}(p) \mid y_N = h(y') \right\},$$

and recall that f_m vanishes in $\overline{\Omega} \setminus B_m$. Then the desired result,

$$\int_\Omega \partial_j f_m = \int_{\partial\Omega} n_j f_m \qquad \text{whenever } j \in \{1, \ldots, N\}, \tag{D.9}$$

is equivalent to the statement

$$\int_{V_+} (\partial_k \tilde{f}_m)(y)\, dy = \int_\Gamma v_k(y') \tilde{f}_m(y)\, d\sigma(y') \quad \text{whenever } k \in \{1, \ldots, N\}; \tag{D.10}$$

here v and $d\sigma$ denote [as in Chapter 0, (viii)] the outward unit normal and element of surface area, respectively, in terms of the local co-ordinates. [To pass from (D.9) to (D.10), multiply (D.9) by $A_{kj}(p)$ and sum over j; to pass from (D.10) to (D.9), multiply (D.10) by $A_{kj}(p)$ and sum over k.] Upon use of the formulae (0.2) and (0.3) for v and $d\sigma$, equation (D.10) becomes

$$\int_{V_+} (\partial_k \tilde{f}_m)(y)\, dy = \begin{cases} \displaystyle\int_{Q'(0,\beta)} (\partial_k h)(y')\, \tilde{f}_m(y', h(y'))\, dy' & \text{if } k \in \{1,\dots,N-1\}, \\[4mm] \displaystyle -\int_{Q'(0,\beta)} \tilde{f}_m(y', h(y'))\, dy' & \text{if } k = N. \end{cases}$$

(D.11)

(iv) In order to prove (D.11), we make the further co-ordinate transformation

$$z' = y', \qquad z_N = y_N - h(y') \qquad (y \in \overline{\mathscr{V}(p)}),$$

which maps $\mathscr{V}(p)$ onto the cube $Q'(0,\beta) \times (-\beta, \beta)$ shown in Figure D.3. Then

$$\frac{\partial}{\partial y_k} = \frac{\partial}{\partial z_k} - (\partial_k h)(y')\frac{\partial}{\partial z_N} \quad \text{if } k \in \{1,\dots,N-1\}, \qquad \frac{\partial}{\partial y_N} = \frac{\partial}{\partial z_N},$$

and the Jacobian determinant is

$$\det\left(\frac{\partial z_i}{\partial y_j}\right) = \begin{vmatrix} 1 & 0 & \dots & 0 \\ 0 & 1 & & 0 \\ \vdots & & & \\ -(\partial_1 h)(y') & -(\partial_2 h)(y') & & 1 \end{vmatrix} = 1.$$

Let $W_+ := Q'(0,\beta) \times (0,\beta)$ and let $\hat{f}_m(z) := \tilde{f}_m(y)$; by the rule for subjecting integrals to co-ordinate transformations (Apostol 1974, p.421; Weir 1973, p.158), the desired result (D.11) becomes

$$\int_{W_+} \left\{ (\partial_k \hat{f}_m)(z) - (\partial_k h)(z')\,(\partial_N \hat{f}_m)(z) \right\} dz$$
$$= \int_{Q'(0,\beta)} (\partial_k h)(z')\,\hat{f}_m(z',0)\, dz' \quad \text{if } k \in \{1,\dots,N-1\}, \quad \text{(D.12a)}$$

$$\int_{W_+} (\partial_N \hat{f}_m)(z)\, dz = -\int_{Q'(0,\beta)} \hat{f}_m(z',0)\, dz' \quad \text{if } k = N. \tag{D.12b}$$

Now in (D.12a) we have

$$\int_{W_+} (\partial_k \hat{f}_m)(z)\, dz = 0 \quad \text{if } k \in \{1,\dots,N-1\}, \tag{D.13}$$

because, very much as in step (ii), Fubini's theorem allows us to replace the integral over W_+ by a repeated integral and to integrate $\partial_k \hat{f}_m$ first

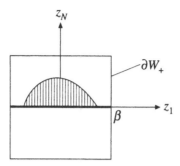

Fig. D.3.

with respect to x_k; moreover, $\hat{f}_m(z) = 0$ on $\partial W_+ \setminus \{ z \mid z_N = 0 \}$. With (D.13) established, (D.12) is seen to be true because

$$\int_0^\beta \left(\partial_N \hat{f}_m\right)(z', z_N)\, \mathrm{d}z_N = -\hat{f}_m(z', 0),$$

and because we may use repeated integrals again, integrating $\partial_N \hat{f}_m$ first with respect to z_N. □

D.2 Extension to some sets with edges and vertices

In many applications one needs a divergence theorem for sets Ω with boundaries $\partial \Omega$ that are not of class C^1 because they have edges and vertices. A general treatment of such boundaries would lead us into deep water, but results for some particular cases are useful and not difficult.

Remark D.4 We shall derive a divergence theorem for the following bounded open sets in $\mathbb{R}^N$ ($N > 2$). As elsewhere, $x'' := (x_2, \ldots, x_N)$.

(i) An open *interval*, say $I := (a_1, b_1) \times (a_2, b_2) \times \cdots \times (a_N, b_N)$.

(ii) A *circular cylinder*, say $C := \{ x \in \mathbb{R}^N \mid a < x_1 < b, |x''| < \rho \}$.

(iii) A *segment of a ball*, by which we mean the (non-empty) intersection of a ball and a half-space, say

$$D := \mathscr{B}(0, b) \cap \{ x \in \mathbb{R}^N \mid x_1 > b \cos \alpha \}, \quad \text{where } \alpha \in (0, \pi).$$

(iv) A finite *sector* in $\mathbb{R}^2$, or *circular cone* in $\mathbb{R}^N$ with $N \geq 3$, possibly truncated near its vertex: in terms of $r := |x|$ and θ_1 such that $x_1 =$

$r\cos\theta_1$ and $|x''| = r\sin\theta_1$, say

$$K := \left\{\, x \in \mathbb{R}^N \mid a < r < b,\ 0 \le \theta_1 < \gamma \,\right\},$$
$$\text{where } 0 \le a < b \quad \text{and} \quad \gamma \in (0,\pi).$$

(v) A *set bounded by two intersecting spheres*, say

$$E := \mathscr{B}(0,R) \setminus \overline{\mathscr{B}(c,\rho)} \qquad \text{with } R - \rho \le |c| < R + \rho,$$

or

$$E' := \mathscr{B}(c,\rho) \setminus \overline{\mathscr{B}(0,R)} \qquad \text{with } R - \rho < |c| < R + \rho,$$

where $0 < \rho \le R$ in both cases.

(vi) Sets resulting from those in (i) to (v) when the co-ordinate frame is translated and rotated.

(vii) *Complements* of the foregoing in a larger admissible set: $\Omega = \Omega_2 \setminus \overline{\Omega}_1$, where $\overline{\Omega}_1 \subset \Omega_2$ and Ω_1, Ω_2 are listed in (i) to (vi), or else one of them is listed while the other is a ball. $\qquad\Box$

The reader may wish to lengthen the list in Remark D.4; the method that will be used to prove Theorem D.5 can be applied to other configurations. Alternatively, the reader may prefer to proceed to more advanced forms of the divergence theorem (Evans & Gariepy 1992, p.209; Maz'ja 1985, p.304; Ziemer 1969, p.248) which allow sets Ω and functions f that are far less restricted than those considered here.

A first question is: how should one write the normal $n(x)$ and element $dS(x)$ of surface area on boundaries with edges and vertices? But this is hardly a difficulty for the catalogue in Remark D.4; for these sets, a direct application of Cartesian, cylindrical or spherical co-ordinates is possible; alternatively, one can apply the definitions in Chapter 0, (viii), to each smooth piece of the boundary. We shall exhibit both methods.

Consider as an example the segment D of a ball in Remark D.4, (iii). Partition ∂D into a *flat part* $\{x \mid x_1 = b\cos\alpha, |x''| < b\sin\alpha\}$, a *spherical part* $\{x \mid |x| = b, x_1 > b\cos\alpha\}$ and an *edge* $\{x \mid x_1 = b\cos\alpha, |x''| = b\sin\alpha\}$. The edge does not contribute to the boundary integral because it has zero surface area (has $(N-1)$-dimensional measure zero). On the flat part of ∂D,

$$n(x) = (-1, 0, \ldots, 0), \qquad dS(x) = dx'' := dx_2\, dx_3 \ldots dx_N. \qquad (\text{D}.14)$$

For the spherical part of ∂D we use spherical co-ordinates as in Exercise D.18; then on the spherical part of ∂D,

$$\left.\begin{array}{c} n(x) = x/|x|, \qquad dS(x) = b^{N-1}A(\theta)\, d\theta, \\[4pt] \text{where} \quad A(\theta) := (\sin\theta_1)^{N-2}(\sin\theta_2)^{N-3}\ldots(\sin\theta_{N-2}), \\[4pt] d\theta := d\theta_1\, d\theta_2\ldots d\theta_{N-1}. \end{array}\right\} \qquad \text{(D.15)}$$

This formula for $dS(x)$ is not invalidated at $x = (b,0,\ldots,0)$ by the fact that the Jacobian determinant $\det \partial(x_i)/\partial(r,\theta_j)$ vanishes there. [Consider the contribution to an integral over ∂D of any piece of surface that contains the point $x = (b,0,\ldots,0)$ and has diameter ε.]

In order to illustrate how the formulae in Chapter 0, (viii), extend to a boundary with an edge, we now represent ∂D, near an edge point p, as the graph of a function that fails to be C^1 only at the projection of the edge onto the domain of the function. Let

$$\beta := \frac{\alpha}{2} \in \left(0, \frac{\pi}{2}\right), \qquad p := (b\cos 2\beta, 0, \ldots, 0, -b\sin 2\beta),$$

and let the transformation $y = Y_p(x)$ to local co-ordinates (Figure D.4) be

$$\left.\begin{array}{rcl} y_1 &=& (\cos\beta)(x_1 - p_1) - (\sin\beta)(x_N - p_N), \\[3pt] y_2 &=& x_2, \\ &\vdots& \\ y_{N-1} &=& x_{N-1}, \\ y_N &=& (\sin\beta)(x_1 - p_1) + (\cos\beta)(x_N - p_N). \end{array}\right\}$$

The flat part of ∂D has radius $R := b\sin 2\beta = -p_N$, and its projection onto the hyperplane $\{y \mid y_N = 0\}$ is the elliptic set

$$E := \left\{ y' \in \mathbb{R}^{N-1} \ \middle|\ \left(\frac{y_1 + R\sin\beta}{R\sin\beta}\right)^2 + \frac{\eta^2}{R^2} < 1 \right\},$$

where

$$\eta^2 := y_2^2 + \cdots + y_{N-1}^2.$$

Our local representation of ∂D is restricted to the cylinder

$$V := \left\{ y \in \mathbb{R}^N \ \middle|\ |y'| < \rho, \quad -b\sin\beta < y_N < b\sin\beta \right\}$$

having radius $\rho \in (0, b - b\cos\beta)$ and cross-section

$$B := \left\{ y' \in \mathbb{R}^{N-1} \ \middle|\ y_1^2 + \eta^2 < \rho^2 \right\}.$$

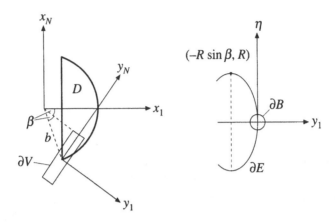

Fig. D.4.

If $1 - \cos\beta \leq \sin 2\beta$ (as is the case in Figure D.4), then $\rho < R$, so that a part of E is outside B; however, if β is so near $\pi/2$ that $1 - \cos\beta > \sin 2\beta$, then it can happen that $\overline{E} \subset B$.

A calculation now yields the following representation of $Y_p(\partial D) \cap V$:

$$y_N = h(y') := \begin{cases} -y_1 \cot\beta & \text{if } y' \in B \cap \overline{E}, \\ b\sin\beta - \sqrt{(b\sin\beta)^2 - 2by_1\cos\beta - y_1^2 - \eta^2} & \text{if } y' \in B \setminus \overline{E}; \end{cases}$$
(D.16)

it is easy to check that these two expressions are equal on $B \cap \partial E$. The formulae (0.2) and (0.3) for the outward unit normal $v(y')$ and element $d\sigma(y')$ of surface area are valid in $B \setminus \partial E$ for this function h.

Theorem D.5 *If G is one of the (open, bounded) sets listed in Remark D.4, then*

$$\int_G \partial_j f = \int_{\partial G} n_j f \tag{D.17}$$

whenever $f \in C^1(\overline{G})$ and $j \in \{1, \ldots, N\}$. The boundary integral is over those parts of ∂G on which the outward unit normal n is defined and continuous.

Proof (i) We shall introduce a sequence (G_m) of open subsets of G, each having closure $\overline{G}_m \subset G$ and boundary ∂G_m of class C^1, such that

$$\int_{G_m} \partial_j f \rightarrow \int_G \partial_j f \quad \text{and} \quad \int_{\partial G_m} n_j f \rightarrow \int_{\partial G} n_j f \quad \text{as } m \rightarrow \infty, \tag{D.18}$$

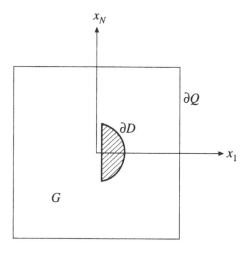

Fig. D.5.

whenever $f \in C^1(\overline{G})$ and $j \in \{1, \ldots, N\}$. Since $\int_{G_m} \partial_j f = \int_{\partial G_m} n_j f$ for each m, by Theorem D.3, this will prove the present theorem.

(ii) Consider the particular set $G := Q \setminus \overline{D}$ (Figure D.5), where Q is the cube $(-a, a)^N$ and D is the segment of a ball in Remark D.4, (iii), with $b < a$. It will suffice to display G_m for this configuration, because a similar construction is possible for all cases of our list. (The symmetry of Q and the alignment of Q and D save words, but are not necessary for the construction. Exercise D.12 deals with another set in the list.) Our method is simply to round off all edges and vertices; in order to make the resulting set a subset of G, we must first enlarge D slightly. At the same time we shall shrink Q slightly, in order to have $\overline{G}_m \subset G$; the virtue of this condition will be seen in §D.3.

The approximating sets $G_m := Q_m \setminus \overline{D}_m$ are formed as follows. Choose and fix $\delta_1 \in \left(0, \frac{1}{3}(a - b)\right)$ and set $\delta_m := \delta_1/m$ for $m \in \mathbb{N}$. The rounded set D_m (Figure D.6) is

$$D_m := \bigcup_{c \in \overline{D}} \mathscr{B}(c, \delta_m) = \left\{ x \in \mathbb{R}^N \mid \operatorname{dist}(x, \overline{D}) < \delta_m \right\},$$

and its boundary ∂D_m is partitioned into a flat part F_m, a spherical part S_m and a toroidal part T_m by the definitions

$$F_m := \partial D_m \cap \left\{ x \mid x_1 = b \cos \alpha - \delta_m \right\},$$

$$S_m := \partial D_m \cap \partial \mathscr{B}(0, b + \delta_m) \quad \text{and} \quad T_m := \partial D_m \setminus \left(F_m \cup S_m \right).$$

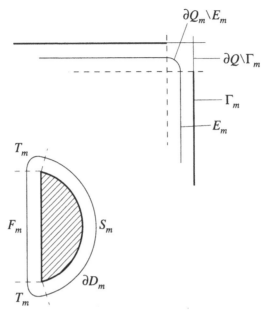

Fig. D.6.

The toroidal part has surface area $|T_m| \leq \text{const.}\delta_m$, where the constant is independent of m (and may be taken to be $(\pi - \alpha) |\partial \mathscr{B}_{N-1}(0, 1)| \times (b \sin \alpha + \delta_1)^{N-2}$). Since ∂D_m is a surface of revolution, one checks without difficulty that it is of class C^1.

Now consider the approximation Q_m to Q. First form the smaller cube $M_m := (-a + 2\delta_m, a - 2\delta_m)^N$, then define

$$Q_m := \bigcup_{c \in \overline{M}_m} \mathscr{B}(c, \delta_m).$$

Defining an intermediate cube by $P_m := (-a + \delta_m, a - \delta_m)^N$, we set $E_m := \partial Q_m \cap \partial P_m$ and so partition ∂Q_m into a flat part E_m and a curved part $\partial Q_m \setminus E_m$. Note that E_m consists of $2N$ squares (that is, $(N-1)$-dimensional cubes) like

$$\left\{ a - \delta_m \right\} \times (-a + 2\delta_m, a - 2\delta_m)^{N-1}.$$

Correspondingly, the original boundary ∂Q is partitioned into a part Γ_m consisting of $2N$ squares like $\{a\} \times (-a + 2\delta_m, a - 2\delta_m)^{N-1}$ and a part $\partial Q \setminus \Gamma_m$ containing the vertices of Q. The curved part of ∂Q_m and the corner part of ∂Q have surface areas

$$|\partial Q_m \setminus E_m| \leq \text{const.}\delta_m, \qquad |\partial Q \setminus \Gamma_m| \leq \text{const.}\delta_m,$$

where the constants are independent of m. That ∂Q_m is of class C^1 is easily seen for $N = 2$ or 3 with the help of figures, and is the subject of Exercise D.13 for arbitrary $N \geq 2$.

(iii) To establish the first limit in (D.18), we fix j and write $h := \partial_j f$, $h_m := \chi_m h$, where χ_m is the characteristic function of G_m [equal to 1 on G_m, and to 0 on $\mathbb{R}^N \setminus G_m$]. Then $h_m(x) \to h(x)$ at each fixed $x \in G$ as $m \to \infty$; also $|h_m(x)| \leq |h(x)|$ and $|h| \in C(\overline{G})$. By the Lebesgue dominated convergence theorem,

$$\int_{G_m} h = \int_G h_m \to \int_G h \qquad \text{as } m \to \infty.$$

(iv) Consider the second limit in (D.18). Let

$$\|f\| := \sup_{x \in G} \{ |f(x)| + |\nabla f(x)| \};$$

we have $\|f\| < \infty$ because $f \in C^1(\overline{G})$. In view of remarks made in (ii) about surface area,

$$\left| \int_{T_m} n_j f \right| + \left| \int_{\partial Q_m \setminus E_m} n_j f \right| + \left| \int_{\partial Q \setminus \Gamma_m} n_j f \right|$$

$$\leq \|f\| \{ |T_m| + |\partial Q_m \setminus E_m| + |\partial Q \setminus \Gamma_m| \} \to 0 \quad \text{as } m \to \infty. \quad \text{(D.19)}$$

Let F and S denote, respectively, the flat and spherical parts of ∂D. It remains to prove that (as $m \to \infty$)

$$\int_{F_m} n_j f - \int_F n_j f \to 0, \quad \int_{S_m} n_j f - \int_S n_j f \to 0, \quad \int_{E_m} n_j f - \int_{\Gamma_m} n_j f \to 0.$$
$$\text{(D.20)}$$

We shall prove the second of these; the first and third are similar but easier, because only Cartesian co-ordinates are required.

In order to have a convenient pairing of points of S_m and points of S, we use spherical co-ordinates $r, \theta_1, \ldots, \theta_{N-1}$ as in Exercise D.18; we also use the notation in (D.15) and write $\theta := (\theta_1, \ldots, \theta_{N-1})$, $\tilde{n}_j(\theta) := n_j(x)$ and $\tilde{f}(r, \theta) :- f(x)$. The labelling $\theta_1 < \alpha$ of the domain of integration means $-\alpha < \theta_1 < \alpha$ if $N = 2$ or $0 < \theta_1 < \alpha$ if $N \geq 3$. Then

$$\left| \int_{S_m} n_j f - \int_S n_j f \right|$$

$$= \left| \int_{\theta_1 < \alpha} \tilde{n}_j(\theta) \{ \tilde{f}(b + \delta_m, \theta) - \tilde{f}(b, \theta) \} (b + \delta_m)^{N-1} A(\theta) \, d\theta \right.$$

$$\left. + \int_{\theta_1 < \alpha} \tilde{n}_j(\theta) \, \tilde{f}(b, \theta) \{ (b + \delta_m)^{N-1} - b^{N-1} \} A(\theta) \, d\theta \right|$$

$$\leq \text{const.} \, \|f\| \delta_m, \qquad \qquad \text{(D.21)}$$

where the constant is independent of f and m because

$$\left|\tilde{f}(b+\delta_m,\theta) - \tilde{f}(b,\theta)\right| = \left|\int_b^{b+\delta_m} \frac{\partial \tilde{f}}{\partial r}(r,\theta)\,\mathrm{d}r\right| \leq \|f\|\delta_m,$$

$$(b+\delta_m)^{N-1} \leq (b+\delta_1)^{N-1}, \quad (b+\delta_m)^{N-1} - b^{N-1} \leq (N-1)(b+\delta_1)^{N-2}\delta_m.$$

The inequality (D.21) establishes the second limit in (D.20), and we have noted that the first and third are proved similarly but more easily. $\quad\square$

Remark D.6 The result (D.18) can be improved. We note again that $\overline{G}_m \subset G$ and that ∂G_m is of class C^1 for each $m \in \mathbb{N}$, and now claim that, as $m \to \infty$,

$$\int_{G_m} u \to \int_G u \quad \text{whenever } u \in L_1(G), \tag{D.22a}$$

$$\int_{\partial G_m} n_j v \to \int_{\partial G} n_j v \quad \text{whenever } v \in C(\overline{G}) \text{ and } j \in \{1,\dots,N\}. \tag{D.22b}$$

[Recall that the condition $u \in L_1(G)$ means merely that u is integrable on G.] As a description of how well G_m approximates G, the statement (D.22) is more satisfactory than (D.18), because the condition $f \in C^1(\overline{G})$ is not natural for such a description.

The proof of (D.22a) is virtually the same as step (iii) of the foregoing proof: $\chi_m(x)u(x) \to u(x)$ at each fixed $x \in G$, and the dominant function is now $|u| \in L_1(G)$.

To prove (D.22b), we note that, since $\overline{G}$ is compact, v is bounded and uniformly continuous on $\overline{G}$: there are positive numbers K and γ_ε (depending on v) such that $|v(x)| \leq K$ for all $x \in \overline{G}$ and such that, for every $\varepsilon > 0$,

$$|v(x) - v(\xi)| < \varepsilon \quad \text{whenever } x, \xi \in \overline{G} \text{ and } |x - \xi| < \gamma_\varepsilon.$$

Accordingly, K replaces $\|f\|$ in (D.19) when v replaces f; thereafter, we use the fact that, for every $\varepsilon > 0$ and for the relevant (x_1, x'') and (r, θ),

$$|v(x_1 - \delta_m, x'') - v(x_1, x'')| < \varepsilon \quad \text{and} \quad |\tilde{v}(r + \delta_m, \theta) - \tilde{v}(r,\theta)| < \varepsilon \tag{D.23}$$

if m is so large that $\delta_m < \gamma_\varepsilon$. $\quad\square$

D.3 Interior approximations to the boundary $\partial\Omega$

Not infrequently the condition $f \in C^1(\overline{\Omega})$ in Theorem D.3 is an embarrassment; here is an example. If in (D.6) or in (D.17) we set $f = u\partial_j u$ and sum over j, there results the identity

$$\int_\Omega \left\{ |\nabla u|^2 + u\triangle u \right\} = \int_{\partial\Omega} u\frac{\partial u}{\partial n} \qquad \left(\frac{\partial}{\partial n} := n \cdot \nabla \right), \qquad \text{(D.24)}$$

for admissible Ω, *provided that* $u \in C^2(\overline{\Omega})$. In the uniqueness theorems for which this identity is commonly used, it is natural to assume that $u \in C^1(\overline{\Omega}) \cap C^2(\Omega)$ and that $\triangle u = 0$ in Ω, but artificial to assume that $u \in C^2(\overline{\Omega})$. As far as we know, an individual second derivative such as $\partial_1^2 u(x)$ might misbehave as x approaches some boundary point. To overcome this difficulty, we shall develop not a third divergence theorem but, rather, approximations Ω_m to Ω. These sets Ω_m satisfy $\overline{\Omega}_m \subset \Omega$, have boundaries $\partial\Omega_m$ of class C^1 and have the approximation properties for integrals that are displayed in (D.22) for the sets G_m considered there.

With such approximations Ω_m in hand, we can apply (D.24) to Ω_m rather than to Ω, noting that $u \in C^2(\overline{\Omega}_m)$ when $u \in C^2(\Omega)$. In fact, if $u \in C^1(\overline{\Omega}) \cap C^2(\Omega)$ and $\triangle u = 0$ in Ω, we shall have

$$\begin{aligned} \int_\Omega |\nabla u|^2 &= \lim_{m\to\infty} \int_{\Omega_m} |\nabla u|^2 \\ &= \lim_{m\to\infty} \int_{\partial\Omega_m} u\frac{\partial u}{\partial n} = \int_{\partial\Omega} u\frac{\partial u}{\partial n}, \end{aligned}$$

which vanishes, as desired in uniqueness proofs, if $u = 0$ on $\partial\Omega$ or $\partial u/\partial n = 0$ on $\partial\Omega$.

If $\partial\Omega$ were of class C^2, then the approximations Ω_m could be formed, for sufficiently small numbers $\delta_m > 0$, by defining $\partial\Omega_m$ to be the set consisting of all points

$$q(p, \delta_m) := p - \delta_m n(p), \qquad p \in \partial\Omega. \qquad \text{(D.25)}$$

It can be shown that $\partial\Omega_m$ is then a boundary of class C^2. This is the conventional method in geometry for the construction of what are called *tubular neighbourhoods* of a surface (the word 'tubular' actually referring to layers on both sides of the surface). When $\partial\Omega$ is merely of class C^1, however, points of Ω arbitrarily close to $\partial\Omega$ may be on more than one line normal to $\partial\Omega$ (Exercise D.17) and the set of points $q(p, \delta_m)$ need not form a boundary of class C^1, however small δ_m may be. Therefore we use a different method and proceed with due caution.

Notation We continue to write $\partial/\partial n := n \cdot \nabla$, where $n(x)$ denotes the outward unit normal at $x \in \partial\Omega$.

Lemma D.7 *If Ω is bounded and $\partial\Omega$ is of class C^1, there exists a function $w \in C^1(\overline{\Omega})$ such that*

$$w\big|_{\partial\Omega} = 0, \quad -\frac{\partial w}{\partial n}\bigg|_{\partial\Omega} \geq 1$$

and $w \geq 0$ in Ω.

Proof We shall use the local representations $y_N = h(y', p)$ of $\partial\Omega$ in Definition D.1; the transformations (D.2) to local co-ordinates will now be written

$$y = Y(x, p), \quad \text{where } Y = (Y', Y_N) \text{ and } Y' := (Y_1, \ldots, Y_{N-1}),$$

and the y_N-axis points into Ω. We shall also use the partition functions $\psi_K, \ldots, \psi_M$ of the second kind in Lemma D.2. Since the functions $\psi_1, \ldots, \psi_{K-1}$ of the first kind all vanish on $\partial\Omega$, those of the second kind form a partition of unity on $\partial\Omega$:

$$\sum_{k=K}^{M} \psi_k(x) = 1 \quad \text{if } x \in \partial\Omega;$$

also, $\psi_k \in C^\infty(\overline{\Omega})$, $\operatorname{supp} \psi_k \subset \overline{\Omega} \cap \mathscr{U}(p^k)$ and $0 \leq \psi_k(x) \leq 1$ for each $k \in \{K, \ldots, M\}$ and all $x \in \overline{\Omega}$. We define

$$w(x) := \sum_{k=K}^{M} \psi_k(x)\eta_k(x), \quad x \in \overline{\Omega}, \tag{D.26a}$$

where

$$\eta_k(x) := \begin{cases} Y_N(x, p^k) - h(Y'(x, p^k), p^k) & \text{if } x \in \overline{\Omega} \cap \mathscr{U}(p^k), \\ 0 & \text{elsewhere in } \overline{\Omega}. \end{cases} \tag{D.26b}$$

[The definition of $\eta_k(x)$ for $x \notin \operatorname{supp} \psi_k$ is unimportant, because $\psi_k(x)\eta_k(x) = 0$ outside $\operatorname{supp}\psi_k$ under any definition of $\eta_k : \overline{\Omega} \to \mathbb{R}$.] Note that, if $x \in \overline{\Omega} \cap \mathscr{U}(p^k)$ and $y = Y(x, p^k)$, then

$$\eta_k(x) = y_N - h(y', p^k);$$

therefore, at points $x \in \partial\Omega \cap \mathcal{U}(p^k)$ and with the abbreviation $h(y') :=$ $h(y', p^k)$,

$$
\begin{aligned}
-\frac{\partial\eta_k}{\partial n}(x) &= \sum_{j=1}^{N} v_j(y')\frac{\partial}{\partial y_j}\{h(y') - y_N\} \\
&= \sum_{j=1}^{N-1} v_j(y')\frac{\partial h}{\partial y_j}(y') - v_N(y') \\
&= \left\{ (\partial_1 h)(y')^2 + \cdots + (\partial_{N-1}h)(y')^2 + 1 \right\}^{1/2} \quad \text{(D.27)}
\end{aligned}
$$

upon application of the formula (0.2) for $v(y')$.

The function $w \in C^1(\overline{\Omega})$ because each $\psi_k \in C^\infty(\overline{\Omega})$, each $h(.,p^k) \in$ $C^1(\overline{Q'_k})$, where $Q'_k := Q'(0, \beta(p^k))$, and each $Y(.,p^k)$ is an affine function (that is, a linear function plus a constant). Moreover, $w|_{\partial\Omega} = 0$ because $\eta_k|_{\partial\Omega} = 0$ for all k; and the definition (D.26) also shows that $w \geq 0$ in Ω. Finally, for each k and all $x \in \partial\Omega \cap \mathcal{U}(p^k)$ we have $-\partial\eta_k(x)/\partial n \geq 1$ by (D.27); since also $\eta_k|_{\partial\Omega} = 0$, this yields

$$
-\frac{\partial w}{\partial n}\bigg|_{\partial\Omega} = \sum_{k=K}^{M} \psi_k \left(-\frac{\partial\eta_k}{\partial n}\right)\bigg|_{\partial\Omega} \geq \sum_{k=K}^{M} \psi_k\bigg|_{\partial\Omega} = 1,
$$

as desired. $\qquad\square$

To construct Ω_m, we shall choose a suitably small number $\alpha_1 > 0$, set $\alpha_m := \alpha_1/m$ for $m \in \mathbb{N}$, and define $\partial\Omega_m$ to be the part of the level set $\{x \in \Omega \mid w(x) = \alpha_m\}$ that lies close to $\partial\Omega$. It will save trouble later to record here some implications of Lemma D.7.

Remark D.8 (i) Since ∇w is continuous on the compact set $\overline{\Omega}$, it is uniformly continuous: for every $\varepsilon > 0$ there is a number $\delta_\varepsilon > 0$ such that

$$
x, \xi \in \overline{\Omega} \quad \text{and} \quad |x - \xi| < \delta_\varepsilon \;\Rightarrow\; |(\nabla w)(x) - (\nabla w)(\xi)| < \varepsilon. \quad \text{(D.28)}
$$

(ii) On $\partial\Omega$ the vector field ∇w is normal to $\partial\Omega$ because $w|_{\partial\Omega} = 0$, and it points inwards to Ω because $-\partial w/\partial n > 0$. Therefore $|\nabla w| = -\partial w/\partial n \geq 1$ on $\partial\Omega$, whence

$$
x \in \overline{\Omega}, \; p \in \partial\Omega \quad \text{and} \quad |x - p| < \delta_\varepsilon \;\Rightarrow\; |(\nabla w)(x)| > 1 - \varepsilon. \quad \text{(D.29)}
$$

(iii) Define

$$
\left.\begin{aligned}
d(x) &:= \text{dist}(x, \partial\Omega), \\
e(x) &:= \nabla w(x)/|\nabla w(x)| \quad \text{if } x \in \overline{\Omega} \text{ and } d(x) < \delta_{1/2}.
\end{aligned}\right\} \quad \text{(D.30)}
$$

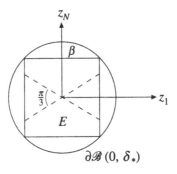

Fig. D.7.

[This last ensures, in view of (D.29), that $|\nabla w(x)| > \frac{1}{2}$.] The direction field (or unit-vector field) e is a continuous extension of the inward unit normal field $-n$ on $\partial\Omega$; that is, $|e(x)| = 1$ and $e\big|_{\partial\Omega} = -n$. Condition (D.28) now implies the existence of a number $\delta_* > 0$ as follows: if $x \in \overline{\Omega}$, $p \in \partial\Omega$ and $|x - p| < \delta_*$, then

$$e(p) \cdot e(x) > \frac{\sqrt{3}}{2} = \cos\frac{\pi}{6} \quad \text{and} \quad e(p) \cdot \nabla w(x) > \frac{3}{5}. \tag{D.31}$$

(A calculation, outlined in Exercise D.14, shows that $\delta_{1/4}$ can serve as δ_*.)

(iv) We use (D.31) to construct a useful cylinder $D(p)$ about any boundary point $p \in \partial\Omega$. Take local co-ordinates z_j such that

$$\left.\begin{array}{l} z := Z_p(x) := C(p)(x - p), \\[2mm] \text{where} \quad \nabla_x z_N = \big(C_{Nj}(p)\big)_{j=1}^{N} = e(p) = -n(p), \end{array}\right\} \tag{D.32}$$

and where $C(p)$ is an orthogonal $N \times N$ matrix depending on p. Define (Figure D.7)

$$\beta := \frac{\delta_*}{\sqrt{2}}, \quad E := \big\{ z \in \mathbb{R}^N \mid |z'| < \beta, \ -\beta < z_N < \beta \big\}, \quad D(p) := Z_p^{-1}(E), \tag{D.33}$$

observing that E is independent of p, so that a translation and rotation of the same cylinder E gives $D(p)$ for all $p \in \partial\Omega$.

The inward normal $-n(x)$ at any point $x \in \partial\Omega \cap D(p)$ makes an angle with $e(p)$ that is less than $\pi/6$, by (D.31); hence

$$z \in Z_p(\partial\Omega) \cap E \quad \Rightarrow \quad |z_N| \le \left(\tan\frac{\pi}{6}\right)|z'| < \frac{\beta}{\sqrt{3}}, \tag{D.34}$$

with equality only if $z' = 0$. Thus $Z_p(\partial\Omega) \setminus \{0\}$ lies between the two conical surfaces shown by broken lines in Figure D.7.

(v) Let $\hat{w}(z) := w(x)$ under the transformation $z = Z_p(x)$, so that $\partial\hat{w}(z)/\partial z_N = e(p) \cdot \nabla w(x)$. Integrating $\partial\hat{w}/\partial z_N$ along a vertical line $\{z \mid z' = \text{const.}\}$ in E from a point of $Z_p(\partial\Omega)$ to the point above it on the ceiling of E, and applying (D.31) and (D.34), we obtain

$$\hat{w}(z',\beta) > \frac{3}{5}\left(\beta - \frac{\beta}{\sqrt{3}}\right) > \frac{1}{4}\beta \qquad (|z'| < \beta). \tag{D.35}$$

It follows from (D.31) and (D.35) that *every vertical line segment* $\{z \mid z' = \text{const.}\}$ *in* $E \cap Z_p(\overline{\Omega})$ *contains exactly one point of* $Z_p(\partial\Omega)$, *at which* $\hat{w}(z) = 0$, *and exactly one point of* $Z_p(\partial\Omega_m)$, *at which* $\hat{w}(z) = \alpha_m$, *if* $0 < \alpha_1 < \frac{1}{4}\beta$ *and* $\alpha_m := \alpha_1/m$ *for* $m \in \mathbb{N}$. We label these points $(z', g(z', p))$ in the case of $Z_p(\partial\Omega)$, and $(z', g_m(z', p))$ in the case of $Z_p(\partial\Omega_m)$. $\qquad\square$

Theorem D.9 *Suppose that* Ω *is bounded and that either* $\partial\Omega$ *is of class* C^1 *or* Ω *is one of the sets listed in Remark* D.4. *Then there exists a sequence* (Ω_m) *of open sets such that*
  (a) *$\overline{\Omega}_m \subset \Omega$ for each m,*
  (b) *each boundary $\partial\Omega_m$ is of class C^1,*
  (c) *$\int_{\Omega_m} u \to \int_\Omega u$ as $m \to \infty$ whenever $u \in L_1(\Omega)$,*
  (d) *$\int_{\partial\Omega_m} n_j v \to \int_{\partial\Omega} n_j v$ as $m \to \infty$ whenever $v \in C(\overline{\Omega})$ and $j \in \{1,\dots,N\}$.*

Proof (i) If Ω is one of the sets listed in Remark D.4, then the present theorem merely repeats the second sentence of Remark D.6. The truth of that sentence rests on the construction of G_m in the proof of Theorem D.5, step (ii); on analogous constructions, like that in Exercise D.12, for other sets in the list; on the proof in Remark 6 of (D.22a, b); and on slight variants of (D.23) for sets other than $Q \setminus \overline{D}$.

(ii) For the rest of this proof, let $\partial\Omega$ be of class C^1 and let w be as in Lemma D.7. We retain the notation in Remark D.8 [in particular, $d(x) := \text{dist}(x, \partial\Omega)$] and shall use the estimates there. The graph of w on the inward normal line from any point $p \in \partial\Omega$ is qualitatively as in Figure D.8.

We define Ω_m by setting $\alpha_1 = \frac{1}{5}\beta$, $\alpha_m := \alpha_1/m$ for $m \in \mathbb{N}$ and

$$\Omega_m := \left\{ x \in \Omega \mid w(x) > \alpha_m \quad \text{or} \quad d(x) > \beta \right\},$$

equivalently,

$$\Omega \setminus \Omega_m = \left\{ x \in \Omega \mid w(x) \le \alpha_m \quad \text{and} \quad d(x) \le \beta \right\}.$$

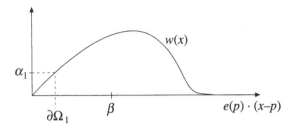

Fig. D.8.

In the definition of Ω_m, the alternative condition $d(x) > \beta$ merely ensures that Ω_m contains those points x relatively far from $\partial\Omega$ at which $w(x) \leq \alpha_m$. We now show that, in fact, $d(x) < \frac{1}{3}\beta$ at all $x \in \Omega \setminus \Omega_m$; therefore $\partial\Omega_m$ lies well inside the set $\{\, x \in \Omega \mid d(x) \leq \beta \,\}$ and is determined there by the condition $w(x) = \alpha_m$.

Let $x_0 \in \Omega \setminus \Omega_m$ and let $p_0 \in \partial\Omega$ with $|x_0 - p_0| = d(x_0)$. There is at least one such nearest boundary point p_0 for x_0. The line segment from p_0 to x_0 has direction $e(p_0)$; let $d\ell(x)$ denote the element of length at a point x of this line segment. Then, by (D.31),

$$\alpha_m \geq w(x_0) = \int_{p_0}^{x_0} e(p_0) \cdot \nabla w(x)\, d\ell(x) > \tfrac{3}{5}|x_0 - p_0| = \tfrac{3}{5}\, d(x_0),$$

whence

$$d(x_0) < \tfrac{5}{3}\alpha_m \leq \tfrac{5}{3}\alpha_1 = \tfrac{1}{3}\beta \qquad (x_0 \in \Omega \setminus \Omega_m). \tag{D.36}$$

(iii) We must prove that $\partial\Omega_m$ is of class C^1. Let a point $q \in \partial\Omega_m$ be given; again there is at least one point $p \in \partial\Omega$ such that $|q - p| = d(q)$, and we take local co-ordinates z_j at p as in (D.32). The final observation of Remark D.8 states that the set $Z_p(\partial\Omega_m) \cap E$ has a representation $z_N = g_m(z', p)$ [and $Z_p(q)$ must satisfy this equation because it is the only point on the line $\{z \mid z' = 0\}$ in E at which $\hat{w}(z) = \alpha_m$]. The implicit-function theorem ensures that $g_m(., p)$ is a C^1 function on some set open in $\mathbb{R}^{N-1}$ and containing $0 \in \mathbb{R}^{N-1}$; we may take this set to be $\mathscr{B}_{N-1}(0, \beta)$, again by the final observation of Remark D.8. Thus $g_m(., p) \in C^1(\overline{G})$ whenever $\overline{G} \subset \mathscr{B}_{N-1}(0, \beta)$.

The implicit-function theorem also ensures existence of a set $U(q, m)$, open in $\mathbb{R}^N$ and containing q, such that

$$x \in U(q, m) \quad \text{and} \quad w(x) = \alpha_m \iff z_N = g_m(z', p), \tag{D.37}$$

under the co-ordinate transformation $z = Z_p(x)$. The uniqueness part ($\Rightarrow$) of this result deserves emphasis here. Suppose that q has another nearest boundary point: there exists a point $p_* \in \partial\Omega \setminus \{p\}$ such that $|q - p_*| = d(q)$. Taking local co-ordinates ζ_j at p_*, again of the type (D.32), we obtain another local representation $\zeta_N = \gamma_m(\zeta', p_*)$ of $\partial\Omega_m$ near q. Suppose that under the affine transformation $\zeta \mapsto z$ this second representation becomes $z_N = \hat{\gamma}_m(z', p_*)$. Then $\hat{\gamma}_m(., p_*) = g_m(., p)$ wherever both are defined, otherwise (D.37) is contradicted.

(iv) The proof that $\int_{\Omega_m} u \to \int_\Omega u$ whenever $u \in L_1(\Omega)$ is like step (iii) in the proof of Theorem D.5 and the extension of that step in Remark D.6.

(v) The proof of condition (d) of the theorem involves a partition of unity on $\overline{\Omega} \setminus \Omega_1$ very reminiscent of that in Lemma D.2. The family $\{ \mathscr{B}(p, \frac{2}{3}\beta) \mid p \in \partial\Omega \}$ is an open cover of the compact set $\overline{\Omega} \setminus \Omega_1$, because of (D.36). Hence there is a finite subcover, say $\{ \mathscr{B}(p^k, \frac{2}{3}\beta) \mid k = 1, \ldots, K \}$ with each $p^k \in \partial\Omega$, and we define, for each $k \in \{1, \ldots, K\}$,

$$\varphi_k(x) := \begin{cases} \exp\left(-\dfrac{1}{1-s^2}\right) & \text{if } s := \dfrac{3|x - p^k|}{2\beta} < 1, \\ 0 & \text{if } s \geq 1, \end{cases}$$

$$\psi_k(x) := \varphi_k(x) \Big/ \sum_{n=1}^{K} \varphi_n(x) \quad \text{if } x \in \overline{\Omega} \setminus \Omega_1.$$

Each partition function ψ_k has support well within the cylinder $D(p^k)$ defined by (D.33).

(vi) It remains to prove condition (d). Given $v \in C(\overline{\Omega})$, we define $v_k := v\psi_k$. It is sufficient to prove that

$$\int_{\partial\Omega_m} n_j v_k \to \int_{\partial\Omega} n_j v_k \qquad \text{as } m \to \infty, \tag{D.38}$$

for $j \in \{1, \ldots, N\}$ and $k \in \{1, \ldots, K\}$; then summation over k yields the result for v because $v = \sum v_k$ on $\overline{\Omega} \setminus \Omega_1$.

We fix k, abbreviate p^k to p, make the transformation $z = Z_p(x)$ to local co-ordinates as in (D.32) and write $\hat{v}(z) := v_k(x)$, *with the understanding that $\hat{v}$ is continuous on $Z_p(\overline{\Omega})$ and that* $\operatorname{supp} \hat{v} \subset E$. The local representations of $\partial\Omega$ and of $\partial\Omega_m$ are abbreviated to $z_N = g(z')$ and to $z_N = g_m(z')$, respectively, and $B := \mathscr{B}_{N-1}(0, \beta)$ denotes the cross-section of E.

In view of the formulae (0.2) and (0.3) for the unit normal and element of surface area, we must prove that, if $j \in \{1, \ldots, N-1\}$,

$$\int_B (\partial_j g_m) \, \hat{v}(z', g_m) \, \mathrm{d}z' \to \int_B (\partial_j g) \, \hat{v}(z', g) \, \mathrm{d}z', \qquad (D.39)$$

where $g_m = g_m(z')$ and $g = g(z')$, and that, if $j = N$,

$$\int_B \hat{v}(z', g_m) \, \mathrm{d}z' \to \int_B \hat{v}(z', g) \, \mathrm{d}z'. \qquad (D.40)$$

For (D.40) it is enough that $\hat{v}$ is uniformly continuous and that

$$0 < g_m(z') - g(z') < \frac{5}{3}\alpha_m = \frac{\beta}{3m} \quad \text{for all } z' \in B, \qquad (D.41)$$

which is proved as (D.36) was. For (D.39) we must also compare $\nabla' g_m$ and $\nabla' g$, where $\nabla' := (\partial_1, \ldots, \partial_{N-1})$; these gradients are calculated by differentiation of the equations

$$\hat{w}\big(z', g_m(z')\big) = \alpha_m \quad \text{and} \quad \hat{w}\big(z', g(z')\big) = 0.$$

Accordingly, for every $\varepsilon > 0$,

$$|\nabla' g_m(z') - \nabla' g(z')| = \left| \frac{\big(\nabla' \hat{w}\big)\big(z', g(z')\big)}{\big(\partial_N \hat{w}\big)\big(z', g(z')\big)} - \frac{\big(\nabla' \hat{w}\big)\big(z', g_m(z')\big)}{\big(\partial_N \hat{w}\big)\big(z', g_m(z')\big)} \right|$$

$$< \text{const.}\,\varepsilon \quad \text{if} \quad \frac{\beta}{3m} < \delta_\varepsilon \quad \text{and} \quad z' \in B, \qquad (D.42)$$

where we have used (D.41) and the basic inequality (D.28). A calculation based on (D.31) shows that the constant may be taken to be $(5/3)\big(1 + 3^{-1/2}\big)$. □

D.4 Exercises

Exercise D.10 Prove the divergence theorem,

$$\int_G \partial_j f = \int_{\partial G} n_j f \quad \text{if } f \in C^1(\overline{G}) \quad \text{and} \quad j \in \{1, 2\}, \qquad (D.43)$$

for an interval in $\mathbb{R}^2$, say $G := (a_1, b_1) \times (a_2, b_2)$, by a method that is more direct for this case than the proof of Theorem D.5.

Exercise D.11 Prove the divergence theorem (D.43) for the truncated sector

$$G := \{(r\cos\theta, r\sin\theta) \in \mathbb{R}^2 \mid a < r < b, \ \lambda < \theta < \mu\},$$

where $a > 0$ and $\mu - \lambda < 2\pi$, by writing $f(x_1, x_2) =: \tilde{f}(r, \theta)$, observing that

$$\iint_G \partial_1 f \, dx_1 \, dx_2 = \iint_{(a,b)\times(\lambda,\mu)} \left\{ r \cos\theta \frac{\partial \tilde{f}}{\partial r} - \sin\theta \frac{\partial \tilde{f}}{\partial \theta} \right\} \, dr \, d\theta,$$

$$\iint_G \partial_2 f \, dx_1 \, dx_2 = \iint_{(a,b)\times(\lambda,\mu)} \left\{ r \sin\theta \frac{\partial \tilde{f}}{\partial r} + \cos\theta \frac{\partial \tilde{f}}{\partial \theta} \right\} \, dr \, d\theta,$$

and integrating by parts along lines of constant θ and lines of constant r.

Exercise D.12 This exercise concerns verification of Theorem D.5 for the set

$$W := \mathcal{B}(0, R) \setminus \overline{\mathcal{B}(c, \rho)}, \quad \text{where} \quad c := (R - \rho, 0, \ldots, 0) \quad \text{and} \quad 0 < \rho < R,$$

which perhaps is the worst of those in Remark D.4 because ∂W is not of class C.

Choose $\delta_1 \in \left(0, \frac{1}{2}(R - \rho)\right)$, set $\delta_m := \delta_1/m$ for $m \in \mathbb{N}$, and define

$$V_m := \mathcal{B}(0, R - 2\delta_m) \setminus \overline{\mathcal{B}(c, \rho + 2\delta_m)}, \qquad W_m := \bigcup_{z \in \overline{V}_m} \mathcal{B}(z, \delta_m).$$

Prove that $\overline{W}_m \subset W$, that ∂W_m is of class C^1, that $\int_{W_m} u \to \int_W u$ as $m \to \infty$ whenever $u \in L_1(W)$ and that $\int_{\partial W_m} n_j v \to \int_{\partial W} n_j v$ as $m \to \infty$ whenever $v \in C(\overline{\Omega})$ and $j \in \{1, \ldots, N\}$.

Exercise D.13 The following result implies that, in the proof of Theorem D.5, the boundary ∂Q_m is of class C^1 for each dimension $N \geq 2$. Let A denote the closed 'negative octant' in $\mathbb{R}^N$, that is, $A := (-\infty, 0]^N$, and let $B := \bigcup_{a \in A} \mathcal{B}(a, \rho)$ for some $\rho > 0$. Prove that ∂B is of class C^1.

[The boundary ∂B is the union of $2^N - 1$ pieces as follows. Call a boundary piece *of type* k iff, for some subset $J := \{j_1, j_2, \ldots, j_k\}$ of $\{1, 2, \ldots, N\}$, the piece is

$$\left\{ x \in \mathbb{R}^N \mid x_{j_1}^2 + \cdots + x_{j_k}^2 = \rho^2, \quad x_j \geq 0 \text{ if } j \in J, \quad x_m \leq 0 \text{ if } m \notin J \right\}.$$

The outward unit normal n on this piece has components $n_j(x) = x_j/\rho$ if $j \in J$, $n_m(x) = 0$ if $m \notin J$. There are $N(N-1) \cdots (N-k+1)/k!$ pieces of type k; the N pieces of type 1 are flat; the single piece of type N is spherical.

Since (0.2) states that $\partial_j h = -\nu_j/\nu_N$ in the notation used there, it suffices here to prove that n is continuous on $S \cup T$ whenever S and T are boundary pieces that intersect.]

Exercise D.14 Prove for the vector field $e := \nabla w / |\nabla w|$ in (D.30) that, if $x \in \overline{\Omega}$, $p \in \partial\Omega$, $|x - p| < \delta_\varepsilon$ and $0 < \varepsilon \leq \frac{1}{2}$, then

$$|e(x) - e(p)| < 2\varepsilon, \quad e(p) \cdot e(x) > 1 - 2\varepsilon^2, \quad e(p) \cdot \nabla w(x) > \left(1 - 2\varepsilon^2\right)(1 - \varepsilon).$$

Infer that $\delta_{1/4}$ can serve as δ_* in (D.31).

[If $a := (\nabla w)(x)$ and $b := (\nabla w)(p)$, then

$$e(x) - e(p) = \frac{(|b| - |a|)a + |a|(a - b)}{|a|\,|b|} \quad \text{and} \quad |b| \geq 1.]$$

Exercise D.15 Let $G(\delta) := \left\{ x \in \Omega \mid \text{dist}(x, \partial\Omega) > \delta \right\}$ for a given (open) set Ω and given number $\delta > 0$. Prove that

(a) $G(\delta) = \left\{ x \in \mathbb{R}^N \mid \overline{\mathscr{B}(x, \delta)} \subset \Omega \right\}$;

(b) if Ω is convex [Apostol 1974, p.66; Rudin 1976, p.31], then $G(\delta)$ is also convex.

Exercise D.16 Let $\mathbf{x} = (x, y)$ and $\boldsymbol{\xi} = (\xi, \eta)$ denote points of $\mathbb{R}^2$. Given $g \in C^1(\mathbb{R})$, define

$$\Omega := \left\{ \mathbf{x} \in \mathbb{R}^2 \mid x \in \mathbb{R}, \quad y > g(x) \right\},$$

$$\boldsymbol{\xi}(t, \delta) := \left(t, g(t)\right) + \delta \frac{\left(-g'(t), 1\right)}{\sqrt{g'(t)^2 + 1}}, \quad t \in \mathbb{R}, \quad \delta = \text{const.} > 0,$$

and let $P(\delta) := \left\{ \boldsymbol{\xi}(t, \delta) \mid t \in \mathbb{R} \right\}$. Note that $P(\delta)$ is the set of points called $q(., \delta)$ in (D.25); also that $P(\delta)$, as the range of the continuous function $\boldsymbol{\xi}(., \delta) : \mathbb{R} \to \mathbb{R}^2$, may be called a path (or curve or arc).

(i) Show that $P(\delta)$ is parallel to $\partial\Omega$ in the sense that

$$\left.\frac{dy}{dx}\right|_{P(\delta)} := \frac{\eta_t(t, \delta)}{\xi_t(t, \delta)} = g'(t)$$

at any point t such that $g''(t)$ happens to exist and $\xi_t(t, \delta) \neq 0$.

(ii) Prove that, if $G(\delta)$ is as in Exercise D.15, then $\partial G(\delta) \subset P(\delta)$.

Exercise D.17 Consider the particular case of Exercise D.16 in which $g(x) = \frac{2}{3}|x|^{3/2}$ for all $x \in \mathbb{R}$. Show that (a) each point $(0, y) \in \Omega$ is the point of intersection of three straight lines normal to $\partial\Omega$; (b) the path $P(\delta)$ is qualitatively as in Figure D.9, the numbers $t_0 = t_0(\delta)$ and $t_1 = t_1(\delta)$ being uniquely determined by $\xi(t_0, \delta) = 0$, $t_0 > 0$ and by $\xi_t(t_1, \delta) = 0$, $t_1 > 0$; (c) neither $P(\delta)$ nor $\partial G(\delta)$ is a boundary of class C^1.

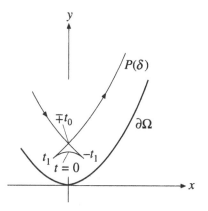

Fig. D.9.

Exercise D.18 Spherical co-ordinates $r, \theta_1, \ldots, \theta_{N-1}$ for $\mathbb{R}^N, N \geq 2$, are defined as follows. If $N = 2$, $x_1 = r \cos \theta_1$, $x_2 = r \sin \theta_1$;

$$\text{if } N = 3, \quad \left. \begin{aligned} x_1 &= r \cos \theta_1, \\ x_2 &= r \sin \theta_1 . \cos \theta_2, \\ x_3 &= r \sin \theta_1 . \sin \theta_2; \end{aligned} \right\}$$

$$\text{if } N = 4, \quad \left. \begin{aligned} x_1 &= r \cos \theta_1, \\ x_2 &= r \sin \theta_1 . \cos \theta_2, \\ x_3 &= r \sin \theta_1 . \sin \theta_2 . \cos \theta_3, \\ x_4 &= r \sin \theta_1 . \sin \theta_2 . \sin \theta_3; \end{aligned} \right\}$$

$$\text{if } N \geq 5, \quad \left. \begin{aligned} x_1 &= r \cos \theta_1, \\ x_2 &= r \sin \theta_1 . \cos \theta_2, \\ &\vdots \\ x_{N-1} &= r \sin \theta_1 \ldots \sin \theta_{N-2} . \cos \theta_{N-1}, \\ x_N &= r \sin \theta_1 \ldots \sin \theta_{N-2} . \sin \theta_{N-1}. \end{aligned} \right\}$$

Write this transformation as $x = f(v)$, $v \in \overline{E}$, where $v := (r, \theta_1, \ldots, \theta_{N-1})$ and $E := (0, \infty) \times (0, \pi)^{N-2} \times (-\pi, \pi)$, with the understanding that $E := (0, \infty) \times (-\pi, \pi)$ if $N = 2$.

Prove the following.

(i) $x_1^2 + \cdots + x_N^2 = r^2$ for all $v \in \overline{E}$. [Begin with $x_N^2 + x_{N-1}^2$.]

(ii) $x_N^2 + x_{N-1}^2 > 0$ for $N \geq 3$ if and only if $r > 0$ and $0 < \theta_j < \pi$ for $j = 1, \ldots, N-2$.

(iii) If f is restricted to $\hat{E} := (0, \infty) \times (0, \pi)^{N-2} \times (-\pi, \pi]$, then its range is

$$ f(\hat{E}) = \left\{ x \in \mathbb{R}^N \mid x_N^2 + x_{N-1}^2 > 0 \right\}, $$

and $f|_{\hat{E}}$ is injective (one-to-one).

(iv) Spherical co-ordinates are orthogonal co-ordinates: for $v \in \hat{E}$,

$$ \frac{\partial x}{\partial v_i} \cdot \frac{\partial x}{\partial v_j} = 0 \quad \text{if } i \neq j, $$

and the arc-length functions $h_j := |\partial x / \partial v_j|$ are given by

$$ h_1(v) = 1, \quad h_2(v) = r, \quad h_3(v) = r \sin \theta_1, \ldots, h_N(v) = r \sin \theta_1 \ldots \sin \theta_{N-2}. $$

[This can be proved by induction, since $y \in \mathbb{R}^{N+1}$ can be represented as

$$ y = (x_1, \ldots, x_{N-1}, x_N \cos \theta_N, x_N \sin \theta_N), $$

where $x \in \mathbb{R}^N$ and $x_j = f_j(r, \theta_1, \ldots, \theta_{N-1})$ for $j = 1, \ldots, N$.]

(v) The Jacobian determinant of the transformation is

$$ \begin{aligned} J(v) \quad := \quad & \det \left(\frac{\partial x_i}{\partial v_j} \right) = h_1 h_2 \cdots h_N \\ = \quad & r^{N-1} (\sin \theta_1)^{N-2} (\sin \theta_2)^{N-3} \cdots (\sin \theta_{N-2}). \end{aligned} $$

(vi) The Laplace operator becomes, for $v \in \hat{E}$,

$$ \begin{aligned} \Delta = \quad & \frac{1}{r^{N-1}} \frac{\partial}{\partial r} \left(r^{N-1} \frac{\partial}{\partial r} \right) + \frac{1}{J(v)} \left\{ \frac{\partial}{\partial \theta_1} \left(\frac{h_1 h_3 \cdots h_N}{h_2} \frac{\partial}{\partial \theta_1} \right) + \cdots \right. \\ & \left. + \frac{\partial}{\partial \theta_{N-1}} \left(\frac{h_1 h_2 \cdots h_{N-1}}{h_N} \frac{\partial}{\partial \theta_{N-1}} \right) \right\}. \end{aligned} $$

[See, for example, Kellogg 1929, p.183; or Spiegel 1959, p.151.]

Appendix E. The Edge-Point Lemma

E.1 Preliminaries

In this appendix we shall try to maintain a certain similarity between the boundary-point lemma (Lemma 2.12 and Theorem 2.15) and the edge-point lemma (Theorems E.8 and E.9 below), but a greater complexity of the present situation cannot be avoided. Example E.5 will illustrate this, after we have established an appropriate terminology. The edge-point lemma will be described loosely after Example E.5.

Definition E.1 By a *neighbourhood* of a point $p \in \mathbb{R}^N$ we mean a set that contains p and is *open in* $\mathbb{R}^N$.

A point $p \in \partial\Omega$ will be called an *edge point* of Ω iff, for some neighbourhood U of p,

(a) $\Omega \cap U = (\Omega_\mathrm{I} \cap \Omega_\mathrm{II}) \cap U$, where Ω_I and Ω_II are open subsets of $\mathbb{R}^N$ having boundaries of class C^2 and outward unit normals n^I and n^II, respectively;

(b)

$$p \in \partial\Omega_\mathrm{I} \cap \partial\Omega_\mathrm{II} \quad \text{and} \quad -1 < n^\mathrm{I}(p) \cdot n^\mathrm{II}(p) < 1, \qquad \text{(E.1)}$$

as is shown in Figure E.1.

A connected set of edge points of Ω will be called an *edge* of Ω. □

If, for example, $\Omega := \mathbb{R}^3 \setminus [0,\infty) \times [0,\infty) \times \mathbb{R}$ (the set obtained by removing two octants from $\mathbb{R}^3$), then the origin is *not* an edge point, even though there is a corner there. This set Ω is a union of sets with smooth boundaries, rather than an intersection. Also, this set Ω has the interior-ball property at the origin (Definition 2.14), whereas there is never an interior ball at an edge point.

305

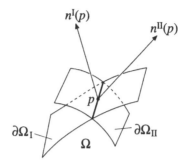

Fig. E.1.

Definitions E.1 to E.3 would still have meaning if $\partial\Omega_{\mathrm{I}}$ and $\partial\Omega_{\mathrm{II}}$ were only of class C^1 and if φ_{I} and φ_{II} were only in $C^1(U)$. However, C^2 smoothness of all these objects will be required in Theorems E.8 and E.9, and it seems simpler to adopt this smoothness *ab initio*.

Definition E.2 We shall say that φ_{I} is an *admissible function describing* Ω_{I} *near* p iff, for some neighbourhood U of p,

(a) $\varphi_{\mathrm{I}} \in C^2(U)$;

(b) $|\nabla\varphi_{\mathrm{I}}(x)| > 0$ for all $x \in U$;

(c) for $x \in U$, the value $\varphi_{\mathrm{I}}(x)$ is negative, zero or positive according as $x \in \Omega_{\mathrm{I}}$, $x \in \partial\Omega_{\mathrm{I}}$ or $x \notin \overline{\Omega}_{\mathrm{I}}$.

Admissible functions describing Ω_{II} near p are defined similarly.   □

Definition E.3 Let p be an edge point of Ω (with $\Omega = \Omega_{\mathrm{I}} \cap \Omega_{\mathrm{II}}$ near p); let φ_{I} and φ_{II} be admissible functions describing Ω_{I} and Ω_{II}, respectively, in a neighbourhood U of p; and let $a = (a_{ij})$, now defined on $\overline{\Omega}$, be the matrix of leading coefficients of an elliptic operator L of order two. Then the function B defined by

$$
\begin{aligned}
B(x) &:= \nabla\varphi_{\mathrm{I}}(x) \cdot a(x) \cdot \nabla\varphi_{\mathrm{II}}(x) \\
&= \sum_{i,j=1}^{N} \partial_i\varphi_{\mathrm{I}}(x)\, a_{ij}(x)\, \partial_j\varphi_{\mathrm{II}}(x), \qquad x \in \overline{\Omega} \cap U, \qquad \text{(E.2)}
\end{aligned}
$$

will be called a *bluntness function* for the edge $\partial\Omega_{\mathrm{I}} \cap \partial\Omega_{\mathrm{II}}$ relative to the operator L.   □

We use the word bluntness because at an edge point $B(x)$ is a weighted form of $n^{\mathrm{I}}(x) \cdot n^{\mathrm{II}}(x)$, and this latter varies from -1 at a perfectly sharp edge to 1 at a perfectly blunt edge. Of course, both extremes are excluded by (E.1).

In §E.3 we shall encounter the following alternative bluntness conditions at an edge point $p \in \partial\Omega_I \cap \partial\Omega_{II}$.

$$B(p) > 0. \tag{E.3}$$

$$\left.\begin{array}{l} B(p) = 0 \quad \text{and} \quad (\partial B/\partial\tau)(p) = 0 \quad \text{for every differential} \\ \text{operator} \quad \partial/\partial\tau \quad \text{tangential to} \quad \partial\Omega_I \cap \partial\Omega_{II} \quad \text{at} \quad p. \end{array}\right\} \tag{E.4}$$

The coefficients a_{ij} are to be sufficiently smooth for continuity of B at p in the case of (E.3), and for continuity of ∇B in $\overline{\Omega} \cap U$ in the case of (E.4). The derivatives $(\partial B/\partial\tau)(p)$ may be evaluated either as

(a) $\lim_{x\to p}(\tau \cdot \nabla B)(x)$, where $x \in \Omega \cap U$ and τ is a non-zero vector satisfying $\tau \cdot n^I(p) = 0$ and $\tau \cdot n^{II}(p) = 0$ [equivalently, $(\tau \cdot \nabla\varphi_I)(p) = 0$ and $(\tau \cdot \nabla\varphi_{II})(p) = 0$]; or as

(b) $dB(\xi(t))/dt$ at $t = t_0$, where $\{\xi(t) \mid 0 \leq t \leq 1\}$ is an arc in $\partial\Omega_I \cap \partial\Omega_{II}$ for which $\xi \in C^1[0,1]$, $|\xi'(t)| > 0$ on $[0,1]$ and $\xi(t_0) = p$ for some $t_0 \in [0,1]$.

Remark E.4 Each of the bluntness conditions (E.3) and (E.4) has a meaning *independent of the choice of* φ_I *and* φ_{II}, provided that these are admissible in the sense of Definition E.2.

To demonstrate this, we define

$$\beta(x) := n^I(x) \cdot a(x) \cdot n^{II}(x) \quad \text{and} \quad g(x) := |\nabla\varphi_I(x)| \, |\nabla\varphi_{II}(x)|$$

for $x \in (\partial\Omega_I \cap \partial\Omega_{II}) \cap U$, and adopt the evaluation (b) of $(\partial B/\partial\tau)(p)$. Then

$$B(x) = \beta(x)g(x),$$

with β independent of the choice of φ_I and φ_{II}, while $g(x) > 0$. Thus (E.3) is equivalent to $\beta(p) > 0$, and the pair of conditions (E.4) is equivalent to the pair $\beta(p) = 0$ and $(\partial\beta/\partial\tau)(p) = 0$. $\square$

Example E.5 Consider the sector

$$\Omega := \left\{ (r\cos\theta, r\sin\theta) \in \mathbb{R}^2 \;\middle|\; 0 < r < R, \; -\frac{\alpha}{2} < \theta < \frac{\alpha}{2} \right\}, \quad \alpha \in (0,\pi).$$

The points $(0,0)$, $\left(R\cos\frac{\alpha}{2}, -R\sin\frac{\alpha}{2}\right)$ and $\left(R\cos\frac{\alpha}{2}, R\sin\frac{\alpha}{2}\right)$ are all edge points, but only the origin will be considered. Defining half-planes

$$\Omega_I := \left\{ x \in \mathbb{R}^2 \;\middle|\; \varphi_I(x) := -x_1 \sin\frac{\alpha}{2} - x_2 \cos\frac{\alpha}{2} < 0 \right\},$$

$$\Omega_{\mathrm{II}} := \left\{ x \in \mathbb{R}^2 \ \Big| \ \varphi_{\mathrm{II}}(x) := -x_1 \sin\frac{\alpha}{2} + x_2 \cos\frac{\alpha}{2} < 0 \right\},$$

we have $\Omega \cap \mathscr{B}(0,\rho) = (\Omega_{\mathrm{I}} \cap \Omega_{\mathrm{II}}) \cap \mathscr{B}(0,\rho)$ whenever $\rho < R$. Relative to the Laplace operator $\triangle$ [for which $a_{ij}(x) = \delta_{ij}$], a suitable bluntness function is

$$B(x) = \nabla\varphi_{\mathrm{I}}(x) \cdot \nabla\varphi_{\mathrm{II}}(x) = -\cos\alpha \qquad (|x| < R).$$

We restrict attention to harmonic functions u having a maximum at the origin:

$$u \in C^1(\overline{\Omega}) \cap C^2(\Omega), \qquad \triangle u = 0 \ \text{ in } \ \Omega, \quad u(0) = 0 = \max_{\overline{\Omega}} u.$$

In fact, at this stage it is sufficient to contemplate [cf. Exercise 2.41]

$$u(x;\beta) := -r^{\pi/\beta} \cos\frac{\pi\theta}{\beta}, \qquad \beta \in [\alpha,\pi], \qquad -\frac{\alpha}{2} \le \theta \le \frac{\alpha}{2}.$$

Throughout this appendix we pursue mainly first and second outward derivatives at an edge point; in the present case we merely calculate limits, as $r \downarrow 0$ with θ fixed in $(-\alpha/2, \alpha/2)$, of

$$-u_r(x;\beta) = \frac{\pi}{\beta} r^{(\pi-\beta)/\beta} \cos\frac{\pi\theta}{\beta} \qquad (x \in \Omega),$$

$$u_{rr}(x;\beta) = -\frac{\pi(\pi-\beta)}{\beta^2} r^{(\pi-2\beta)/\beta} \cos\frac{\pi\theta}{\beta} \qquad (x \in \Omega).$$

It is instructive to list four cases.

(i) $\beta = \pi$, $u(x;\pi) = -x_1$. For each $\alpha \in (0,\pi)$ and all $\theta \in (-\alpha/2, \alpha/2)$,

$$-u_r(x;\pi) \to \cos\theta > 0 \qquad \text{as } r \downarrow 0 \ \text{ with } \ \theta \ \text{ fixed.}$$

The bluntness function is irrelevant.

(ii) $\alpha \le \beta < \pi$ and $\beta > \pi/2$. For each $\alpha \in (0,\pi)$ and all $\theta \in (-\alpha/2, \alpha/2)$,

$$-u_r(0;\beta) = 0,$$

$$u_{rr}(x;\beta) \to -\infty \qquad \text{as } r \downarrow 0 \ \text{ with } \ \theta \ \text{ fixed.}$$

The bluntness function has a little significance: if $\alpha = \beta$, then $B(0) > 0$.

(iii) $\alpha \le \beta = \pi/2$, so that $B(0) \le 0$; $u(x;\pi/2) = -x_1^2 + x_2^2$. For each $\alpha \in (0,\pi/2]$ and all $\theta \in (-\alpha/2, \alpha/2)$,

$$-u_r(0;\pi/2) = 0,$$

$$u_{rr}(x;\pi/2) \to -2\cos 2\theta < 0 \qquad \text{as } r \downarrow 0 \ \text{ with } \ \theta \ \text{ fixed.}$$

(iv) $\alpha \leq \beta < \pi/2$, so that $B(0) < 0$. For each $\alpha \in (0, \pi/2)$ and all $\theta \in (-\alpha/2, \alpha/2)$,

$$-u_r(0; \beta) = 0, \qquad u_{rr}(0; \beta) = 0.$$

$\square$

In §E.3, Theorems E.8 and E.9 will exclude the analogue of (iv), in Example E.5, by means of the bluntness condition (E.3) in Theorem E.8 and (E.4) in Theorem E.9. Both theorems establish that, *under these bluntness conditions, first and second outward derivatives of a subsolution u, at an edge point where* $\sup_\Omega u$ *is attained, must be essentially as in* (i) *or* (ii) *or* (iii) *of Example* E.5.

Second outward derivatives at an edge point, of a sufficiently smooth function $u : \overline{\Omega} \to \mathbb{R}$, result from setting $v = \partial u / \partial m$ in the following definition.

Definition E.6 Let p be an edge point as in Definition E.1. A constant vector m is *outward from* Ω *at* p iff it is outward from Ω_{I} and from Ω_{II} at p : that is, $m \cdot n^{\mathrm{I}}(p) > 0$ and $m \cdot n^{\mathrm{II}}(p) > 0$. In that case, $\partial/\partial m := m \cdot \nabla$ at points $x \in \Omega$ near p, while

$$\frac{\partial v}{\partial m}(p) := \lim_{t \downarrow 0} \frac{v(p) - v(p - tm)}{t}$$

whenever this limit exists [cf. (2.21)]. $\square$

E.2 Bluntness and ellipticity under co-ordinate transformations

The proofs of Theorems E.8 and E.9 depend on the conservation of various conditions under co-ordinate transformations more general than the translations and rotations of axes considered previously. By a C^ℓ *co-ordinate transformation* or C^ℓ *diffeomorphism*, where $\ell \in \{1, 2, 3, \ldots\}$, we mean a bijection f of an open set U in $\mathbb{R}^N$ onto an open set $f(U)$ in $\mathbb{R}^N$ such that $f \in C^\ell(\overline{U}, \mathbb{R}^N)$ and $f^{-1} \in C^\ell(\overline{f(U)}, \mathbb{R}^N)$. In the present context, U will always be *bounded*. It is often convenient to use the archaic notation

$$y = f(x), \qquad \frac{\partial y_r}{\partial x_i} := (\partial_i f_r)(x), \qquad \frac{\partial x_j}{\partial y_s} := (\partial_s f_j^{-1})(y), \qquad \text{(E.5)}$$

where $x \in U$, $y \in f(U)$ and $f_j^{-1} := (f^{-1})_j$. We specify the following transformation rules.

(a) Scalar-valued functions transform in the obvious way:

$$\tilde{\varphi}(y) := \varphi(f^{-1}(y)) = \varphi(x). \qquad \text{(E.6)}$$

(b) Any vector field k used to define a directional derivative $\partial/\partial k := k \cdot \nabla$ (for example, the τ tangential to $\partial\Omega_{\mathrm{I}} \cap \partial\Omega_{\mathrm{II}}$ in (E.4), or the m outward from Ω in Definition E.6) transforms *contravariantly*; so does the matrix (a_{ij}) of leading coefficients of an elliptic operator L. This means that, for $r, s \in \{1, \dots, N\}$,

$$\tilde{k}_r(y) := \sum_{i=1}^{N} \frac{\partial y_r}{\partial x_i} k_i(x), \qquad x = f^{-1}(y), \tag{E.7a}$$

$$\tilde{a}_{rs}(y) := \sum_{i,j=1}^{N} \frac{\partial y_r}{\partial x_i} a_{ij}(x) \frac{\partial y_s}{\partial x_j}, \qquad x = f^{-1}(y). \tag{E.7b}$$

(c) Test vectors ξ for the ellipticity condition $[\sum_{i,j} a_{ij}(x)\xi_i\xi_j \geq \lambda(x)|\xi|^2]$ transform *covariantly*. This means that, for $r \in \{1, \dots, N\}$,

$$\tilde{\xi}_r := \sum_{i=1}^{N} \frac{\partial x_i}{\partial y_r} \xi_i. \tag{E.8}$$

All these transformations are invertible. For example, multiplying (E.7a) by $\partial x_j/\partial y_r$ and summing over r, we obtain

$$k_j(x) = \sum_{r=1}^{N} \frac{\partial x_j}{\partial y_r} \tilde{k}_r(y).$$

The reason for transforming (a_{ij}) as in (E.7b) is essentially the same as in Exercise 2.2: contravariance implies that

$$\sum_{i,j} a_{ij}(x) \frac{\partial}{\partial x_i} \frac{\partial}{\partial x_j} = \sum_{r,s} \tilde{a}_{rs}(y) \frac{\partial}{\partial y_r} \frac{\partial}{\partial y_s} + \quad \text{a lower-order operator.}$$

Remark E.7 Definitions (E.5) to (E.8) and the chain rule imply the following invariance or conservation results. It is to be understood that $y = f(x)$ and $x = f^{-1}(y)$ throughout the list.

(i) Directional derivatives are invariant: with $\partial/\partial\tilde{k} := \tilde{k} \cdot \nabla_y$ and $\partial/\partial k := k \cdot \nabla_x$,

$$\frac{\partial\tilde{\varphi}}{\partial\tilde{k}}(y) = \frac{\partial\varphi}{\partial k}(x). \tag{E.9}$$

(ii) $\tilde{m}$ is outward from $f(\Omega_{\mathrm{I}})$ at $f(p)$ if and only if m is outward from Ω_{I} at p. [In other words, if φ_{I} is an admissible function describing Ω_{I} near p, then

$$\frac{\partial\tilde{\varphi}_{\mathrm{I}}}{\partial\tilde{m}}(f(p)) > 0 \iff \frac{\partial\varphi_{\mathrm{I}}}{\partial m}(p) > 0,$$

which is immediate from (E.9).]

(iii) Let p be an edge point; say $p \in E := \partial\Omega_{\mathrm{I}} \cap \partial\Omega_{\mathrm{II}}$. Then $\tilde{\tau}$ is tangential to $f(E)$ at $f(p)$ if and only if τ is tangential to E at p. [In other words, $(\partial\tilde{\varphi}_J/\partial\tilde{\tau})(f(p)) = 0$ for $J = \mathrm{I}, \mathrm{II}$ if and only if $(\partial\varphi_J/\partial\tau)(p) = 0$ for $J = \mathrm{I}, \mathrm{II}$, which is immediate from (E.9).]

(iv) Bluntness functions are invariant:

$$\tilde{B}(y) := (\nabla\tilde{\varphi}_{\mathrm{I}})(y) \cdot \tilde{a}(y) \cdot (\nabla\tilde{\varphi}_{\mathrm{II}})(y) = (\nabla\varphi_{\mathrm{I}})(x) \cdot a(x) \cdot (\nabla\varphi_{\mathrm{II}})(x) =: B(x).$$
(E.10)

(v) Uniform ellipticity is conserved when U is bounded, although moduli of ellipticity are changed in general. Let λ_0 be the uniform modulus of ellipticity of the operator having leading coefficients $a_{ij}(x)$, $x \in U$. Then, for all $\tilde{\xi} \in \mathbb{R}^N$,

$$\sum_{r,s} \tilde{a}_{rs}(y)\,\tilde{\xi}_r\tilde{\xi}_s = \sum_{i,j} a_{ij}(x)\,\xi_i\xi_j \geq \lambda_0\,|\xi|^2 \geq \lambda_0\,M^{-2}|\tilde{\xi}|^2,$$
(E.11)

where

$$M := \sup_{y \in f(U), |\xi|=1} |A(y)\xi|$$

if we define

$$A(y)\xi := \left(\sum_{i=1}^{N} (\partial_r f_i^{-1})(y)\,\xi_i\right)^N_{r=1}$$

for all $y \in f(U)$ and $\xi \in \mathbb{R}^N$. □

E.3 Two stages of the edge-point lemma

In the following theorems, the operators L_0, L and L_1 are as in Definition 2.3 except for the greater smoothness of the coefficients a_{ij} that is stated in the theorems. The words C^2-*subsolution* and *generalized subsolution* retain the meanings in Definitions 2.4 and 2.10, respectively. The set $C^1(\Omega \cup \{p\})$ consists of those functions $f \in C(\Omega \cup \{p\}) \cap C^1(\Omega)$ for which all first derivatives $\partial_j f$ have extensions continuous on $\Omega \cup \{p\}$.

Theorem E.8 *Suppose that*

(a) *p is an edge point of Ω (the relevant edge being $\partial\Omega_{\mathrm{I}} \cap \partial\Omega_{\mathrm{II}}$);*

(b) *u is a C^2-subsolution relative to L_0 or L and Ω, or a generalized subsolution relative to L_1 and Ω;*

(c) *$u \in C^1(\Omega \cup \{p\})$ and $u(x) < u(p)$ for all $x \in \Omega$, with $u(p) \geq 0$ when the coefficient c is not the zero function;*

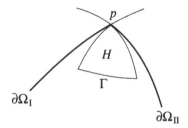

Fig. E.2.

(d) *a bluntness function B, for the edge* $\partial\Omega_I \cap \partial\Omega_{II}$ *relative to the operator considered in* (b), *satisfies* $B(p) > 0$;

(e) *all coefficients* a_{ij} *are continuous at* p.

Let m be a unit vector outward from Ω *at* p. *Then either*

(α) $\dfrac{\partial u}{\partial m}(p) > 0$; *or*

(β) $\dfrac{\partial u}{\partial m}(p) = 0$ *and* $\limsup_{t\downarrow 0} \dfrac{1}{t}\dfrac{\partial u}{\partial m}(p - tm) > 0$,

which implies that $(\partial^2 u/\partial m^2)(p) < 0$ *whenever this derivative exists.*

Proof (i) Assume that we can find a set H and a function v as follows. The set H (Figure E.2) is a bounded open subset of Ω such that $\overline{H} \subset \Omega \cup \{p\}$ and p is an edge point of H; moreover, the representation $H_I \cap H_{II}$ of H near p is such that the outward unit normals at p to ∂H_I and ∂H_{II} coincide with those to $\partial\Omega_I$ and $\partial\Omega_{II}$, respectively.

The function $v \in C^2(\overline{H})$ and

$$v(p) = 0; \tag{I}$$

$$\frac{\partial v}{\partial m}(p) = 0 \quad \text{and} \quad \frac{\partial^2 v}{\partial m^2}(p) > 0 \tag{IIa,b}$$

for every vector m outward from H (equally, from Ω) at p;

$$u + v \leq M \quad \text{on} \quad \overline{H}, \tag{III}$$

where $M := u(p) = \sup_\Omega u$.

With such H and v in hand, we can prove the theorem as follows. Let $w := u + v$ and let a vector m as in (II) be given. By (I) and (III),

$w(p) = M = \sup_H w$; hence $(\partial w/\partial m)(p) \geq 0$ [otherwise, $w(x) > M$ at certain x in H]. Consequently,

$$\frac{\partial u}{\partial m}(p) \geq -\frac{\partial v}{\partial m}(p) = 0.$$

If $(\partial u/\partial m)(p) > 0$, then the theorem is true. Suppose therefore that $(\partial u/\partial m)(p) = 0$, and consider the difference quotient

$$Q(t) := \frac{1}{t}\left\{ \frac{\partial w}{\partial m}(p) - \frac{\partial w}{\partial m}(p - tm) \right\} = -\frac{1}{t}\frac{\partial w}{\partial m}(p - tm) \qquad (t > 0).$$

There must exist a sequence (t_n) such that $t_n \downarrow 0$ and $Q(t_n) \leq 0$, because

$$0 \leq w(p) - w(p - tm) = \int_t^0 \frac{\mathrm{d}}{\mathrm{d}s} w(p - sm) \, \mathrm{d}s = \int_0^t \frac{\partial w}{\partial m}(p - sm) \, \mathrm{d}s,$$

and this is contradicted if there is a number $\varepsilon > 0$ such that $Q(s) > 0$ [hence $(\partial w/\partial m)(p - sm) < 0$] for all $s \in (0, \varepsilon)$. Now consider

$$P(t) := \frac{1}{t}\left\{ \frac{\partial v}{\partial m}(p) - \frac{\partial v}{\partial m}(p - tm) \right\} = -\frac{1}{t}\frac{\partial v}{\partial m}(p - tm) \qquad (t > 0).$$

Since $\lim_{t\downarrow 0} P(t) > 0$, by (IIb), we have $P(t_n) \geq c$ for some constant $c > 0$ whenever t_n is sufficiently small. For all such t_n,

$$\frac{1}{t_n}\frac{\partial u}{\partial m}(p - t_n m) = P(t_n) - Q(t_n) \geq c,$$

and this proves the theorem for case (β).

(ii) The function v will be of the form δV, where δ is a small, positive constant to be chosen presently, while $V \in C^2(\overline{H})$ and satisfies (IIa,b); in addition,

$$V = 0 \quad \text{on} \quad \partial H \setminus \Gamma, \qquad\qquad (\mathrm{I}')$$

where Γ is a compact subset of ∂H that is disjoint from $\{p\}$, and

$$LV \geq 0 \quad \text{in} \quad H, \qquad\qquad (\mathrm{III}')$$

where L denotes whichever of L_0, L and L_1 is considered in hypothesis (b).

Consider the values of $u + \delta V$ on ∂H. On $\partial H \setminus \Gamma$ we have $u \leq M, V = 0$ and hence $u + \delta V \leq M$, with equality at p. On Γ, which is a subset of Ω, we have $u < M$ by hypothesis (c); since Γ is compact and u continuous, $u \leq M - \alpha$ on Γ for some constant $\alpha > 0$. Choose $\delta = \frac{1}{2}\alpha / \max_\Gamma V$; then $u + \delta V < M$ on Γ, so that

$$\max_{\partial H} (u + \delta V) = M.$$

Condition (III) now follows from (III′) and one of our three versions of the weak maximum principle, in precisely the way that was spelled out in the proof of Lemma 2.12.

(iii) In order to construct the set H and the function V, we simplify the geometry by means of co-ordinate transformations. First, let X_j be local co-ordinates of the usual kind (resulting from a translation and rotation of axes as in Exercise 5.5) such that

(a) $X = 0$ at $x = p$;

(b) $\nabla_x X_N = \dfrac{n^{\mathrm{I}}(p) + n^{\mathrm{II}}(p)}{\left| n^{\mathrm{I}}(p) + n^{\mathrm{II}}(p) \right|}$;

(c) the (two-dimensional) plane spanned by $\nabla_x X_1$ and $\nabla_x X_N$ is that spanned by $n^{\mathrm{I}}(p)$ and $n^{\mathrm{II}}(p)$.

Conditions (E.1) and (b) imply that $n^J(p) \cdot \nabla_x X_N > 0$ for $J = \mathrm{I}, \mathrm{II}$; then, by Exercise 5.5, $\partial \Omega_{\mathrm{I}}$ and $\partial \Omega_{\mathrm{II}}$ have local representations $X_N = g_J(X')$ (for $J = \mathrm{I}, \mathrm{II}$) such that $X_N < g_J(X')$ when x is near p and $x \in \Omega_J$. Moreover, conditions (a) to (c) and the C^2 nature of the boundaries $\partial \Omega_J$ (Definition E.1) allow us to write

$$\left. \begin{array}{ll} g_{\mathrm{I}}(X') = \gamma X_1 + r_{\mathrm{I}}(X'), & r_{\mathrm{I}} \in C^2(\overline{S}), \\ g_{\mathrm{II}}(X') = -\gamma X_1 + r_{\mathrm{II}}(X'), & r_{\mathrm{II}} \in C^2(\overline{S}), \end{array} \right\} \tag{E.12}$$

where the constant $\gamma > 0$; the set S is a convex neighbourhood of 0 in $\mathbb{R}^{N-1}$; and, as $X' \to 0$, $r_J(X') = O\big(|X'|^2\big)$ and $\nabla r_J(X') = O\big(|X'|\big)$ for $J = \mathrm{I}, \mathrm{II}$.

The next co-ordinate transformation is

$$\left. \begin{array}{rcl} y_1 & = & X_1 + \dfrac{1}{2\gamma}\{r_{\mathrm{I}}(X') - r_{\mathrm{II}}(X')\}, \\[2mm] y_\lambda & = & X_\lambda \quad \text{for each } \lambda \in \{2, \ldots, N-1\}, \\[2mm] y_N & = & X_N - \tfrac{1}{2}\big\{ r_{\mathrm{I}}(X') + r_{\mathrm{II}}(X') \big\}. \end{array} \right\} \tag{E.13}$$

If we write this as $y = f(X)$, then the Fréchet or total or linear derivative $f'(0) = I$, the identity operator $\mathbb{R}^N \to \mathbb{R}^N$. Thus $f'(0)$ is certainly invertible (is certainly a linear homeomorphism) and the inverse-function theorem ensures that f is a C^2 diffeomorphism on some neighbourhood of 0 in $\mathbb{R}^N$. [For a form of the inverse-function theorem that yields the C^2 property of f^{-1}, see Dieudonné 1969, p.273.]

Define open half-spaces

$$\left. \begin{array}{rcl} \tilde{\Omega}_{\mathrm{I}} & := & \big\{ y \in \mathbb{R}^N \mid y_N < \gamma y_1 \big\}, \\[2mm] \tilde{\Omega}_{\mathrm{II}} & := & \big\{ y \in \mathbb{R}^N \mid y_N < -\gamma y_1 \big\}, \end{array} \right\} \tag{E.14}$$

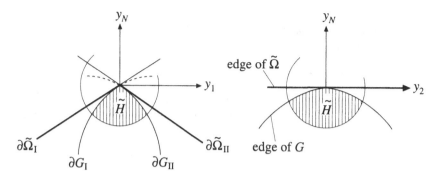

Fig. E.3.

and set $\tilde{\Omega} := \tilde{\Omega}_{\mathrm{I}} \cap \tilde{\Omega}_{\mathrm{II}}$. Then, *for sufficiently small* $|x - p|$ *and* $|y|$, *we have* $x \in \Omega$ *if and only if* $y \in \tilde{\Omega}$, because (E.13) transforms $X_N - g_{\mathrm{I}}(X')$ to $y_N - \gamma y_1$ and transforms $X_N - g_{\mathrm{II}}(X')$ to $y_N + \gamma y_1$.

(iv) Now choose at pleasure a constant $\kappa > 0$ (which will be a curvature), let

$$G_{\mathrm{I}} \; := \; \left\{ y \in \mathbb{R}^N \; \middle| \; y_N < \gamma y_1 - \frac{\kappa}{2}|y'|^2 \right\},$$

$$G_{\mathrm{II}} \; := \; \left\{ y \in \mathbb{R}^N \; \middle| \; y_N < -\gamma y_1 - \frac{\kappa}{2}|y'|^2 \right\},$$

set $G := G_{\mathrm{I}} \cap G_{\mathrm{II}}$ (Figure E.3) and define

$$\tilde{H} := G \cap \mathscr{B}(0, \rho) \tag{E.15}$$

for some small $\rho > 0$ that remains to be chosen. The inverse image of $\tilde{H}$ under the C^2 diffeomorphism $x \mapsto y$ is to be the set H specified in step (i).

(v) To construct the comparison function V, we first define

$$\left. \begin{aligned} \varphi(y) := y_N - \gamma y_1 + \frac{\kappa}{2}|y'|^2, \qquad P(y) := e^{-K\varphi(y)} - 1, \\[2mm] \psi(y) := y_N + \gamma y_1 + \frac{\kappa}{2}|y'|^2, \qquad Q(y) := e^{-K\psi(y)} - 1, \end{aligned} \right\} \tag{E.16}$$

for all $y \in \mathbb{R}^N$ and for some large $K > 0$ that also remains to be chosen. Observe that φ and $-P$ are admissible functions describing G_{I} (Definition E.2) and that ψ and $-Q$ are admissible functions describing G_{II}; in particular, $P(y) > 0$ and $Q(y) > 0$ if $y \in G$, and P, Q vanish on $\partial G_{\mathrm{I}}, \partial G_{\mathrm{II}}$ respectively.

Now define $\tilde{V} \in C^{\infty}(\mathbb{R}^N)$ by $\tilde{V} := PQ$, and let $y = Y(x)$ denote the C^2 diffeomorphism that we have constructed on a neighbourhood of p. Denoting this neighbourhood by $\mathscr{M}(p)$, we set

$$V(x) := \tilde{V}\big(Y(x)\big), \qquad x \in \mathscr{M}(p). \tag{E.17}$$

Then $V \in C^2(\overline{H})$ if $\overline{H} \subset \mathscr{M}(p)$, and we proceed to show that V satisfies (I'), (IIa,b) and (III') if ρ and K are chosen suitably. This will complete the proof.

(vi) Condition (I') holds if we define Γ to be the inverse image of $\overline{G} \cap \partial\mathscr{B}(0,\rho)$ under the map Y [because PQ vanishes on ∂G]. For (IIa,b), let a unit vector m, outward from H at p, be given; define, for $r \in \{1,\dots,N\}$,

$$\tilde{m}_r(y) := \sum_{i=1}^{N} \frac{\partial Y_r}{\partial x_i}(x)\, m_i, \qquad \frac{\partial}{\partial\tilde{m}} = \sum_{r=1}^{N} \tilde{m}_r(y)\frac{\partial}{\partial y_r}.$$

Then $(\partial V/\partial m)(p) = (\partial\tilde{V}/\partial\tilde{m})(0)$ and $\tilde{m}(0)$ is outward from $\tilde{H}$ at 0, by Remark E.7, (i) and (ii). Now $\nabla\tilde{V} = (\nabla P)Q + P(\nabla Q)$ and $P(0) = Q(0) = 0$, so that certainly $(\partial V/\partial m)(p) = \big(\partial\tilde{V}/\partial\tilde{m}\big)(0) = 0$. Next, using once more the condition $\varphi(0) = \psi(0) = P(0) = Q(0) = 0$, we obtain

$$\frac{\partial^2 V}{\partial m^2}(p) = \frac{\partial^2 \tilde{V}}{\partial\tilde{m}^2}(0) = 2\frac{\partial P}{\partial\tilde{m}}(0)\frac{\partial Q}{\partial\tilde{m}}(0) = 2K^2 \frac{\partial\varphi}{\partial\tilde{m}}(0)\frac{\partial\psi}{\partial\tilde{m}}(0) > 0,$$

since $\tilde{m}$ is outward from G_{I} and from G_{II} at 0.

(vii) It remains to prove that V satisfies (III') if ρ and K are chosen suitably; equivalently, that $\tilde{L}\tilde{V} \geq 0$ in $\tilde{H}$, where $\tilde{L}$ is the operator to which L is transformed (§E.2) by the map $y = Y(x)$. Let $\tilde{a}_{ij}(y)$, $\tilde{b}_j(y)$ and $\tilde{c}(y)$ denote the coefficients of $\tilde{L}$. Then

$$\tilde{L}\tilde{V} = (\tilde{L}P)Q + 2\nabla P \cdot \tilde{a} \cdot \nabla Q + P(\tilde{L}Q) - \tilde{c}PQ. \tag{E.18}$$

Since $P > 0$ and $Q > 0$ in $\tilde{H}$, the essentially new term relative to Lemma 2.12 is

$$2\nabla P \cdot \tilde{a} \cdot \nabla Q = 2e^{-K\varphi - K\psi}K^2 \nabla\varphi \cdot \tilde{a} \cdot \nabla\psi,$$

where $\nabla\varphi \cdot \tilde{a} \cdot \nabla\psi =: B_G$, say, is a bluntness function for the edge $\partial G_{\mathrm{I}} \cap \partial G_{\mathrm{II}}$ relative to $\tilde{L}$. By hypothesis (d) and the two kinds of invariance of the condition $B(p) > 0$ [Remark E.4 and (E.10)], we have $B_G(0) > 0$, and B_G is continuous at 0 by hypothesis (e) and the definitions of φ and ψ. Therefore, if ρ is chosen sufficiently small in the definition (E.15) of $\tilde{H}$, then

$$2\nabla P \cdot \tilde{a} \cdot \nabla Q > 0 \qquad \text{in } \tilde{H}.$$

Consider the remaining terms in (E.18). We have $-\tilde{c}PQ \geq 0$ in $\tilde{H}$, and make $\tilde{L}P > 0$ and $\tilde{L}Q > 0$ in $\tilde{H}$ by choosing K sufficiently large, very much as in the proof of Lemma 2.12. In fact,

$$\tilde{L}P = e^{-K\varphi}\left\{ K^2 \sum_{i,j} \tilde{a}_{ij}\,(\partial_i\varphi)(\partial_j\varphi) - K\sum_{ij}\tilde{a}_{ij}\,\partial_i\partial_j\varphi \right.$$

$$\left. -K\sum_j \tilde{b}_j\,\partial_j\varphi + \tilde{c} \right\} - \tilde{c}, \qquad (E.19)$$

where $-\tilde{c} \geq 0$ and

$$\left(\partial_j\varphi(y)\right)_{j=1}^{N} = (-\gamma + \kappa y_1,\ \kappa y_2,\ \ldots,\ \kappa y_{N-1},\ 1), \qquad (E.20a)$$

$$\partial_i\partial_j\varphi(y) = \begin{cases} \kappa & \text{if } i = j \in \{1,\ldots,N-1\}, \\ 0 & \text{otherwise.} \end{cases} \qquad (E.20b)$$

Since uniform ellipticity, say with modulus $\tilde{\lambda}_0 > 0$ [Remark E.7, (v)], implies that

$$\sum_{i,j} \tilde{a}_{ij}\,(\partial_i\varphi)(\partial_j\varphi) \geq \tilde{\lambda}_0|\nabla\varphi|^2 \geq \tilde{\lambda}_0 \qquad \text{in } \tilde{H}, \qquad (E.21)$$

we can certainly choose K so large that $\tilde{L}P > 0$ in $\tilde{H}$, and similarly for Q. $\qquad\square$

Theorem E.9 *Suppose that hypotheses* (a) *to* (c) *of Theorem* E.8 *stand, while* (d) *and* (e) *are replaced by*

(d′) *a bluntness function B, for the edge* $\partial\Omega_{\mathrm{I}} \cap \partial\Omega_{\mathrm{II}}$ *relative to the operator considered in* (b), *satisfies* $B(p) = 0$ *and* $(\partial B/\partial\tau)(p) = 0$ *for every differential operator* $\partial/\partial\tau$ *tangential to* $\partial\Omega_{\mathrm{I}} \cap \partial\Omega_{\mathrm{II}}$ *at* p;

(e′) *all coefficients* $a_{ij} \in C^2\left(\overline{\Omega} \cap \mathscr{B}(p,\sigma)\right)$ *for some* $\sigma > 0$.

Then the conclusion of Theorem E.8 *remains true : we have either* (α) *or* (β) *there.*

Proof Steps (i) to (iii) in the proof of Theorem E.8 are not changed. However, in order to have $LV \geq 0$ in H, we introduce here: (a) a further co-ordinate transformation, which will simplify the matrix $\tilde{a}(y)$ of leading coefficients; (b) a further parameter, in the definitions of H and V, such that $LV \geq 0$ in H when the new parameter is sufficiently large.

(iv′) Proceeding from the co-ordinates y defined by (E.13) and the set $\tilde{\Omega}$ defined by (E.14) and $\tilde{\Omega} := \tilde{\Omega}_{\mathrm{I}} \cap \tilde{\Omega}_{\mathrm{II}}$, we transform linearly to co-ordinates

z, obtaining a set $\hat{\Omega}$ and coefficients $\hat{a}_{ij}(z), \hat{b}_j(z), \hat{c}(z)$ such that $\hat{\Omega} = \tilde{\Omega}$ and $\hat{a}_{\lambda N}(0) = 0$ for $\lambda \in \{2, \ldots, N-1\}$.

Notation In this proof, Greek subscripts take only the values $2, \ldots, N-1$; 'for each λ' will mean for each $\lambda \in \{2, \ldots, N-1\}$; and $\sum_\lambda := \sum_{\lambda=2}^{N-1}$.

Define, for constants C_λ that we are about to determine,

$$z_1 = y_1, \qquad z_\lambda = y_\lambda + C_\lambda y_N \quad \text{for each } \lambda, \qquad z_N = y_N; \qquad \text{(E.22)}$$

then

$$\frac{\partial z_\lambda}{\partial y_i} = \delta_{\lambda i} + C_\lambda \delta_{Ni}, \qquad \frac{\partial z_N}{\partial y_j} = \delta_{Nj},$$

so that

$$\hat{a}_{\lambda N}(z) = \sum_{i,j} \frac{\partial z_\lambda}{\partial y_i} \tilde{a}_{ij}(y) \frac{\partial z_N}{\partial y_j} = \tilde{a}_{\lambda N}(y) + C_\lambda \tilde{a}_{NN}(y).$$

Here $\tilde{a}_{NN}(0) \geq \tilde{\lambda}_0 > 0$ by the uniform ellipticity of $\tilde{L}$ with modulus $\tilde{\lambda}_0$ and with test vector $(0, \ldots, 0, 1)$; therefore we choose $C_\lambda = -\tilde{a}_{\lambda N}(0)/\tilde{a}_{NN}(0)$. The transformation (E.22) is invertible [since $y_\lambda = z_\lambda - C_\lambda z_N$ for each λ]; its only effect is that

$$\hat{a}_{\lambda N}(0) = 0 \qquad \text{for each } \lambda; \qquad \text{(E.23)}$$

by (E.14) and (E.22), $z \in \hat{\Omega} := \tilde{\Omega}$ if and only if $y \in \tilde{\Omega}$.

Now define a new set $W := W_{\mathrm{I}} \cap W_{\mathrm{II}}$ (Figure E.4) by

$$W_{\mathrm{I}} := \left\{ z \in \mathbb{R}^N \;\middle|\; z_N < \gamma z_1 - \frac{\kappa}{2} z_1^2 - \frac{A}{2} \sum_\lambda z_\lambda^2 \right\},$$

$$W_{\mathrm{II}} := \left\{ z \in \mathbb{R}^N \;\middle|\; z_N < -\gamma z_1 - \frac{\kappa}{2} z_1^2 - \frac{A}{2} \sum_\lambda z_\lambda^2 \right\},$$

where $\kappa > 0$ and $A \geq \kappa/\gamma$. Any $\kappa > 0$ will serve, as before, but the main choice of the large parameter A has yet to be made. Next, let

$$\hat{H} := W \cap \mathscr{B}(0, \rho) \qquad \text{(E.24)}$$

for some small $\rho > 0$ that also remains to be chosen. The inverse image of $\hat{H}$ under the C^2 diffeomorphism $x \mapsto z$ is to be the set H specified in step (i).

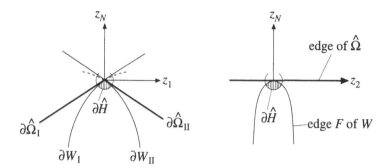

Fig. E.4.

(v′) To construct the comparison function V, we first define [cf. (E.16)]

$$\left.\begin{aligned}
f(z) &:= z_N - \gamma z_1 + \frac{\kappa}{2}z_1^2 + \frac{A}{2}\sum_\lambda z_\lambda^2, \quad \Phi(z) := e^{-Kf(z)} - 1, \\
g(z) &:= z_N + \gamma z_1 + \frac{\kappa}{2}z_1^2 + \frac{A}{2}\sum_\lambda z_\lambda^2, \quad \Psi(z) := e^{-Kg(z)} - 1,
\end{aligned}\right\}$$
(E.25)

for all $z \in \mathbb{R}^N$ and for some large $K > 0$ that will be chosen after A has been fixed. We note that f and $-\Phi$ are admissible functions describing W_I, while g and $-\Psi$ are admissible functions describing W_II.

Now define $\hat{V} \in C^\infty(\mathbb{R}^N)$ by $\hat{V} := \Phi\Psi$, and let $z = Z(x)$ denote the C^2 diffeomorphism that we have constructed on a neighbourhood, say $\mathcal{N}(p)$, of p. We set

$$V(x) := \hat{V}\big(Z(x)\big), \qquad x \in \mathcal{N}(p);$$
(E.26)

then $V \in C^2(\overline{H})$ if $\overline{H} \subset \mathcal{N}(p)$.

(vi′) The argument in step (vi) of the earlier proof shows that the function V in (E.26) satisfies conditions (I′) and (IIa,b) [stated in steps (ii) and (i), respectively]. Condition (III′), or equivalently that $\hat{L}\hat{V} \geq 0$ in $\hat{H}$, requires further labour. Here L denotes whichever of L_0, L and L_1 is considered in hypothesis (b); $\hat{L}$ is the operator to which L is transformed by the map $z = Z(x)$; and the coefficients of $\hat{L}$ are $\hat{a}_{ij}(z), \hat{b}_j(z)$ and $\hat{c}(z)$. Accordingly,

$$\hat{L}\hat{V} = (\hat{L}\Phi)\Psi + 2\nabla\Phi \cdot \hat{a} \cdot \nabla\Psi + \Phi(\hat{L}\Psi) - \hat{c}\Phi\Psi,$$
(E.27)

and the troublesome term is

$$\left.\begin{aligned}
2\,\nabla\Phi \cdot \hat{a} \cdot \nabla\Psi &= 2\,e^{-Kf-Kg}\,K^2\,B_W, \\
\text{where} \qquad B_W &:= \nabla f \cdot \hat{a} \cdot \nabla g \quad \text{on } \overset{\frown}{\hat{H}}.
\end{aligned}\right\}$$
(E.28)

The bluntness hypothesis (d′) concerns not the edge of W and the function B_W, but [in view of Remark E.4 and the transformation law (E.10)] the edge of $\hat{\Omega}$ and the function

$$
\begin{aligned}
B_{\hat{\Omega}}(z) &:= (-\gamma, 0, \ldots, 0, 1) \cdot \hat{a}(z) \cdot (\gamma, 0, \ldots, 0, 1) \\
&= -\gamma^2 \hat{a}_{11}(z) + \hat{a}_{NN}(z) \qquad (z \in \overline{\hat{\Omega}}, \ |z| \ \text{small}),
\end{aligned}
$$

in which the $\hat{a}_{1N}$-terms have cancelled each other. Thus hypothesis (d′) implies that

$$
\left.
\begin{aligned}
&\qquad -\gamma^2 \hat{a}_{11}(0) + \hat{a}_{NN}(0) = 0, \\
\text{and} \qquad &-\gamma^2 \big(\partial_\lambda \hat{a}_{11}\big)(0) + \big(\partial_\lambda \hat{a}_{NN}\big)(0) = 0 \quad \text{for each} \ \lambda.
\end{aligned}
\right\} \quad (\text{E.29})
$$

(vii′) We begin with the restriction of B_W to the edge $F := \partial W_{\mathrm{I}} \cap \partial W_{\mathrm{II}}$ of W; more precisely, to $F \cap \overline{\mathscr{B}(0, \rho_*)}$, the radius ρ_* begin so small that

$$
\overline{\mathscr{B}(0, \rho_*)} \subset Z\big(\mathcal{N}(p) \cap \mathscr{B}(p, \sigma)\big), \qquad (\text{E.30})
$$

where $\mathcal{N}(p)$ is the domain of Z and σ is as in hypothesis (e′). The edge F of W has a representation

$$
z = z_F(\zeta) := \Big(0, \zeta, -\frac{A}{2}|\zeta|^2\Big), \qquad \zeta := (\zeta_2, \ldots, \zeta_{N-1}) \in \mathbb{R}^{N-2}.
$$

Let $\alpha_{ij}(\zeta) := \hat{a}_{ij}\big(z_F(\zeta)\big)$ and let

$$
\begin{aligned}
B_F(\zeta) &:= B_W\big(z_F(\zeta)\big) \\
&= (-\gamma, A\zeta, 1) \cdot \alpha(\zeta) \cdot (\gamma, A\zeta, 1) \quad \text{for} \ |\zeta| \le \frac{1}{A} \le \frac{2}{\sqrt{5}}\rho_*; \quad (\text{E.31})
\end{aligned}
$$

this last condition ensures that $|z_F(\zeta)| \le \rho_*$. We shall see that hypothesis (d′) and uniform ellipticity make $B_F(\zeta) \ge 0$ if A is sufficiently large. First, because α_{1j}-terms cancel for $j \ne 1$,

$$
B_F(\zeta) = A^2 \sum_{\lambda,\mu} \alpha_{\lambda\mu}(\zeta)\,\zeta_\lambda\zeta_\mu + 2A \sum_\lambda \alpha_{\lambda N}(\zeta)\,\zeta_\lambda - \gamma^2 \alpha_{11}(\zeta) + \alpha_{NN}(\zeta).
$$

Second, $B_F(\zeta)$ is of the second order in ζ: by (E.23) and (E.29),

$$
\begin{aligned}
B_F(\zeta) = \ & A^2 \sum_{\lambda,\mu} \alpha_{\lambda\mu}(\zeta)\,\zeta_\lambda\zeta_\mu + 2A \sum_\lambda \big\{\alpha_{\lambda N}(\zeta) - \alpha_{\lambda N}(0)\big\}\zeta_\lambda \\
& - \gamma^2 \left\{ \alpha_{11}(\zeta) - \alpha_{11}(0) - \sum_\lambda \big(\partial_\lambda \alpha_{11}\big)(0)\,\zeta_\lambda \right\} \\
& + \left\{ \alpha_{NN}(\zeta) - \alpha_{NN}(0) - \sum_\lambda \big(\partial_\lambda \alpha_{NN}\big)(0)\,\zeta_\lambda \right\}.
\end{aligned}
$$

By the uniform ellipticity of $\hat{L}$, say with modulus $\hat{\lambda}_0$, and the C^2 smoothness of the α_{ij}, we have

$$B_F(\zeta) \geq \left(A^2\hat{\lambda}_0 - \text{const.}\,A - \text{const.}\right)|\zeta|^2 \qquad \left(|\zeta| \leq \frac{1}{A} \leq \frac{2}{\sqrt{5}}\rho_*\right),$$

where the constants are independent of ζ and A. *We choose and fix A so large that*

$$A \geq \frac{\kappa}{\gamma}, \quad \frac{1}{A} \leq \frac{2}{\sqrt{5}}\rho_* \quad \text{and} \quad B_F(\zeta) \geq 0 \;\; \text{for} \;\; |\zeta| \leq \frac{1}{A}. \qquad \text{(E.32)}$$

(viii') In order to estimate B_W effectively in $\hat{H}$, define

$$d_F(z) := \text{dist}(z, F),$$

$$h(z) := -\frac{1}{2}\{f(z) + g(z)\} = -z_N - \frac{\kappa}{2}z_1^2 - \frac{A}{2}\sum_\lambda z_\lambda^2,$$

so that $h(z) > 0$ in W and $h(z) = 0$ on F; we establish an inequality

$$0 < c_1 \leq \frac{d_F(z)}{h(z)} \leq C_1 \qquad \text{for} \;\; z \in W \cap \mathscr{B}(0, 1/A), \qquad \text{(E.33)}$$

in which c_1 and C_1 are independent of z.

First, consider points in $W_0 := W \cap \{z \mid z_1 = 0\}$. Both d_F and h are continuous and positive in W_0; so, therefore, is d_F/h. Moreover, if $z \in W_0$ and $z \to z_0 \in F$, then

$$h(z) = |\nabla h(z)|\,d_F(z) + O\big(d_F(z)^2\big),$$

whence

$$\frac{d_F(z)}{h(z)} \to \left\{1 + A^2\sum_\lambda z_{0\lambda}^2\right\}^{-1/2}.$$

Thus d_F/h, extended to $\overline{W}_0$ by these limiting values, is continuous and positive on $\overline{W}_0$, hence is bounded between positive constants on any compact subset of $\overline{W}_0$. In particular, there are constants c_0 and C_0 such that

$$0 < c_0 \leq \frac{d_F(0, z'')}{h(0, z'')} \leq C_0 \qquad \text{for} \;\; (0, z'') \in \overline{W \cap \mathscr{B}(0, 1/A)}; \qquad \text{(E.34)}$$

here $z'' := (z_2, \ldots, z_N)$, so that $(0, z'') \in \overline{W}_0$. In addition, we have

$$\gamma|z_1| < -z_N - \frac{A}{2}\sum_\lambda z_\lambda^2 = h(0, z'') \qquad \text{for all} \;\; z \in W,$$

from which it follows that, for $z \in W \cap \mathscr{B}(0, 1/A)$,

$$1 \leq \frac{h(0, z'')}{h(z)} \leq \left(1 - \frac{\kappa}{2A\gamma} \right)^{-1}, \qquad \text{(E.35)}$$

$$1 \leq \frac{d_F(z)}{d_F(0, z'')} \leq \left(1 + \frac{1}{\gamma^2 c_0^2} \right)^{1/2}, \qquad \text{(E.36)}$$

where $\kappa/2A\gamma \leq \frac{1}{2}$ by (E.32). The inequalities (E.34), (E.35) and (E.36) imply (E.33).

(ix') We now estimate B_W in $\hat{H} := W \cap \mathscr{B}(0, \rho)$. If ρ is sufficiently small, then to each point $z \in \hat{H}$ there corresponds a point $z_0 \in F$ such that $|z - z_0| = d_F(z)$ and $z_0 \in \mathscr{B}(0, 1/A)$. In fact, the line segment from z to z_0 is then contained in $\mathscr{B}(0, 1/A)$. A calculation shows that $\rho \leq 1/2A$ is quite small enough for this condition; we adopt this bound for definiteness. Then $\rho \leq \rho_*/\sqrt{5}$, where ρ_* is as in (E.30), so that $B_W \in C^2(\hat{\overline{H}})$. Since $B_W(z_0) \geq 0$ when $z_0 \in F \cap \mathscr{B}(0, 1/A)$, by (E.32), there is a constant $C_2 > 0$ such that

$$B_W(z) \geq B_W(z) - B_W(z_0) \geq -C_2|z - z_0| = -C_2 d_F(z)$$

for all $z \in \hat{H}$. By the inequality (E.33),

$$B_W(z) \geq -C_1 C_2 h(z) = \tfrac{1}{2} C_1 C_2 \{f(z) + g(z)\} \qquad \text{for all } z \in \hat{H}, \quad \text{(E.37)}$$

provided that $\rho \leq 1/2A$.

(x') It remains to choose K, and to make a final choice of ρ, in order that $\hat{L}\hat{V} \geq 0$ in $\hat{H}$. First we observe [cf. (E.19) to (E.21)] that

$$\hat{L}\Phi \geq e^{-Kf} \left\{ K^2 \hat{\lambda}_0 |\nabla f|^2 - K \sum_{i,j} \hat{a}_{ij} \, \partial_i \partial_j f - K \sum_j b_j \, \partial_j f + \hat{c} \right\}$$

$$\geq \tfrac{1}{2} e^{-Kf} K^2 \hat{\lambda}_0 \qquad \text{in } \hat{H}$$

if we make a first, sufficiently large choice of K and if $\rho \leq 1/2A$. The function $\hat{L}\Psi$ is treated similarly. The formula (E.27) for $\hat{L}\hat{V}$ now yields

$$\hat{L}\hat{V} \geq \tfrac{1}{2} e^{-Kf} K^2 \hat{\lambda}_0 \left\{ e^{-Kg} - 1 \right\} + 2 e^{-Kf - Kg} K^2 B_W$$

$$\qquad + \tfrac{1}{2} e^{-Kg} K^2 \hat{\lambda}_0 \left\{ e^{-Kf} - 1 \right\}$$

$$= \tfrac{1}{2} e^{-Kf - Kg} K^2 \hat{\lambda}_0 \left\{ 2 - e^{Kf} - e^{Kg} + (4/\hat{\lambda}_0) B_W \right\}$$

$$\geq \tfrac{1}{2} e^{-Kf - Kg} K^2 \hat{\lambda}_0 \left\{ 2 - e^{Kf} - e^{Kg} + C_3 f + C_3 g \right\}, \quad \text{(E.38)}$$

where $C_3 := (2/\hat{\lambda}_0)C_1C_2$ and we have used (E.37). The result (E.38) holds in $\hat{H}$ if $\rho \leq 1/2A$. Now define

$$q(t) := 1 - e^{-Kt} - C_3t \qquad \text{for} \ 0 \leq t \leq \frac{1}{K};$$

whence

$$q(0) = 0 \quad \text{and} \quad q'(t) = Ke^{-Kt} - C_3 \geq Ke^{-1} - C_3.$$

If necessary, increase K beyond our choice for $\hat{L}\Phi$ and $\hat{L}\Psi$ so that $K \geq C_3e$; then $q(t) > 0$ for $0 < t \leq 1/K$. If necessary, decrease ρ below $1/2A$ so that $-f(z) < 1/K$ and $-g(z) < 1/K$ when $z \in \hat{H}$. Then (E.38) implies that, in $\hat{H}$,

$$\hat{L}\hat{V} \geq \tfrac{1}{2}e^{-Kf-Kg} K^2 \hat{\lambda}_0 \{q(-f) + q(-g)\} > 0,$$

as desired. □

Notes on Sources

I apologize for the inadequacy of these notes. They describe material that I used, rather than all the material that I should have used. The notes also point to further theorems that I happen to have encountered; they do not amount to a sketch of the present state of the subject.

Chapter 1 It is worth mention that Aharonov, Schiffer & Zalcman (1981) obtained the conclusion of Theorem 1.1 from quite different hypotheses. Their first result concerns a compact set P in $\mathbb{R}^N$ which, *inter alia*, is connected and such that $\mathbb{R}^N \setminus P$ is connected. They proved that, if in a (non-empty) open set outside P the Newtonian potential of constant density in P is equal to the potential of a point mass, then P is a ball.

In using a flower to illustrate a symmetry group, I have followed Weyl (1952, p.66), who used *vinca herbacea* for the cyclic group C_5 and used a geranium for the dihedral group D_5. Weyl's book is full of charming examples; every page reflects the erudition and humanity of a master from an age less barbarous than ours.

For problems that allow symmetry of solutions and that can be formulated as exercises in the calculus of variations, there is an approach to proofs of symmetry that is older than, and quite different from, the methods set out in this book. This approach is based on various geometrical operations (devised by J. Steiner and H.A. Schwarz in the nineteenth century) called *symmetrization with respect to a plane, a line or a point*. Typically, such an operation assigns to a given set $A \subset \mathbb{R}^3$ (which may be the set below the graph of a function of two variables) a set A^* that has a specified symmetry and has the same cross-sectional area as A for each of a family of cross-sections. In particular, symmetrization with respect to a hyperplane, also called *Steiner symmetrization*, yields sets of the form called Steiner symmetric in Definition 3.2. The books of Pólya and Szegö (1951), of Bandle (1980) and of Kawohl (1985) explain these procedures fully.

Chapter 2 In §§2.2 and 2.3 the weak maximum principle, the boundary-point lemma and the strong maximum principle for C^2-subsolutions are taken from Gilbarg & Trudinger's book (1983, Chapter 3); of course, with expanded proofs and some differences of detail. The boundary-point lemma in this form is due to E. Hopf (1952b) and to Oleinik (1952); the others are somewhat older. Theorem 2.11 (the weak maximum principle for distributional subsolutions relative to the operator L_1, which has constant coefficients) is a very particular case of maximum principles due to Littman (1959, 1963). The elementary proof of Theorem 2.11 seems (to me) to be new, but may well occur somewhere in the literature of the 1950s or 1960s; my colleague Buffoni suggested the use of (2.16) after I had used (2.17).

The maximum principle for thin sets (Theorem 2.19) is a very restricted version, although not quite a particular case, of theorems due to Berestycki & Nirenberg (1991, Proposition 1.1), to Dancer (1992, Lemma 1) and to Healey, Kielhöfer & Stuart (1994, Theorem 3.2). It seems that, for theorems of this kind, the basic observation appeared first in work of Bakelman preparatory to publication of his book (1994) and in unpublished remarks of S.R.S. Varadhan. Theorem 2.19 is more naive than its counterparts elsewhere because the constant coefficients of the operator L_{10} make it easy to avoid derivatives of the subsolution (and thus to avoid Sobolev spaces).

The Phragmén–Lindelöf theory in §§2.4 and 2.6 is a mixture based on my reading of Protter & Weinberger (1984, §2.9) and of Stein & Weiss (1971, §II.5.2), but with different comparison functions. As is mentioned in §2.6, the primary comparison function g, described by Theorem 2.27, is an extension to all dimensions $N \geq 2$ of a function introduced by D. Siegel (1988) for $N = 2$. Recently, Siegel has abandoned this function in favour of a more elaborate one, defined on polytopes for all $N \geq 2$ (Siegel & Talvila 1996, Theorem 4.1), that allows larger rates of growth of a subharmonic function when $|x| \to \infty$ and $x_N/|x| \to 0$ in the half-space $\{ x \in \mathbb{R}^N \mid x_N > 0 \}$. I have not seen previously the comparison functions g_e and g_2, constructed from g in §2.6 by means of the Kelvin transformation; the trio g, g_e and g_2 is intended to be a compromise between the simplest comparison functions and difficult ones that yield results close to best possible.

Chapter 2 provides only an introduction to the subject of maximum

principles for elliptic equations. For example, information about positive
solutions of $\triangle u + f(u) = 0$ in a bounded set Ω can be gleaned by means
of P-functions; these are of form $|\nabla u|^2 g(u) + h(u)$ and, for suitable choices
of g and h, attain their supremum either on $\partial\Omega$ or at a stationary point of
u. The construction and use of such functions are the main topics of the
book by Sperb (1981). In another direction, Bakelman (1994) considers
fully non-linear equations, such as those of Monge–Ampère type, and
derives for such equations maximum principles (Chapter 8 of his book)
that make those in our Chapter 2 look like toys.

Chapter 3 For boundaries $\partial\Omega$ of class C^2, monotonicity and symmetry
results like Theorem 3.3 and its corollaries, but for more general elliptic
equations, were proved by Gidas, Ni & Nirenberg (1979), building on
work of Alexandrov (see H. Hopf 1956, p.147) and of Serrin (1971). A
maximum principle for narrow sets Ω, and use of the compact set called
F in the proof of Theorem 3.3, were introduced into problems of this
kind by Berestycki & Nirenberg (1991); thereby the need for smoothness
of $\partial\Omega$ was removed and proofs were shortened. Parallel steps were taken
by Dancer (1992), who made more use of Sobolev spaces and proved
monotonicity results like those in Theorem 3.3 under hypotheses on u
that are weaker than those of Berestycki & Nirenberg (and considerably
weaker than those in Theorem 3.3).

   Theorem 3.6 and Corollary 3.9 have their roots in the paper of Gidas,
Ni & Nirenberg (1979), in the work of Berestycki & Nirenberg (1991)
and in an extension of two results of the 1979 paper to the case of discontinuous f (Amick & Fraenkel 1986). The combination of discontinuous
f and a set Ω that is merely Steiner symmetric seems to be new.

   Exercises 3.18 and 3.19 (which show that one cannot relax significantly
the conditions on f used in most of the book) extend a construction of
Gidas, Ni & Nirenberg (1979, p.220) in a way demonstrated for ordinary
differential equations by work of C.A. Stuart (1976) and by unpublished
examples of J.F. Toland.

   Gidas, Ni & Nirenberg (1979, p.220) raised the question of whether
results like those of Theorem 3.3 remain true if the hypothesis $u > 0$ in
Ω is changed to $u \geq 0$ in Ω with $u > 0$ somewhere in Ω. They observed
that for $N = 1$ (for $\Omega \subset \mathbb{R}$) the answer is No because

$$u(x) = 1 + \cos \pi x \implies u'' + \pi^2(u - 1) = 0 \text{ in } (-3, 3), \quad u(\pm 3) = 0,$$

and u' has the wrong sign in $(1, 2)$; they proved that the answer is Yes
for all $N \in \mathbb{N}$ if $f(0) \geq 0$. Castro & Shivaji (1989) have shown that for

$f(0) < 0$ the wider hypothesis is allowable in Theorem 3.3 if Ω is a ball and $N \geq 2$. Note that extensions of the functions u in Exercises 3.10 and 3.11 are consistent with all this.

Allegretto & Siegel (1995) have considered the problem

$$\triangle u + \lambda\, p(x)\, g(u) = 0 \quad \text{in } \Omega, \quad u\big|_{\partial\Omega} = 0,$$

where $\Omega := (-1, 1) \times \Omega'$ is a cylinder, Ω' being open and bounded in $\mathbb{R}^{N-1}$, and $p(-x_1, x'') = p(x_1, x'')$. Using Sobolev-space methods, a family of non-negative eigenfunctions defined on caps of Ω, and corresponding hypotheses on p and g, they avoid pointwise estimates near $\partial\Omega$ and prove symmetry under hypotheses on u slightly weaker than those of Berestycki & Nirenberg (1991).

Chapter 4 Regarding the behaviour of $u(x)$ as $r := |x| \to \infty$: by now there are many different sets of conditions that lead to symmetry for solutions u of suitable equations in $\mathbb{R}^N$. Gidas, Ni & Nirenberg (1979, Theorem 4) considered positive solutions of rather general equations and introduced hypothesis (C) of Definition 4.1, but with three explicit terms before a remainder. Caffarelli & Friedman (1980, Theorem 7.1) observed for the equation $\triangle u + f(u) = 0$ in $\mathbb{R}^2$ that positivity of u could be abandoned, that logarithmic growth at infinity could be tolerated and that the asymptotic approximation could be shortened. Definition 4.1, and the sufficiency of its conditions for Theorem 4.2, evolved from these two sources in the course of my lectures at Bath, given in the years 1989–1992.

The approach of Gidas, Ni & Nirenberg in their 1981 paper is quite different: for positive C^2-solutions of $\triangle u + f(u) = 0$ in $\mathbb{R}^N$, with $N \geq 2$ in some cases and $N \geq 3$ in others, f is specified rather precisely but very little is demanded of u as $r \to \infty$; for example, merely that $u(x) = O(r^{-m})$ with $m > 0$. More detailed asymptotic behaviour is then *deduced*. Li & Ni (1993) have taken this further, even for fully non-linear equations, by demanding only that $u(x) \to 0$ as $r \to \infty$; no rate of decay is prescribed. Of course, the function f or its fully non-linear counterpart must be specified carefully.

The reader who is bewildered but attracted by remarks in the text about Sobolev spaces and variational principles (for example, after Theorem 4.13 and in Exercises 4.22, 5.29 and 5.30) will find lucid and satisfying explanations in the books by Berger (1977), by Brezis (1983) and by Friedman (1982).

Regarding steady vortex motions. Turkington (1983), considering various steady vortex flows in two dimensions, used reflection in hyperplanes (in this case reflection in lines) to prove that intense vortices of small diameter are approximately circular and, suitably scaled, become exactly circular in a certain limit. Concerning Hill's vortex and Hicks's vortex: the solutions of Exercises 4.22 and 4.23 can be found, to a large extent, in papers by Amick & Fraenkel (1986) and by Fraenkel (1992). Uniqueness of the vortex pair in $\mathbb{R}^2$ that corresponds to Hill's vortex has been proved by Burton (1996).

Two conspicuous omissions from this book are (a) *overdetermined problems*, by which we mean here that two boundary conditions

$$u\big|_{\partial\Omega} = \text{constant} \quad \text{and} \quad \frac{\partial u}{\partial n}\bigg|_{\partial\Omega} = \text{constant}$$

accompany an elliptic equation of order two; (b) problems on an *exterior domain* Ω, by which we mean a connected set in $\mathbb{R}^N$ the complement of which is compact. The first of these omissions is unfortunate in that overdetermined problems in a bounded set Ω were the subject of Serrin's early and fruitful use of reflection in hyperplanes (Serrin 1971); the conclusion was that Ω must be a ball and the solution u spherically symmetric and monotonic.

Reichel (1996, 1997) has considered overdetermined problems on exterior domains for the equation $\triangle u + f(u, |\nabla u|) = 0$ and for certain fully non-linear equations, recovering Serrin's conclusions under suitable hypotheses on $f(u, |\nabla u|)$ or on its fully non-linear counterpart. These hypotheses, however, exclude the case $f(u, |\nabla u|) = f_0(u)$ with f_0 positive and increasing on $(0, \infty)$. This defect has been made good by Aftalion & Busca (1998), who combined skilful use of the Kelvin transformation with a maximum principle for thin sets.

Chapter 5 Most authors seem to regard results like Theorem 5.7 (two pleasant properties of boundaries of class C^1) and Proposition 5.20 (resolution of a topological question) as obvious; however, colleagues attending my lectures at Bath insisted on proof of such matters. Theorem 5.7 is taken from Amick & Fraenkel (1986, Lemma A.1) and from Keady & Kloeden (1987, Lemma 2.1); Proposition 5.20 appears to be new. Rhomboid neighbourhoods (Definition 5.4) have a long history in the theory of Sobolev spaces, although they are seldom given a name; their treatment in Chapter 5 follows that of Fraenkel (1979). References to their earlier use are given in that paper.

The monotonicity results in §5.3 are due to Gidas, Ni & Nirenberg (1979). In our version, somewhat more attention is given to the components of a cap, and less smoothness is required of $\partial\Omega$, f and u. On the other hand, our equations (5.1) and (5.2) are far less general than those considered by Gidas, Ni & Nirenberg.

The result in Exercise 5.30, that symmetry results for a ball do not extend to an annulus or spherical shell A, is not new; examples of positive solutions in A that lack spherical symmetry have been given by Brezis & Nirenberg (1983, p.453) and by Coffman (1984). Possibly the example in Exercise 5.30 is simpler than previous ones; it is a modified form of an example suggested to me by G. Keady.

The annulus or spherical shell A has also been used by Dancer (1997) to prove that positive solutions $u \in C^2(\overline{\Omega})$, of $\triangle u + f(u) = 0$ in Ω with $u = 0$ on $\partial\Omega$, can have stationary points arbitrarily close to $\partial\Omega$. Dancer used $\Omega = A$ and a family of functions $f = f(.,\varepsilon)$ to construct spherically symmetric, positive solutions $u(.,\varepsilon)$ having a stationary point that approaches the inner part of ∂A as $\varepsilon \to 0$. Thereby he answered a question of Gidas, Ni & Nirenberg (1979, p.223), who had gathered evidence for the opposite conclusion.

Keady & Kloeden (1987) applied monotonicity results like those in Theorem 5.10 to discuss solutions of

$$\triangle u + \lambda f_H(u - k) = 0 \quad \text{in a bounded set } \Omega \subset \mathbb{R}^2, \quad u\big|_{\partial\Omega} = 0,$$

where λ and k are positive constants and f_H denotes the Heaviside function. They showed that, for convex Ω with smooth boundary, the solutions resemble those when Ω is a ball, provided that the set $C := \{x \in \Omega \mid u(x) > k\}$, which they called the *core*, is connected. In particular, there is a branch of solutions along which $\text{diam}\, C \to 0$ and $\lambda \to \infty$. (When Ω is a ball, all solutions are known explicitly; see Exercises 3.15 and 3.16, in which the same equation occurs and Ω is a ball in $\mathbb{R}$ or in $\mathbb{R}^5$.)

Two important techniques, beyond those in Chapters 3 to 5, were introduced by Berestycki & Nirenberg (1988, 1990) and used to prove monotonicity, symmetry or anti-symmetry in a variety of problems with boundary conditions more general than $u\big|_{\partial\Omega} = 0$. These techniques were also simplified and extended in their 1991 paper.

First, equations more general than $\triangle u + f(u) = 0$ may lead to

$$\triangle w + b(x,\mu) \cdot \nabla w + c(x,\mu)w - \frac{1}{2}\beta(x,\mu)\frac{\partial w}{\partial \mu} \geq 0 \quad \text{for } x \in Z(\mu), \quad (\dagger)$$

330 — *Notes on Sources*

in place of our (3.6); here $\beta(x,\mu) \geq 0$ and the identity

$$\frac{\partial}{\partial x_1} u(2\mu - x_1, x'') = \frac{1}{2}\frac{\partial}{\partial \mu}\left\{ u(x) - u(2\mu - x_1, x'') \right\} = \frac{1}{2}\frac{\partial w}{\partial \mu}(x,\mu)$$

has been used. Now, if μ is regarded as a time-like variable, then (†) is an inequality for a *parabolic* partial differential operator (although a degenerate one wherever $\beta(x,\mu) = 0$); somewhat surprisingly, boundary conditions that suit such an inequality also arise in the (basically elliptic) problems considered by Berestycki & Nirenberg. Thus maximum principles for parabolic operators play an important part.

Second, the method of *sliding domains* was introduced and used to good effect. In this method, one compares the graph of a function not with a reflection of that graph, but with a translation of it, defining, for example,

$$\Omega_\tau := \left\{ x \in \mathbb{R}^N \mid x - \tau e^1 \in \Omega \right\},$$

$$w(x,\tau) := u(x) - u(x - \tau e^1) \quad \text{for } \tau > 0 \text{ and } x \in \Omega \cap \Omega_\tau.$$

When the boundary conditions suggest it, monotonicity in the whole of Ω is often established by an application of maximum principles that is not unlike that used for reflection and symmetry.

In the same direction and exhibiting further developments, the work of Craig & Sternberg (1988, 1992) established the symmetry of certain gravity waves in hydrodynamics; in particular, the symmetry of the classical solitary wave. This is of interest because, when surface tension is included, there is no shortage of asymmetric solitary waves.

Appendix A Little need be said about this material except to acknowledge that I first learned the main results, some thirty years ago, from the book of Günter (1967). However, the resemblance of Appendix A to Günter's book now seems rather faint; for example, in the context of Theorems A.15 and A.16, Günter (1967, p.308) was willing to accept $|h|^\mu \log(1/|h|)$ in place of the $|h|^\mu$ for which we work so hard.

Exercise A.25 is a fragment of generalized axially symmetric potential theory (Weinstein 1953).

Appendix B In the proof of Theorem B.6, the choice of test function is that in Sobolev's monograph (1963, p.92). Theorem B.9 is taken from Gilbarg & Trudinger (1983, p.23). Theorems B.29 and B.30, on the Poisson integral for a ball, should really be accompanied by a description of how boundary values are attained when the boundary-value function g is merely in $L_p(\partial B)$; Folland (1976, p.124) demonstrates this beautifully.

Of course, my hope is that Appendices A and B may prepare beginners for the more concise and extensive material in, say, Dautray & Lions (1990), Folland (1976) and Gilbarg & Trudinger (1983).

Appendix E As was remarked in the Preface, the edge-point lemma (or boundary-point lemma at a corner) is due to Serrin (1971) and was extended by Gidas, Ni & Nirenberg (1979, pp.214 and 237–243), who called it Lemma S. The version in Appendix E is essentially Lemma S, with slight changes intended to increase the resemblance to our boundary-point lemma, and with a vastly expanded proof.

References

Aftalion, A. & Busca, J. (1998). Radial symmetry of overdetermined boundary value problems in exterior domains. *Archive for Rational Mechanics and Analysis*, **143**, 195–206.

Aharonov, D., Schiffer, M.M. & Zalcman, L. (1981). Potato Kugel. *Israel Journal of Mathematics*, **40**, 331–339.

Allegretto, W. & Siegel, D. (1995). Picone's identity and the moving plane procedure. *Electronic Journal of Differential Equations*, **1995**, 1–13.

Amick, C.J. & Fraenkel, L.E. (1986). The uniqueness of Hill's spherical vortex. *Archive for Rational Mechanics and Analysis*, **92**, 91–119.

Apostol, T.M. (1974). *Mathematical analysis*. 2nd ed. Reading, Massachusetts: Addison-Wesley.

Arnold, V.I. (1990). *Huygens and Barrow, Newton and Hooke*. Basel: Birkhäuser Verlag.

Bakelman, I.J. (1994). *Convex analysis and nonlinear geometric elliptic equations*. Berlin: Springer.

Bandle, C. (1980). *Isoperimetric inequalities and applications*. London: Pitman.

Batchelor, G.K. (1967). *An introduction to fluid dynamics*. Cambridge University Press.

Berestycki, H. & Nirenberg, L. (1988). Monotonicity, symmetry and antisymmetry of solutions of semilinear elliptic equations. *Journal of Geometry and Physics*, **5**, 237–275.

Berestycki, H. & Nirenberg, L. (1990). Some qualitative properties of solutions of semilinear elliptic equations in cylindrical domains. In *Analysis, et cetera*, edited by P.H. Rabinowitz & E. Zehnder. New York: Academic Press, pp. 115–164.

Berestycki, H. & Nirenberg, L. (1991). On the method of moving planes and the sliding method. *Boletim Sociedade Brasileira de Matemática*, **22**, 1–37.

Berger, M.S. (1977). *Nonlinearity and functional analysis*. New York: Academic Press.

Brezis, H. (1983). *Analyse fonctionnelle, théorie et applications*. Paris: Masson.

Brezis, H. & Nirenberg, L. (1983). Positive solutions of nonlinear elliptic equations involving critical Sobolev exponents. *Communications on Pure and Applied Mathematics*, **36**, 437–477.

Burkill, J.C. (1975). *The theory of ordinary differential equations.* 3rd ed. London: Longman.

Burkill, J.C. & Burkill, H. (1970). *A second course in mathematical analysis.* Cambridge University Press.

Burton, G.R. (1996). Uniqueness for the circular vortex-pair in a uniform flow. *Proceedings of the Royal Society of London*, A **452**, 2343–2350.

Caffarelli, L.A. & Friedman, A. (1980). Asymptotic estimates for the plasma problem. *Duke Mathematical Journal*, **47**, 705–742.

Cartan, H. (1971). *Differential calculus.* Paris: Hermann; also Boston: Houghton Mifflin.

Castro, A. & Shivaji, R. (1989). Nonnegative solutions to a semilinear Dirichlet problem in a ball are positive and radially symmetric. *Communications in Partial Differential Equations*, **14**, 1091–1100.

Coffman, C.V. (1984). A non-linear boundary value problem with many positive solutions. *Journal of Differential Equations*, **54**, 429–437.

Coxeter, H.S.M. (1963). *Regular polytopes.* 2nd ed. New York: Macmillan.

Coxeter, H.S.M. (1981). *Introduction to geometry.* 2nd ed. New York: Wiley.

Craig, W. & Sternberg, P. (1988). Symmetry of solitary waves. *Communications in Partial Differential Equations*, **13**, 603–633.

Craig, W. & Sternberg, P. (1992). Symmetry of free-surface flows. *Archive for Rational Mechanics and Analysis*, **118**, 1–36.

Dancer, E.N. (1992). Some notes on the method of moving planes. *Bulletin of the Australian Mathematical Society*, **46**, 425–434.

Dancer, E.N. (1997). Some singularly perturbed problems on annuli and a counterexample to a problem of Gidas, Ni and Nirenberg. *Bulletin of the London Mathematical Society*, **29**, 322–326.

Dautray, R. & Lions, J.-L. (1990). *Mathematical analysis and numerical methods for science and technology.* Vol. 1, *Physical origins and classical methods.* Berlin: Springer.

Dieudonné, J. (1969). *Foundations of modern analysis.* New York: Academic Press.

Evans, L.C. & Gariepy, R.F. (1992). *Measure theory and fine properties of functions.* Boca Raton, Florida: CRC Press.

Fauvel, J., Flood, R., Shorland, M. & Wilson, R. (1988). *Let Newton be!* Oxford University Press.

Ferraro, V.C.A. & Plumpton, C. (1966). *An introduction to magneto-fluid mechanics.* 2nd ed. Oxford University Press.

Fleming, W.H. (1965). *Functions of several variables.* Reading, Massachusetts: Addison-Wesley.

Folland, G.B. (1976). *Introduction to partial differential equations.* Princeton University Press.

Fraenkel, L.E. (1979). On regularity of the boundary in the theory of Sobolev spaces. *Proceedings of the London Mathematical Society* (3), **39**, 385–427.

Fraenkel, L.E. (1992). On steady vortex rings with swirl and a Sobolev inequality. In *Progress in partial differential equations: calculus of variations, applications*, edited by C. Bandle *et al.* Harlow, Essex: Longman; also New York: Wiley, pp. 13–26.

References

Friedman, A. (1982). *Variational principles and free-boundary problems.* New York: Wiley.

Gidas, B., Ni, W.-M. & Nirenberg, L. (1979). Symmetry and related properties via the maximum principle. *Communications in Mathematical Physics,* **68**, 209–243.

Gidas, B., Ni, W.-M. & Nirenberg, L. (1981). Symmetry of positive solutions of nonlinear elliptic equations in R^n. In *Mathematical analysis and applications,* part A, *Advances in mathematical supplementary studies,* **7A**, edited by L. Nachbin. New York: Academic Press, pp. 369–402.

Gilbarg, D. (1952). The Phragmén–Lindelöf theorem for elliptic partial differential equations. *Journal of Rational Mechanics and Analysis,* **1**, 411–417.

Gilbarg, D. & Trudinger, N.S. (1983). *Elliptic partial differential equations of second order,* 2nd ed. Berlin: Springer.

Günter, N.M. (1967). *Potential theory.* New York: Ungar.

Healey, T.J., Kielhöfer, H. & Stuart, C.A. (1994). Global branches of positive weak solutions of semilinear elliptic problems over nonsmooth domains. *Proceedings of the Royal Society of Edinburgh,* **124A**, 371–388.

Hicks, W.M. (1899). Researches in vortex motion. Part III. On spiral or gyrostatic vortex aggregates. *Philosophical Transactions of the Royal Society of London,* **A 192**, 33–99.

Hill, M.J.M. (1894). On a spherical vortex. *Philosophical Transactions of the Royal Society of London,* **A 185**, 213–245.

Hille, E. (1973). *Analytic function theory,* vols. I and II. 2nd ed. New York: Chelsea.

Hopf, E. (1952a). Remarks on the preceding paper by D. Gilbarg. *Journal of Rational Mechanics and Analysis,* **1**, 419–424.

Hopf, E. (1952b). A remark on linear elliptic differential equations of second order. *Proceedings of the American Mathematical Society,* **3**, 791–793.

Hopf, H. (1946, 1956). *Differential geometry in the large.* Seminar lectures New York University 1946 and Stanford University 1956. Re-issued as Lecture Notes in Mathematics, **1000**, by Springer, Berlin, 1983.

Kawohl, B. (1985). *Rearrangements and convexity of level sets in PDE.* Lecture Notes in Mathematics, **1150**. Berlin: Springer.

Keady, G. & Kloeden, P.E. (1987). An elliptic boundary-value problem with a discontinuous nonlinearity, II. *Proceedings of the Royal Society of Edinburgh,* **105A**, 23–36.

Kellogg, O.D. (1929). *Foundations of potential theory.* Berlin: Springer. Re-issued by Dover, New York, 1954.

Kingman, J.F.C. & Taylor, S.J. (1966). *Introduction to measure and probability.* Cambridge University Press.

Lamb, H. (1932). *Hydrodynamics.* 6th ed. Cambridge University Press.

Li, Y. & Ni, W.-M. (1993). Radial symmetry of positive solutions of nonlinear elliptic equations in R^n. *Communications in Partial Differential Equations,* **18**, 1043–1054.

Littman, W. (1959). A strong maximum principle for weakly L-subharmonic functions. *Journal of Mathematics and Mechanics,* **8**, 761–770.

Littman, W. (1963). Generalized subharmonic functions: monotonic approximations and an improved maximum principle. *Annali della Scuola Normale*

Superiore di Pisa (3), **17**, 207–222.

Maz'ja, V.G. (1985). *Sobolev spaces.* Berlin: Springer.

Moffat, H.K. (1969). The degree of knottedness of tangled vortex lines. *Journal of Fluid Mechanics,* **35**, 117–129.

Oleinik, O.A. (1952). On properties of solutions of certain boundary problems for equations of elliptic type. *Matematichesky Sbornik* (N.S.), **30**, 695–702.

Pólya, G. & Szegö, G. (1951). *Isoperimetric inequalities in mathematical physics.* Princeton University Press.

Protter, M.H. & Weinberger, H.F. (1984). *Maximum principles in differential equations.* 2nd ed. Berlin: Springer.

Rees, E.G. (1983). *Notes on geometry.* Berlin: Springer.

Reichel, W. (1996). Radial symmetry for an electrostatic, a capillarity and some fully nonlinear overdetermined problems on exterior domains. *Zeitschrift für Analysis und ihre Anwendungen,* **16**, 619–635.

Reichel, W. (1997). Radial symmetry for elliptic boundary value problems on exterior domains. *Archive for Rational Mechanics and Analysis,* **137**, 381–394.

Rudin, W. (1970). *Real and complex analysis.* New York: McGraw–Hill.

Rudin, W. (1976). *Principles of mathematical analysis.* 3rd ed. New York: McGraw–Hill.

Serrin, J. (1971). A symmetry problem in potential theory. *Archive for Rational Mechanics and Analysis,* **43**, 304–318.

Siegel, D. (1988). The Dirichlet problem in a half-space and a new Phragmén–Lindelöf principle. In *Maximum principles and eigenvalue problems in partial differential equations,* edited by P.W. Schaefer. Harlow, Essex: Longman; also New York: Wiley, pp. 208–217.

Siegel, D. & Talvila, E.O. (1996). Uniqueness for the n-dimensional half space Dirichlet problem. *Pacific Journal of Mathematics,* **175**, 571–587.

Simmons, G.F. (1972). *Differential equations with applications and historical notes.* New York: McGraw–Hill.

Sobolev, S.L. (1963). *Applications of functional analysis in mathematical physics.* Providence: American Mathematical Society.

Sobolev, S.L. (1964). *Partial differential equations of mathematical physics.* Oxford: Pergamon Press. Re-issued by Dover, New York, 1989.

Sperb, R.P. (1981). *Maximum principles and their applications.* New York: Academic Press.

Spiegel, M.R. (1959). *Vector analysis.* Schaum's outline series. New York: McGraw–Hill.

Stein, E.M. & Weiss, G. (1971). *Introduction to Fourier analysis on Euclidean spaces.* Princeton University Press.

Stuart, C.A. (1976). Differential equations with discontinuous non-linearities. *Archive for Rational Mechanics and Analysis,* **63**, 59–75.

Thompson, W.B. (1962). *An introduction to plasma physics.* 2nd ed. Oxford: Pergamon Press.

Titchmarsh, E.C. (1932). *The theory of functions.* Oxford University Press.

Turkington, B. (1983). On steady vortex flow in two dimensions, I and II. *Communications in Partial Differential Equations,* **8**, 999–1030 and 1031–1071.

Weinstein, A. (1953). Generalized axially symmetric potential theory. *Bulletin of the American Mathematical Society*, **59**, 20–38.

Weir, A.J. (1973). *Lebesgue integration and measure.* Cambridge University Press.

Weyl, H. (1952). *Symmetry.* Princeton University Press.

Whittaker, E.T. & Watson, G.N. (1927). *A course of modern analysis.* 4th ed. Cambridge University Press.

Ziemer, W.P. (1989). *Weakly differentiable functions.* Berlin: Springer.

Index